Spaces Speak, Are You Listening?

Spaces Speak, Are You Listening?

Experiencing Aural Architecture

Barry Blesser and Linda-Ruth Salter

The MIT Press
Cambridge, Massachusetts
London, England

MIT Press books may be purchased at special quantity discounts for business or sales promotional use. For information, please email special_sales@mitpress.mit.edu or write to Special Sales Department, The MIT Press, 55 Hayward Street, Cambridge, MA 02142.

This book was set in Stone Serif and Stone Sans on 3B2 by Asco Typesetters, Hong Kong and was printed and bound in the United States of America.

Library of Congress Cataloging-in-Publication Data

Blesser, Barry.
Spaces speak, are you listening? : experiencing aural architecture / by Barry Blesser and Linda-Ruth Salter.
　　p.　cm.
Includes bibliographical references and index.
ISBN-13: 978-0-262-02605-5 (hc : alk. paper)
ISBN-10: 0-262-02605-8 (hc : alk. paper)
1. Space perception. 2. Auditory perception. 3. Architectural acoustics. 4. Psychoacoustics. 5. Sound—Recording and reproducing—Digital techniques. I. Salter, Linda-Ruth. II. Title.

QP443.B585　2006
612.8′5—dc22   2006044046

10　9　8　7　6　5　4　3　2

Contents

Acknowledgments

The book is based on the work of thousands of scientists, engineers, architects, scholars, and researchers who created the intellectual foundation for aural architecture without necessarily realizing that they were laying a foundation for this field. Without their efforts, it would not have been possible to integrate and synthesize the enormous scope of this subject into a coherent exposition.

I am especially grateful to my wife and best friend, Linda-Ruth Salter, for her unwavering encouragement during the four years that it took to write this book. As my editor, critic, reviewer, and debating partner, she broadened and improved the text by providing a social science perspective that complemented my scientific and engineering one. Because of her invaluable contributions throughout the project, I invited her to be the second author. The ideas reflect a joint effort, spanning the 35 years of our life together.

The insights and knowledge that I accumulated over the years came from a diverse group of colleagues who taught me the power of intellectual collaboration, fusing a broad range of abilities, experiences, and information into shared goals. Early in our careers, both of us were influenced by two scholars who exemplified academic intellectuals at their best: Murray Eden and Philip Kubzansky. More specific to the topic of aural architecture, I acquired an appreciation for the subtleties of audio, perception, and the aural arts through years of working with Karl Otto Bäder, Jan Wissmuller, David Griesinger, and Geoff Steadman, all of whom have successfully integrated the aural arts with modern technology.

During the many drafts of the manuscript, I learned that insight and analysis are useless if not clearly developed and well written. Many of my friends and colleagues spent hundreds of hours editing drafts and teaching me the art of writing. For their valuable contribution, I would like to express my appreciation to David Moran, David Moulton, Jay Rose, Jean Sarris, and Shirley Reiss. As my editor at MIT Press, Doug Sery never wavered in his faith that I would eventually convert an ill-defined proposal into a coherent book. Copy editor Jeff Lockridge greatly improved the flow, clarity, and organization of the writing, for which I am deeply grateful. Finally, my children, Andrew and

Sonya, provided emotional encouragement to continue with the project without resenting the time that it took away from our family life. They, too, believed that this project was worthwhile.

Looking back over the last four years, I have enjoyed the opportunity to work with my truly great friends and colleagues, who gave freely of their time and ideas. I thank them for their enthusiasm and their support. Though I served as the scribe, a large community actually created this book.

A Personal Perspective

This book is my adventure story about transforming a narrow topic that has engaged my professional interest for three decades into a set of broader issues. The story began in the 1970s when I had the good fortune to develop and commercialize the first digital signal-processing products for the recording industry: an audio delay and an artificial reverberator. Now, more than a quarter century later, that technology has expanded into a multibillion dollar industry permeating our culture and supporting thousands of innovators making incremental contributions. Looking backward, it is clear that my initial goal of electronically reproducing the auditory experience of a concert hall had a much broader meaning than my initial concept. This book expands that limited engineering goal into an interdisciplinary research project: the experience of space by attentive listening. Like most of the thousands of people who have contributed to the aural architecture of spaces, I had not appreciated the artistic, social, historical, and philosophical context of my isolated activities.

The nature of a discussion depends on the scope of the questions being asked: narrow or broad. I could have framed the discussion solely in terms of the physical and mathematical properties of sound waves that contribute to the aural experience of a concert hall. A vast body of literature already takes this approach. It is relevant mainly for specialists who have the professional interest and necessary background to appreciate the details of that subject. Rather, I have chosen to explore the broad *phenomenon* of auditory spatial awareness without regard to a specific discipline, culture, or time period. For me, the global approach is an intellectual adventure with gratifying surprises.

What does it mean to explore a phenomenon? An explanation is never the phenomenon itself, but only a refracted image of it, like looking at a scene through a prism. Although models, theories, and pictures are not reality, they present aspects of a phenomenon. For example, in dealing with musical space, a composer sees one aspect of the phenomenon, whereas architects, archaeologists, anthropologists, audio engineers, psychophysical scientists, and blind individuals each see other aspects. When we have access to multiple views, each with its own biases and limitations, we acquire greater understanding of the phenomenon. As we explore these views, we must remember that each version of a phenomenon is always constrained by the questions being asked and the answers being offered. On the other hand, the union of diverse viewpoints, like multiple shadows from an object that we cannot see, allows us to form an image of the phenomenon, which by definition always remains inaccessible.

While doing research for this book, I was frequently surprised by the relative ignorance of professionals about the issues and insights of other disciplines. I was also unnerved by my own lack of useful knowledge about other disciplines. Yet with the massive information overload in modern society, interdisciplinary explorations are too inefficient without a guide to help navigate through the mountains of details. Knowing of the existence of other disciplines is very different from extracting relevant information from them. Interdisciplinary explorations of difficult problems are now commonplace, but their varying degrees of success suggest that there are, as yet, no clear answers to the questions of how or when to invest intellectual energy in such activities. However revolutionary its possibilities, crossing into foreign disciplines is still an ad hoc activity that is sometimes productive and sometimes irrelevant. The pioneering work of Julie Thompson Klein (1990, 1996) and others on the properties of cross-disciplinary activities—interdisciplinarity—exemplifies the need for structuring intellectual collaborations across disciplinary boundaries. However, until interdisciplinarity become formalized, with proven predictive utility and methods, cross-disciplinary works such as this one must be considered as experimental. I hope that my efforts will illustrate the intellectual and professional utility that can result from fusing fragmented viewpoints into a composite picture.

Had I been able to write this book decades ago, I would have managed my career from a different perspective. I might have more accurately interpreted the pivotal events that portended major paradigm shifts in my discipline. For example, I would have seen the shift from solving an old engineering problem, artificially reproducing concert hall acoustics, to inventing experiences without constructing buildings. I would have recognized that artificial reverberation was a subset of musical space, which itself is a subset of aural architecture, which depends on auditory spatial awareness. None of these connections was apparent during my career. Unknowingly, I was a member of an expanding generation of aural architects: electroacoustic designers who were liberating auditory space from its physical roots. My own professional history illustrates how I reacted to changes within my discipline without being aware of those events that marked fundamental shifts.

Because, as one of its many parents, I was present at the birth of the new discipline of digital audio, I had the opportunity to observe how it progressed from infancy to adolescence, and then to maturity. During this progression, the original discipline spawned new disciplines. Many died, but a few thrived to nurture a new generation of disciplines. Although the original parents of digital audio are long retired, and although their grandchildren are thriving, the attitudes of the current generation are still strongly influenced by the values of their parents.

Aural architecture belongs within a family tree of disciplines having branches that spread broadly across thousands of generations. The first marriage of visual and auditory art occurred when paleolithic painters discovered that their paintings of hoofed animals were more intense if they were located in caves producing echoes. Most important, like biological evolution, aural architecture has its own rules for survival, mutation, reproduction, and extinction.

Although the generations of artists, scientists, and engineers who contributed to aural architecture built on the legacies of previous innovators, such legacies are often taken as a given. Yet ignoring legacies does not reduce their impact. Indeed, innovative behavior is itself a reaction to these powerful cultural forces. As much as I take pride in having managed my own career, my activities were actually a response to cultural shifts that elevated the importance of computer technology and audio entertainment within the wider society.

Like our prehistoric ancestors who created wall paintings in resonant caves, my colleagues and I use available tools to create a listening experience for some personal and social purpose. The nature of that experience remains rooted in inherited culture and evolutionary biology. Although the supporting technology moves forward, the experience of aural architecture moves sideways in sweeping spirals. Even after having traveled a long distance, we are not far from the core experience of all peoples. We cannot discern the nature of that movement close-up. We need to take an aerial view to see the larger picture.

As an intellectual travelogue, this book is my attempt to overcome an inadequate education. I enjoyed the challenge of integrating and expanding a vast array of intellectual fragments into a single, broad, and coherent theme: the experience of aural architecture. I also included some speculations because the trip could not be completed without also constructing bridges over the uncharted waters of unexplored topics and missing research.

I take complete responsibility for any errors in fact and logic, and for any speculative foolishness that leaked through the review process. Given the scope of this book, I could not become an expert on the dozens of disciplines that are part of auditory spatial awareness and its application to aural architecture. I hope other scholars will clarify discussions that are either incomplete or debatable, thereby improving, correcting, applying, and extending what is necessarily only a beginning.

Spaces Speak, Are You Listening?

1 Introduction to Aural Architecture

We shape our buildings, and afterward our buildings shape us.
—Winston Churchill, 1943

Architecture, which has been called the "mother of all arts," is concerned with the design, arrangement, and manipulation of the physical properties of a space. Unlike other art forms, architecture provides spaces for the daily activities of life; when more than simply utilitarian, it also appeals to our aesthetic sensibilities. By choosing and combining materials, colors, and shapes, architects embed their respective artistic messages in structures that we see, hear, and feel. Like poets with their specialized language, architects communicate their worldview with a vocabulary of spatial elements that often contain symbolic meaning reflecting their culture.

To communicate the artistic, social, emotional, and historical context of a space, however, architects almost exclusively consider the visual aspects of a structure. Only rarely do they consider the acoustic aspects. The native ability of human beings to sense space by listening is rarely recognized; indeed, some people think such an ability is unique to bats and dolphins. But sensing spatial attributes does not require special skills—all human beings do it: a rudimentary spatial ability is a hardwired part of our genetic inheritance. For example, when blindfolded, nearly all of us can approach a wall without touching it just by attending to the way the wall changes the frequency balance of the background noise. Similarly, the sounds of our footsteps hint at the location of stairs, walls, low ceilings, and open doors. To make this more obvious, walk through your home while listening to loud music through headphones; then do it again without the headphones. Notice how the clear sounds of your shoes on uncarpeted stairs provide navigational confidence, especially when your eyes are focused elsewhere. When crawling through underground caves, spelunkers can gauge the depth of a dark passageway by its resonances. But even nonspelunkers have acoustic awareness. It is available to all of us.

Observing that ordinary people can hear passive objects and sense spatial geometry requires an explanation. As a simple illustration of how we hear an object that itself does not produce any sound, consider a flat wall located at some distance. When the sound wave from a hand clap is reflected from that distant wall, we hear the reflection as a discernible echo. The distance to the wall determines the delay for the arrival of the echo, the area of the wall determines the intensity, and the material of the wall's surface determines the frequency content. These physical facts relate only indirectly to perception. Our auditory cortex converts these physical attributes into perceptual cues, which we then use to synthesize an experience of the external world. On the one hand, we can simply hear the echo as an additional sound (sonic perception) in the same way that we hear the original hand clap (sonic event). On the other hand, we can interpret the echo as a wall (passive acoustic object). The echo is the aural means by which we become aware of the wall and its properties, such as size, location, and surface materials. The wall becomes audible, or rather, the wall has an audible manifestation even though it is not itself the original source of sound energy. When our ability to decode spatial attributes is sufficiently developed using a wide range of acoustic cues, we can readily visualize objects and spatial geometry: we can "see" with our ears.

A real environment, such as an urban street, a concert hall, or a dense jungle, is sonically far more complex than a single wall. The composite of numerous surfaces, objects, and geometries in a complicated environment creates an *aural architecture*. As we hear how sounds from multiple sources interact with the various spatial elements, we assign an identifiable personality to the aural architecture, in much the same way we interpret an echo as the aural personality of a wall. To illustrate that we are aware of aural architecture, consider displacing familiar sounds to unfamiliar environments. Transported to an open desert, urban traffic would not have the aural personality of a dense city environment. Moved to a forest, a symphony concert would not have the aural impact, intimacy, and immediacy of a concert hall. Nor could the aural personality of singing in the bathroom, which takes advantage of the resonances of small spaces, be duplicated in a large living room. In each contrasting space, even if the sound sources were to remain unchanged, the aural architecture would change. Every space has an aural architecture, which will be defined more extensively in chapter 2.

In addition to providing acoustic cues that can be interpreted as objects and surfaces, aural architecture can also influence our moods and associations. Although we may not be consciously aware that aural architecture is itself a sensory stimulus, we react to it. We may experience a living room as cold or warm independent of its actual temperature, or a train station as lonely and forbidding independent of its actual appearance. The acoustics of a grand cathedral can create an exalted mood; those of a chapel can enhance the privacy of quiet contemplation; those of an elevator can produce the feeling of encapsulation and, in the extreme, claustrophobia. The acoustics of an open area can produce feelings of either freedom or insecurity.

Aural architecture can also have a social meaning. For example, the bare marble floors and walls of an office lobby loudly announce the arrival of visitors by the resounding echoes of their footsteps. In contrast, thick carpeting, upholstered furniture, and heavy draperies, all of which suppress incident or reflected sounds, would mute that announcement. The aural architecture of the lobby thus determines whether entering is a public or private event. When applied to a living room, those same acoustic attributes convey a different sense: cold, hard, and barren, as contrasted with warm, soft, and intimate. In a musical performance space, acoustic attributes can produce a blending of sequential notes, almost like chords. In certain religious spaces, they can produce a reverberation that conveys a sense of awe and reverence. As with all sensory aspects of architecture, cultural values and social functions determine the experiential consequences of spatial attributes. In different social settings, the same acoustic features have different meanings, which then influence the mood and behavior of the people in those settings.

Aural architecture, with its own beauty, aesthetics, and symbolism, parallels visual architecture. Visual and aural meanings often align and reinforce each other. For example, the visual vastness of a cathedral communicates through the eyes, while its enveloping reverberation communicates through the ears. For those with ardent religious beliefs, both senses create a feeling of being in the earthly home of their deity. Similarly, the visual elegance of a grand opera hall contributes to the artistry of the performance, and the aura of power in a governmental chamber contributes to the importance of speeches presented there. In these examples, because the aural and visual elements in the space are congruent, symbols and associations are shared.

Although we expect the visual and aural experience of a space to be mutually supportive, this is not always the case. Consider dining at an expensive restaurant whose decorations evoke a sense of relaxed and pampered elegance, but whose reverberating clatter produces stress, anxiety, isolation, and psychological tension, undermining the possibility of easy social exchange. The visual and aural attributes produce a conflicting response.[1]

Although multiple senses contribute to the formation of an internal experience of an external reality, the contribution made by listening varies greatly among individuals and cultures (Classen, 1993). Since listening with understanding depends on culture, rather than on the biology of hearing, auditory spatial awareness must be considered the province of sensory anthropology. To evaluate aural architecture in its cultural context, we must ascertain how acoustic attributes are perceived: by whom, under what conditions, for what purposes, and with what meanings. Understanding aural architecture requires an acceptance of the cultural relativism for all sensory experiences.

Sensory anthropologists study how social structures determine the use of the senses and the meaning of the resulting perceptions (Howes, 1991). In our scientific society with its emphasis on physical explanations, the categories for sensing the external world are mostly sorted by the combination of biological organs and physical stimuli: ears are for hearing sound, eyes are for seeing light, and skin is for touching surfaces (Ackerman, 1990). Yet even with this bias toward concrete labeling, our culture takes no notice of the many different kinds of information processing that actually compose a single sensory modality. For example, the tactile modality—touch—includes independent sensors for vibration, texture, temperature, movement, and so on. Our very concept of the senses arises from our cultural biases.

To illustrate the wide range of choices for labeling the senses and for understanding their relationship to social functions, consider a few examples from other cultures and subcultures. The Hausa people recognize only two senses: seeing and experiencing (Ritchie, 1991). In this culture, the vision sense is only a means for navigating the environment, and the experience sense encompasses intuition, emotion, smell, touch, taste, and hearing. The anthropologist Anthony Seeger (1981), in addressing cultural meaning of sensation, commented: "Just as time and space are not perceived by the vast majority of human societies as a regular continuum and grid, so the [sensorium] is rarely thought of in strictly biological terms. . . . The five senses are given different emphasis and different meanings in different societies. A certain sense may be privileged as a sensory mode." For example, Aivilik Eskimo natives do not describe space in visual terms (Carpenter, 1955) because their environment is an open expanse without visual markers. For this group, the nonvisual senses play a stronger role in their experience of space. Similarly, in many religious subcultures, their gods speak to their disciples rather than leave them written messages. Rehabilitation workers often report that blindness is less socially and emotionally burdensome than deafness. Some cultures revere the role of the blind seer who has learned to accentuate the gift of listening as a better means for "seeing" the future.

From this broad perspective, it is clear that hearing contributes to a wide range of experiences and functions. Hearing, together with its active complement, listening, is a means by which we sense the events of life, aurally visualize spatial geometry, propagate cultural symbols, stimulate emotions, communicate aural information, experience the movement of time, build social relationships, and retain a memory of experiences. To a significant but underappreciated degree, aural architecture influences all of these functions.

Let us digress briefly to clarify a few common words and concepts relating to sound. Over the years, some words have acquired meanings and associations that deviate from their dictionary definitions. *Acoustics*, from the Greek *akoustikos* and meaning that which pertains to hearing, now refers mostly to the behavior of sound waves (vibrations) in solids, liquids, or gases. Listening is not required, and may not even be possible, for underwater, ultrasonic, or high-pressure acoustics. Even when listening is expected, acoustic architecture uses the language of physics to describe sonic processes as phenomena that can be measured. To clarify how key terms are used in this book, the adjective *aural*, which parallels *visual*, refers exclusively to the human *experience* of a sonic process; *hearing*, to the detection of sound; and *listening*, to active attention or reaction to the meaning, emotions, and symbolism contained within sound.

Accordingly, *aural architecture* refers to the properties of a space that can be *experienced* by listening. An *aural architect*, acting as both an artist and a social engineer, is therefore someone who selects specific aural attributes of a space based on what is desirable in a particular cultural framework. With skill and knowledge, an aural architect can create a space that induces such feelings as exhilaration, contemplative tranquillity, heightened arousal, or a harmonious and mystical connection to the cosmos. An aural architect can create a space that encourages or discourages social cohesion among its inhabitants. In describing the aural attributes of a space, an aural architect uses a language, sometimes ambiguous, derived from the values, concepts, symbols, and vocabulary of a particular culture.

In contrast, an *acoustic architect* is a builder, engineer, or physical scientist who implements the aural attributes previously selected by an aural architect. Acoustic design manipulates physical objects, spatial geometries, and mathematical equations using the scientific language of physics. Because of differences in their perspectives, acoustic architects focus on the way that the space changes the physical properties of sound waves (*spatial acoustics*), whereas aural architects focus on the way that listeners experience the space (*cultural acoustics*). Although some individuals function as both aural and acoustic architects, the fundamental difference in the two functions is the distinction between choosing aural attributes and implementing a space with previously defined attributes.

We can sometimes identify the aural architect of a space, but far more frequently, aural architecture is the incidental consequence of unrelated sociocultural forces. Ancient cathedrals possess an aural architecture, without having had aural architects. Towns have an aural architecture that arises from their natural geography and topography, as well as from the uncoordinated construction of streets and buildings. Residential dwellings have an aural architecture determined by design traditions and construction budgets. The aural architecture of many modern spaces is created by architects, space planners, and interior designers with little appreciation for the aural impact of their choices. Living rooms, restaurants, and automobiles are examples of such spaces. Aural architecture thus exists regardless of how the acoustic attributes of a space came into existence: naturally, incidentally, unwittingly, or intentionally. For these reasons, the aural architect is most often not an actual person.

Even when the architects *are* actual people, however, aural architecture is not the exclusive domain of a handful of acoustic professionals who have an opportunity to design classrooms, concert halls, or churches. In a very real sense, we are all aural architects. We function as aural architects when we select a seat at a restaurant, organize a living space, or position loudspeakers.

To broaden the concept still further, aural architecture includes the creation of spatial experiences where a physical space does not actually exist, so-called virtual, phantom, and illusory spaces. While listening to recorded music in our homes, we experience a virtual space created by a mixing engineer who manipulated a spatial synthesizer in a recording studio. There never was a performance space. Defined as the design or selection of a spatial experience, without regard to the means of implementing that experience, aural architecture is as old as civilization, embracing the widest diversity of social and artistic examples in cultures that span thousands of years.

Even though aural architects are most often sociocultural forces rather than actual people, we can still examine how these forces influence spatial designs. Over the millennia, a series of progressive changes in the relationship between aural architecture and its social uses resulted from changes in artistic attitudes, in the prevailing theology, and in how the senses were used to experience physical and social environments. The difference between adapting a cave for a religious ceremony and designing a consumer home theater surround-sound system reflects not only advances in technology, but also changes in culture. Those who built cathedrals and those who designed virtual electroacoustic spaces were seldom aware of how their social context influenced their spatial creations.

Thousands of visual artists, civil engineers, architectural historians, and social scientists have created a comprehensive symbolic language and an extensive literature for visual architecture, whose intellectual foundation draws on archaeology, engineering, history, sociology, anthropology, evolution, psychology, and science. In contrast, even though aural architecture shares the same intellectual foundation, its language and literature are sparse, fragmented, and embryonic.

There are four principal reasons why this might be so. First, aural experiences of space are fleeting, and we lack means for storing their cultural and intellectual legacy in museums, journals, and archives. Second, for both cultural and biological reasons, the language for describing sound is weak and inadequate. Third, being fundamentally oriented toward visual communications, modern culture has little appreciation for the emotional importance of hearing, and thus attaches little value to the art of auditory spatial awareness. And fourth, questions about aural architecture are not generally recognized as a legitimate domain for intellectual inquiry; professional schools provide little or no training in physical acoustics, aural aesthetics, or sensory sociology.

Because aural architecture is not a recognized discipline, its concepts are not a significant part of our cultural and intellectual mainstream. When professional architects focus exclusively on the visual and utilitarian attributes of a space, they are reflecting a tradition that devalues listening. More significant, when listeners tolerate an environment whose acoustics damage their ears, their social relations, or both, they, too, are devaluing the aural experience.

There are, however, segments of our culture that take an interest in aural architecture. When given the freedom to choose the aural attributes of a spatial experience, audio engineers, composers, acoustic scientists, and spatial designers function as aural architects. There are conspicuous and representative examples of artists and architects who explicitly focus on aural architecture. The Finnish architect Juhani Pallasmaa (1996), who rejected the assumption of visual dominance, considered sensory architecture as an umbrella theme that explicitly included aural architecture. R. Murray Schafer (1977), in formulating the concept of the soundscape as a mixture of aural architecture and sound sources, created disciples who have passionately extended and applied his initial concept. Ted Sheridan and Karen van Lengen (2003) suggested that architectural schools should intentionally include aural considerations in order "to achieve a richer, more satisfying built environment." In their treatise on spatial acoustics, Hope Bagenal and Alex Wood (1931) recognized the social and cultural aspects of aural architecture.

The aural architecture of musical spaces, unlike that of religious, political, and social spaces, is well recognized and extensively researched. When a musical space is considered to be an extension of musical instruments, rather than an independent manifestation of aural architecture, it becomes a tool to be used by composers, musicians, and conductors. Musical spaces are intentionally designed for specific audiences that have acquired sensitivity and appreciation for spatial acoustics, as these bear on their experience of music and voice. Musical spaces are also an interesting application of aural architecture because music has played a role far beyond that of entertainment, a role anchored in history, culture, evolution, and neurobiology. Like architecture, music is also a language of aesthetics, spirituality, patriotism, and especially the emotions of joy, love, pride, and sorrow. Although they do not identify themselves as such, many aural architects are found within audio and musical subcultures. Fortunately, we can apply our knowledge of musical spaces to other kinds of space as well.

Even within a given culture, listeners are not homogeneous with regard to how they use their sense of hearing. When, however, listeners share a similar relationship to some aspect of aural architecture, they become a relatively homogeneous group, an *auditory subculture*. We find auditory subcultures both within a culture and across cultures. Active users of particular kinds of acoustic space who share goals, motivation, genetic ability, and opportunities often become a unique auditory subculture. They teach themselves to attend to the particular spatial attributes they consider important. From this perspective, those with an active interest in music—performers, composers, and listeners—form an auditory subculture with an enhanced sensitivity to the aspects of aural architecture that apply to their music. Those blind individuals who orient and navigate a space by listening to objects and geometries form another auditory subculture. The experience of aural architecture depends on the individual's subculture.

A related kind of social grouping is the professional subculture whose members study, design, or manipulate spatial attributes for the purpose of creating aural experiences for others. Often these professionals do not realize they are functioning as aural architects. To name but a few, such subcultures include ancient shamans who performed ceremonies in caves, recording engineers who use virtual space simulators as part of the production process, cinema film directors who match or contrast the visual and auditory experience of space, social psychologists who study human behavior, and designers of religious ceremonial spaces who want the congregation to feel a connection with their deities and their heavenly cosmos. Each of these professional subcultures is unique in terms of its educational training, cultural beliefs, specialized goals, economic rewards, and private agendas. Aural architecture is mostly the result of the values and biases in these professional subcultures.

In one sense, the concept of aural architecture is nothing more than an intellectual edifice built from bricks of knowledge, borrowed from dozens of disciplinary subcultures and thousands of scholars and researchers. I did not create these bricks, all of which appear in published papers. When fused together into a single concept, however, the marriage of aural architecture and auditory spatial awareness provides a way to explore our aural connection to the spaces built by humans and to those provided us by nature. This book is the story of that marriage over the centuries in a variety of cultures and subcultures, and today's artists and scientists are its children.

Individuals who use spaces for a particular purpose, and individuals who design spaces for a particular use, often acquire a heightened sensitivity to particular aspects of aural architecture. Auditory spatial awareness is a multiplicity of related but independent abilities. Although evolution provided our species with the basic neurobiology for hearing space, each sensory and professional subculture emphasizes only a subset of this endowment. Conversely, those who are neither users nor designers of aural architecture are unlikely to display more than the basic abilities to hear space. Furthermore, cultures without any appreciation for aural experiences are unlikely to develop and support those subcultures with an interest in aural architecture.

Spaces Speak is written for three types of reader. First, for those professionals who possess an expertise in one of the supporting disciplines, the discussions provide an overview into related, and possibly unfamiliar, areas. Second, for those with a general curiosity, the discussion integrates the collective knowledge of many artists, designers, and scientists into an accessible presentation of aural architecture. And finally, for those with a love of music, the discussions explore aural architecture as an extension of the auditory arts.

As an intellectual mosaic, *Spaces Speak* explores auditory spatial awareness and its relationship to aural architecture. Discussions move from cave acoustics to home theater audio systems, from evolution to neurobiology, from physics to perception, from science to engineering, from physical to virtual spaces, and from physical sound to emotional response. This book does not require expertise in any of the relevant specialties, and it will not make its readers experts. Rather, it is intended to provide a means of capturing and fusing disparate knowledge into a common framework: the human condition as seen through one particular prism, the aural architecture of spaces.

2 Auditory Spatial Awareness

The life that happens in a building or a town is not merely anchored in the space but made up of the space itself.

—Christopher Alexander, 1979

Auditory spatial awareness is more than just the ability to detect that space has changed sounds; it includes as well the emotional and behavioral experience of space. For example, detecting reverberation is different from responding to it. Listeners react both to sound sources and to spatial acoustics because each is an aural stimulus with social, cultural, and personal meaning. To create a foundation for aural architecture, we must explore these meanings. Depending on the physical design and the cultural context, aural architecture can stimulate anxiety, tranquillity, socialization, isolation, frustration, fear, boredom, aesthetic pleasure, and so on. Although there is a vast body of scholarly work both on the physical acoustics of enclosed spaces and on perceiving acoustic parameters, the literature is relatively silent on the subject of how people experience aural space. We know much about measuring acoustic processes and sensory detection, but less about the phenomenology of aural space.

A complex amalgam of spatial attributes, auditory perception, personal history, and cultural values, auditory spatial awareness manifests itself in at least four different ways. First, it influences our social behavior. Some spaces emphasize aural privacy or aggravate loneliness; others reinforce social cohesion. Second, it allows us to orient in, and navigate through, a space. Hearing acoustic objects and surfaces supplements vision or, in the case of darkness or visual disability, actually replaces vision. Third, it affects our aesthetic sense of a space. Devoid of acoustic features, a space is as sterile and boring as barren, gray walls. Just as visual embellishments can make a space aesthetically pleasing to the eye, so aural embellishments can do so for the ear, by adding aural richness to the space. Fourth, auditory spatial awareness enhances our experience of music and voice. The physical acoustics of a musical space merge with sound sources to create a unified aural experience. Space then becomes an extension of the musical or vocal art form performed within it.

These four aspects of auditory spatial awareness correspond to four aspects of aural architecture: social, navigational, aesthetic, and musical spatiality. To some degree, every space manifests all four, even though only one or two aspects typically dominate the design or selection criteria. A space designed for music can be examined for its aesthetic or navigational attributes, and a space designed for navigation can be evaluated for its musical and social attributes. Investigating auditory spatial awareness establishes a foundation for the language of aural architecture. This chapter focuses on the social, navigational, and aesthetic spatiality of aural architecture; chapters 4 and 5 focus on the musical spatiality of real and virtual spaces.

Introduction to Hearing Space

To discuss auditory spatial awareness, we first need to explore the basics of listening. What does it mean to be aware of sound or spatial acoustics? Although *awareness* implies that the listener is conscious of sound, the cognitive process of interpreting sound is highly complex and incompletely understood. We need an intellectual framework that distinguishes the different manifestations of experiencing the environment. Unfortunately, the cognitive language of consciousness is ill defined, ambiguous, philosophical, and subject to continual revision. Rather than becoming mired in the swamp of competing ideas, let us begin by making certain simple, yet functional distinctions. Aural awareness progresses through a series of stages: transforming physical sound waves to neural signals, detecting the sensations they produce, perceiving the sound sources and the acoustic environment, and finally, influencing a listener's affect, emotion, or mood. Notice that this conceptualization provides a continuum from the physical reality of sound to the personal relevance of that reality. Let us examine this continuum.

A Functional Model of Auditory Awareness
Physical sound is a pressure wave that transports both sonic events and the attributes of an acoustic space to the listener, thereby connecting the external world to the listener's ears. Because the physics of sound is complex, transmission includes such processes as reflection, dispersion, refraction, absorption, and so on, all of which depend on the acoustic properties of the space. When arriving at the inner ear, sound waves are converted to neurological signals that are processed by the brain; the external world is connected to inner consciousness.

At one extreme of auditory awareness, there is only raw sensation. It involves detecting an auditory stimulus that has no meaning or affect, as for example, laboratory signals composed of pure tones, transient clicks, or noise bursts. If we ignore minor physiological differences, there is little behavioral variability among individual listeners when detecting such sounds. Cognitive involvement and memory are minimal; neither personality nor culture strongly influences the ability to detect, discriminate, or localize such sounds. They are so pure that psychophysicists find them useful for modeling the neurological properties of the auditory system in all mammalian species. Raw sensation is predominantly a biological property of a species.

Farther along the continuum, the next stage is perception. Cognitive processes, containing the individual listener's personal history, transform raw sensation into an awareness that has meaning. Perception includes cultural influences and personal experiences. For example, understanding speech requires knowledge of the words—meanings and conventions specific to the culture—in order to decode sounds. Similarly, recognizing that a space, not a vibrating string, creates reverberation requires experience with both strings and spaces. When a culture provides consistent exposure to a class of sounds, perception is reasonably consistent among listeners within that culture. Perception does not require the sound to have any relevance to life; a spoken sequence of random numbers can be perceived as linguistic objects, a sequence of musical notes can be perceived as a melody, and a sound source can be localized. Perception is predominantly a property of cultural exposure.

At the far end of the continuum, we find high-impact, emotionally engaged listening. In this case, sounds produce a visceral response, a heightened arousal (Thayer, 1989), and an elevated state of mental and physical alertness. Such sounds have personal meanings and associations for the listener. For example, the sound of a violin in a small space may generate distress in a listener who associates that sound with hours of coerced practice as a child. A Swiss villager might become homesick when listening to the sounds of alphorns echoing through the mountains. In many situations, a listener may not be consciously aware of the affect induced by listening to engaging sounds or spaces. With emotionally active listening, listeners might burst into tears of sadness or feel overwhelmed with ecstatic pleasure. In some cultures, certain kinds of music are so powerful they are used to create trances, altered states of consciousness (Rouget, 1985; Besmer, 1983).

As opposed to exploring sensation or perception in a laboratory context, investigating the affective aspects of aural architecture is relevant to real experience in real life. Unfortunately, affective reactions are difficult to study for many reasons. An individual listener's history and temperament, rather than particular culture and universal biology, govern meaning. Moreover, a listener may not have the linguistic skill to describe affective reactions, and a researcher may not have an objective means for observing neurological responses corresponding to emotions. Nevertheless, we are mostly interested in listening experiences that have the *capacity* to produce either an overt or a subliminal affect. Overt affect corresponds to strong feelings, *emotions*, whereas subliminal affect corresponds to subtle arousal, *moods*.

Even though a listener may clearly perceive and decode the information in a sound, the experience may produce neither overt nor subliminal affect. There are at least two reasons why listening might be experienced as irrelevant. First, the sound and acoustic space may be without meaningful content for a particular listener; there is nothing being communicated. Exposed to "music" generated by a computer from a concatenation of tone oscillators in an empty space, you may find the resulting sound ("music" and "space") sterile and boring. The computer algorithm is not communicating anything of emotional significance to you. Second, the listener may not be paying attention to the sound and space. Even if these are emotionally charged, you may not be engaged in focused listening; indeed, you may have tuned out altogether, ignoring all sounds while attending to daydreams. In both cases, sound is nothing more than background noise, quickly forgotten.

As understood here, auditory spatial awareness includes all parts of aural experience: sensation (detection), perception (recognition), and affect (meaningfulness). From the broadest perspective, auditory awareness means only that there is some neurological reaction to spatial acoustics, including both conscious and unconscious changes to the listener's body state.[1] Thus you are understood to be aware of an acoustic space when listening to its aural architecture raises or lowers your blood pressure, even though you are not consciously aware of that reaction. With this definition, monitoring brain waves may be the only reliable means of observing a listener's reaction to aural architecture.

Making a distinction among sensation, perception, and meaning is especially important because much of the literature confuses or intermingles these concepts. Whereas physical and perceptual scientists emphasize sensation and perception, artists and social scientists emphasize perception and meaning. When interpreting scholarly research and applying the result to real life, ask yourself whether an assertion is addressing *detectability*, *perceptibility*, or *desirability*. Detectable attributes may not contribute to perceptual attributes, and perceptible attributes may not be emotionally or artistically meaningful. Furthermore, affect can be at once meaningful and undesirable. As discussed in chapter 8, neurological research suggests that detection, perception, emotion, and consciousness involve different brain substrates.

To a large degree, manifestations of awareness involve the active participation of the listener—hearing or ignoring spatial acoustics. Earlier, we described the awareness of an echo off a wall as either the perception of an additional sound or the perception of a physical wall. With training, a listener can consciously switch between these two choices. More commonly, there are additional choices. For example, when listening to an oral interchange in an auditorium, you can attend to the informational content, the geographical dialect of the speakers, their emotional attitudes and personal biases, their location relative to you, or the spatial acoustics of the environment. There are at least five distinct channels of information using a single sensory system. Consciously choosing a channel requires practice and motivation. Auditory spatial awareness is just one of many possible aural channels, which itself is composed of multiple channels, comprising numerous subchannels.

Soundscape as Sonic Events and Aural Architecture

When you listen carefully with your eyes closed, when you attend to the feel of a specific acoustic space, be it concert hall, cathedral, restaurant, kitchen, or forest, you engage in *attentive listening*—intensely focusing on the sounds of life in the immediate environment. Take a moment to visualize the world from its sounds: the songs of birds heralding the onset of spring in a forest park, the creaking of a rocking chair on a front porch, the laughter of children at the playground, or the sound of music blaring from an open window. Solely through sound, an entire environment, complete with memories and emotions, comes alive. Indeed, we feel included in the life of the *soundscape*: the auditory equivalent of a landscape.

Sounds signify events taking place: babies crying, brakes screeching, birds singing, people talking, and water falling. All sounds are the result of dynamic action, periodic vibrations, sudden impacts, or oscillatory resonances. Sounds produced by mechanical activities may dominate the personality of a soundscape. Listening is an important human activity just because it creates an intimate connection to the dynamic activities of life, both human and natural. In fact, from a psychological perspective, we do not so much hear sound as perceive sonic events, with sounds transporting events into our consciousness. Whereas landscapes can be comparatively static and sometimes almost lifeless, soundscapes, of necessity, are dynamic: they require animated activities to produce sonic events. In tribal societies where survival is a continuous struggle against hidden events, soundscapes are frequently more relevant than landscapes (Feld, 1996). Thus soundscapes are alive by definition; they can never be static.

Although we usually think of a soundscape as a collection of sonic events, it also includes the aural architecture of the environment. The experience of listening to a sermon in a cathedral is a combination of the minister's passionate articulation and spatial reverberation. A performance of a violin concerto combines the sounds of musical instruments with the acoustics of the concert hall. The soundscape of a forest combines the singing of birds with the acoustic properties of hills, dales, trees, and turbulent air. To use a food metaphor, sonic events are the raw ingredients, aural architecture is the cooking style, and, as an inseparable blend, a soundscape is the resulting dish.

Those who engage in attentive listening rarely separate a soundscape into its components: the sonic events and their modification by the aural architecture. Although, to discuss aural architecture, we must make that separation, this leads us to two contrasting perspectives. On the one hand, just as light sources are required to illuminate visual architecture, so sound sources (sonic events) are required to "illuminate" aural architecture in order to make it aurally perceptible.[2] On the other hand, we can think of aural architecture as simply modifying our experience of sonic events, such as when reverberation of a concert hall elongates musical notes. Both perspectives are accurate. But traditionally, spatial acoustics have been considered in terms of how they modify sound waves, rather than as something to be experienced separately. The opposite is true for visual architecture, where illumination is of secondary importance to spatial objects and their properties.

Aural architecture requires the presence of sound sources to illuminate the space, and a soundscape is also the same combination of space and sources. What then is the difference between them? With a soundscape, the sounds are important in themselves, as for example, birds singing or people talking, whereas with aural architecture, those same sounds serve only to illuminate it. The personality of a soundscape includes the personality of sounds as well as the personality of the aural architecture illuminated by those sounds. Aural architecture emphasizes sound primarily as illumination, whereas a soundscape emphasizes sound in itself. The distinction is subtle and may not always be relevant.

Architecture, like a giant, hollowed-out sculpture, embeds those who find themselves within it; it is to be apprehended from within. But that embedding differs between the aural and visual modalities because human activities produce sound but not light. Musicians make music, blind individuals tap their canes, diners make conversation, and children shout to one another. In each case, the environment responds as if it were a partner in an auditory dialogue. Snap your fingers, and the space responds. Whistle a note, and the space returns one or more echoes. Sing a song, and the space emphasizes particular pitches. Remain silent, and the space remains silent. The listener is immersed in the space's aural response, and there is rarely a discernible location for that response. By responding to human presence, aural architecture is dynamic, reactive, and enveloping. In contrast, because human beings do not possess an intrinsic means for generating light, a space does not react to our visual presence, which manifests itself there only through interrupted or reflected light—as shadows or mirror images.

The duality between aural and visual architecture diverges still further when we consider that sound is actually more complex than light. Although both have a frequency spectrum and amplitude intensity, *time* is central to sound but mostly irrelevant for vision. Sound and light waves have dramatically different velocities: sound waves traverse a space with perceptible speed; light waves move instantaneously. As either echoes or reverberation, the sounds of the past, at least on the timescale of seconds, exist concurrently with the sounds of the present; by encapsulating air, the interior surfaces of the enclosed space preserve sonic energy as it slowly dissipates. In contrast, visual architecture never modifies our experience of time because light illumination dissipates instantaneously regardless of the number of reflections. Turn off a light source, even in a mirrored room, and abruptly the space becomes dark. Turn off a sound source, and the space continues to speak. The time dimension of sound produces a complex response to sonic illumination, and we hear aural architecture by the way that the space changes a sound's spectrum, intensity, and *temporal* sequence. In comparison with vision, hearing is orders of magnitude more sensitive to temporal changes. In a very real sense, sound *is* time.

There are other parallels and contrasts between sonic and visual illumination of aural and visual architecture. Just as you cannot see visual objects without light, so you cannot hear aural objects without sound. Yet the visual details of most spaces are illuminated with sufficient sunlight or artificial lighting to make them readily apparent, whereas the aural details of a space are seldom illuminated with a full range of sounds (the space would be very noisy), and thus are not readily apparent. Indeed, full sonic illumination of aural architecture requires a mixture of continuous and transient energy over a wide range of frequencies, amplitudes, and locations. Spatial objects, surfaces, and geometries require extensive sonic illumination in order to excite such physical processes as interference, reflections, shadowing, dispersion, absorption, diffraction, and reverberation. You cannot hear the presence of a telephone pole or a partly open door unless background (sonic) illumination excites many of those physical processes. Sonic illumination is typically an artifact of some social activity, such as a concert, lecture, or traffic in an urban environment. Yet, when a space is exposed to full sonic illumination and you have sufficient cognitive skill to interpret the multiplicity of acoustic cues, you can *aurally visualize* passive acoustic objects and spatial geometry.[3]

Because experiencing sound involves time and because spatial acoustics are difficult to record, auditory memory plays a large role in acquiring the ability to hear space. Whereas comparing the visual architecture of two spaces through pictures does not place a burden on short-term memory, comparing the aural architecture of two spaces involves both the unreliability of auditory memory and the time required to travel from one space to another. Spatial simulators, which permit ready comparison of the aural architecture of two different spaces, obviate the need to travel, but only a few professionals have access to such tools, and they yield only approximations. Everyone else is burdened with remembering aural architecture over a span of at least minutes and perhaps hours or days, if not longer. There is no aural equivalent to a picture book of visual architecture, which can be studied at leisure. To preserve our experience of aural architecture, most of us depend on long-term memory, which, without extensive training and practice, is even more unreliable than short-term memory. For this reason, few of us accumulate aural experiences of spaces; our culture cannot readily communicate its aural architectural heritage. Furthermore, when we visit a space, our aural experience depends on sonic events, which result from inconsistent human activities producing unpredictable sonic illumination. The personality of a courtyard late at night is not the same as it is at lunchtime. Similarly, a crowd of people in a space alters its spatial acoustics, as when a concert hall is filled with sound-absorbing listeners.

Absent training, our experience of aural architecture is fragile and perishable. Yet, however difficult to recall, describe, reproduce, or even study, aural architecture can elevate or depress our affective responses—it bears directly on our sense of: privacy, intimacy, security, warmth, encapsulation, socialization, and territoriality. It changes our behavior as individuals and influences the social structure of our groups.

Examples of Common and Unusual Spaces

Just as silence gives us a better appreciation for sound, and just as darkness is a prerequisite for understanding light, so "spacelessness" highlights the experience of a real space. Although not readily available, there are real environments that exhibit auditory spacelessness to varying degrees. Being suspended 300 meters (1,000 feet) in the air from an imaginary skyhook would be an obvious example of such an environment. Its acoustic space is without sonic reflections, resonances, or any object to influence sound waves. A more accessible environment that exhibits an approximation to spacelessness is a suburban town after a heavy winter snowstorm. A thick blanket of snow, which absorbs sonic energy, prevents the objects it covers from influencing sound waves. As if hanging in air from a skyhook, an individual in a snowy soundscape only hears direct sounds; the space approaches the conditions of an echo-free (anechoic) environment.

Scientists often use an anechoic chamber to conduct scientific experiments, and many acoustic laboratories have constructed such spaces with varying degrees of absorption and isolation. The highest-quality research chambers are relatively large, perhaps 2,000 cubic meters (72,000 cubic feet), and their six surfaces are covered with fiberglass wedges up to 1 meter (3 feet) in length. The example in figure 2.1 shows a typical chamber, where 99.9 percent of the incident sound waves are absorbed by the wedges. A wire-mesh floor allows for walking but is acoustically transparent, as if aurally absent. A properly designed anechoic chamber permits an experience that is similar to hanging in the sky. In addition, thick concrete walls and a floating foundation prevent external sounds and vibrations from entering the chamber. From an aural perspective, an ideal anechoic chamber is completely silent and entirely "spaceless."

Forty years after entering an anechoic chamber for the first time, I still remember my strange feelings of pressure, discomfort, and disorientation.[4] Some people report an initial feeling of nausea in such an environment. The aural experience of spacelessness in an anechoic chamber sheds light on a number of aspects of spatial awareness. First, spacelessness breaches a perceptual boundary. The combination of sound isolation and absorption reduces background sound to a level that no longer masks the sound of a listener's beating heart or flowing blood. The activity of the organs enclosed within the listener's body thus becomes part of the listener's acoustic space. Second,

Figure 2.1
View of an anechoic chamber with sound-absorbing wedges and a wire-mesh floor. Courtesy of Roger Russell of McIntosh Laboratories, Birmingham, New York.

because the chamber's absorption of incident sound is not 100 percent effective at the lowest frequencies, listeners experience those inaudible spectral components as ill-defined pressure. Third, absent any reflective surface, listeners experience speaking, clicking, and other familiar sounds as dull, strange, and remote. Except for anechoic environments, all normally habitable spaces on earth include at least one reflective surface, the ground, or its equivalent. Fourth, listeners are made immediately uneasy or anxious by the disorienting sensation of the chamber's unexpected acoustics, which produce strong affective responses. Finally, however strong, this disorientation passes with repeated exposure to spacelessness. Although they never forget their initial experience, those who work in an anechoic chamber become accustomed or indifferent to its unique strangeness.

More typically, an open meadow is the most accessible approximation to space-lessness. It is neither totally quiet nor totally lacking in spatial attributes, having, for example, ground reflections. Nevertheless, the absence of other nearby surfaces to reflect sound can exacerbate a feeling of agoraphobia in those who fear open spaces. In an open field, we hear the absence of enclosing boundaries.

At the other extreme, consider the experience of very small spaces, sitting inside a small closet, for example, or having a box over your head. In these cases, even with your eyes closed, you feel the proximity of the walls, the confinement of encapsulation, which in the extreme, seems like lying in a coffin. The perceptible and unmistakable sensation of nearby walls is created by elevated low-frequency sounds, and by the presence of strong resonances.

Consider from both an aural and a visual perspective, two conceptual variants of a small space. The first variant replaces the solid walls with heavy but clear glass such that the visual scene is now open (unobstructed) while the auditory experience remains that of a box. A second variant replaces the walls with an acoustically transparent surface constructed with open-wire mesh and visually opaque cloth, so that sound travels though it as if it were not present. The auditory scene is open (unobstructed), even as the visual experience remains that of a box. Most listeners find that the feeling of encapsulation is weaker with surfaces that are acoustically transparent but visually opaque. Sound transparency removes the sense of solidity, as if you could leave the space at any time. No matter how constructed, an acoustically transparent wall feels insubstantial. Moreover, with acoustic transparency, the auditory channel, which supports voice communications, is always open, whereas visual communications through a glass partition requires the voluntary control of the point of gaze. The size and properties of an aural and visual space need not be consistent.

The experience of extreme spaces such as anechoic chambers and small enclosures demonstrates that we can "hear" space. Take a moment to mentally compare the following familiar spaces in a hypothetical "space-tasting" activity: a bathroom, an old-fashioned telephone booth, a sports arena, an elegant living room, a school auditorium, a Gothic cathedral, a tiny church, an unfurnished house, an airport lounge, a small passageway, an atrium, and a fast-food restaurant. Most of us can readily imagine the aural experience of these spaces, which suggests that we recognize their aural personalities.

Spatial awareness varies widely among listeners. Those with low to average awareness can vividly experience an acoustic space only when it is unfamiliar, contradictory, or unexpected, whereas those with elevated awareness can accurately remember and describe the aural personality of even ordinary spaces.

The Social Components of Aural Architecture

Let us now focus on how acoustic spaces influence our sense of social cohesion by extending the premise advanced by Steen Eiler Rasmussen (1959), R. Murray Schafer

(1977), and Juhani Pallasmaa (1996) that the experience of architecture involves all the senses. Although the idea is not new, only a few studies have explored the way in which multisensory architecture influences the inhabitants of a space. Because of differences both in light and sound and in the neurobiology of seeing and hearing, aural architecture is distinct from visual architecture, and each has the capacity to enhance or diminish social cohesion.

Experiential Attributes of Space

To begin our discussion of social spatiality, let us turn to its basic attributes: the perceived size and boundaries of a space. Rather than focusing on a space as being determined by physical boundaries, we will focus on intangible, experiential boundaries perceived by listening. In our social definition, the boundaries of an aural enclosure acquire their meaning from the social context.

Though size is a property of a space, our senses are not scientific instruments that measure physical parameters. As a rule, vision both decodes size as length, width, and height, and organizes distance by the way objects obscure one another or change their relative size. In contrast, hearing decodes size as the global metric of volume because sound permeates air as a fluid, flowing around objects and into crevices. We cannot see volume, but we can hear it. Aurally, we sense the volume of a large space by its long reverberation time and the volume of a small space by its sharp frequency resonances. Visually, we can sense volume only by mentally multiplying the three dimensions of a space.

A physical boundary is essentially a visual concept. An observer can see a small boundary even at distant locations, but a listener can hear a boundary only when large or nearby. For hearing, volume or area remains primary, and boundaries are secondary; for vision, the opposite is true. When collaborating and reinforcing each other, the aural and visual sensory systems combine their respective experience of size, merging volume and linear extent.

Because visual and aural boundaries are independent means of enclosing a space, our visual and aural experience of size, the space between boundaries, may not be consistent. For example, glass is an auditory partition but not a visual one, and a black curtain is a visual partition but not an aural one. With two kinds of spatial partitions, we also have two kinds of spatial areas—aural and visual. Only physical boundaries impermeable to both light and sound produce a consistent experience. But consistency is more the exception than the rule.

To understand spatial area and spatial boundaries, think of them as *experiential* concepts that are unrelated to physical partitions. Let us consider virtual partitions. Darkness creates a visual demarcation of a space, and background noise creates an auditory demarcation. We can neither see visual objects nor hear sonic events if they are on the other side of a virtual partition. For example, at a cocktail party with many conversations, we hear only conversations that are above the background noise. Other conversations are inaudible, as if in a neighboring room. The area where a conversation is audible is enclosed by a virtual boundary, thereby creating an experiential region.

The concept of virtual sonic boundaries leads to a new abstraction, *acoustic horizon*, the maximum distance between a listener and source of sound where the sonic event can still be heard. Beyond this horizon, the sound of a sonic event is too weak relative to the masking power of other sounds to be audible or intelligible. The acoustic horizon is thus the experiential boundary that delineates which sonic events are included and which are excluded. The acoustic horizon also delineates an *acoustic arena*, a region where listeners are part of a community that shares an ability to hear a sonic event. An acoustic arena is centered at the sound source; listeners are inside or outside the arena of the sonic event. An acoustic horizon is centered at the listener; sonic events are within the horizon of the listener. Every sonic event has an acoustic arena, and every listener an acoustic horizon. Regardless of the viewpoint, the connection between a sonic event and a listener forms an *auditory channel*. A channel shared among listeners provides social cohesion. The concepts of arena, horizon, and channel originated from the language of soundscapes (Truax, 2001), but they are especially relevant to the analysis of aural architecture. Physical boundaries are only one means of delineating a space, and they are not always the most useful for describing social interactions.

With multiple listeners and sonic events, an environment is a composite of multiple auditory channels that compete with each other. Two conversations across the same dinner table, each with its own arena, compete with each other. Arenas collide and intersect with each other, opening and closing channels, including and excluding listeners. For example, the sudden ringing of the telephone shrinks the acoustic arena for television sound, and a cessation of traffic noise enlarges the acoustic arena of chirping crickets. Sound sources engage in a kind of Darwinian combat; loud sounds claim more area for their arenas than soft sounds. Listeners experience this dynamic as enhancing or degrading their auditory channels; an aural architect can conceptualize and manipulate this interplay among changing arenas.

With this foundation, we define *acoustic arena* as the area where listeners can hear a sonic event (target sound) because it has sufficient loudness to overcome the background noise (unwanted sounds). When the target sound is too soft or when unwanted sounds are too loud, the listener is outside the arena of the target, or the target is beyond the horizon of the listener. Except for a shift in viewpoint, acoustic arenas and acoustic horizons are equivalent. Noise is important because it shapes both the target's acoustic arena and the listener's acoustic horizon. As the contemporary composer John Cage (1961) commented after entering an anechoic chamber for the first time, pure silence does not exist naturally. Ever-present background noise, however low, determines the boundary of an acoustic arena. Noise need not be overwhelming or bothersome to have a social impact on the inhabitants within their acoustic arenas.

Aural architecture is a major factor in determining the size of acoustic arenas. By blocking unwanted sounds from remote locations, physical boundaries enlarge an acoustic arena. At the same time, an enclosed space may produce echoes or reverberation, which listeners may experience as unwanted noise. In contrast to producing noise, the spatial design may concentrate the energy of a target sound in specific parts of the space, a form of acoustic amplification that increases the size of the acoustic arena. By changing the ratio of the target sound to unwanted noise, spatial acoustics determine the size and shape of the arena.

Although echoes and reverberation are the space's response to the target sound, we can think of the space as creating its own sonic noise by accumulating old and obsolete target sounds. With speech, for example, reverberation is the accumulation of dozens of previous syllables, often masking the current syllable. For this reason, public address systems in large reverberant spaces, such as older European railroad stations, are notoriously unintelligible. Electronic amplification of announcements simultaneously increases both the target sound and its reverberation without changing the ratio between them. Despite amplification, the station's overall acoustic arena remains unchanged. But loud announcements dramatically shrink acoustic arenas within it, such as the arena of two travelers in conversation.

Spatial acoustics can amplify the target sound without also amplifying noisy reverberation. When strong reflections from nearby surfaces appear at the listener shortly after the direct sound, they perceptually fuse with it, thereby increasing its loudness but not its reverberation. A spatial geometry that produces the necessary intensity in these early reflections increases the acoustic arena, a phenomenon we will explore in chapters 4 & 6. Only the late arriving reflections become arena-shrinking noisy echoes and reverberation. Similarly, by concentrating sound in a particular direction, walls, ceiling, and panels with curved surfaces focus sound on a specific location. A megaphone and a shotgun microphone both have long and narrow acoustic arenas.

Science museums typically demonstrate this effect with two parabolic acoustic mirrors[5] set 100 meters (330 feet) apart, as shown in figure 2.2. A speaker at one focus (region A) easily communicates with a listener at the other (region B). Two widely spaced areas are acoustically fused into a single arena. These two regions, though visually separate, aurally overlap. Distance always depends on the choice of sensory modality. In fact, in the 1930s, attempts were made to aurally connect England with France by constructing very large surfaces that would project sound across the English Channel. Curved surfaces, acting as sonic lenses, can dramatically enlarge the acoustic arena in the direction of the focused sound, making objects sound closer than they actually are. Curves have a strong influence on the size, shape, and location of acoustic arenas.

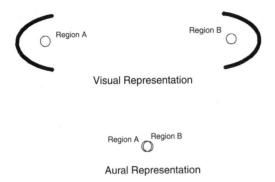

Figure 2.2
Two parabolic acoustic mirrors acoustically fuse physically separate regions.

The definitions of target and unwanted sounds are social concepts determined by those who occupy or live within an acoustic space, rather than abstract concepts determined by aural architects. In a musical space, this definition is explicit and static: sounds produced by musicians, and the space's response to those sounds, are both considered target sounds. When we listen to a musical performance, reverberation confers an aesthetic benefit. But in a different setting, when several groups are independently engaged in conversation, that same reverberation is detrimental. In a social setting, the definition of a target sound is indeterminate: any sound may be experienced as desirable by one listener and undesirable by another. For a mother, her baby's cry is important; to a student nearby, that same cry is noise. The determination of the relevant arena in a social context is complicated because, at any given time, a listener may arbitrarily select from one of many sources, interchanging desirable and undesirable sounds. The social application of aural architecture to acoustic arenas obviously requires a flexible definition of target and unwanted sounds.

The properties of acoustic arenas are determined both by the acoustic designers and by those who occupy or live within these arenas. Aural architecture is thus a social system rather than a simple application of physical science to spatial design. The properties of an arena are obviously influenced by the sonic behavior of the arena's occupants or inhabitants, as well as by the acoustics of the space. When an interior space is properly designed, its acoustics amplify desirable sonic events in *appropriate* areas of the space, while attenuating undesirable sonic events that would otherwise shrink the acoustic arenas within that space. Spatial acoustics are the aural architect's mechanism for changing the size, quality, and behavior of acoustic arenas when their occupants produce sonic events. Once built, the spatial design is relatively static and inflexible. Only the occupants remain free to change their arenas by modifying their social and sonic behavior. In this sense, aural architecture is adaptive and dynamic, even though the physical space may be static.

The following examples illustrate the concept of an acoustic arena. Inside an anechoic space, background noise is so low that biological activities within your body become audible, producing an arena that includes your ears. But when walking down a street with traffic noise amplified by reflections from buildings, you are deaf to the sounds of your footsteps. Your ears are outside the acoustic arena of your footsteps. In a noisy restaurant, dining partners seated across a table may nevertheless be outside each other's acoustic arena. In a public space, the introduction of background music automatically reduces the size of acoustic arenas within it. A concert hall is ideally a single shared acoustic arena for everyone in the audience. In contrast, when listening to music with headphones, you are injected into a recorded arena and simultaneously removed from your immediate social arena, which disappears. Headphones, like high background noise, produce social deafness and isolation from immediate surroundings. Functional deafness, unrelated to biological deafness, is the absence of all acoustic arenas.

The acoustic arena is the experience of a social spatiality, where a listener is connected to the sound-producing activities of other individuals. By manipulating the spatial design, the aural architect influences the relationship among the occupants of a space in a multiplicity of acoustic arenas. Because the occupants also determine the intensity of sonic events, however, spatial attributes are only one component of an acoustic arena. In each situation, both collectively and individually, those who occupy or live within a space have the prerogative to manipulate the size and shape of their acoustic arenas. Open the door, and you are now inside the acoustic arena for the activities taking place in the other room; close the windows, and you are no longer in the arena of children playing on the street. Turn up the volume of your entertainment system, and you are now beyond the acoustic arena of your telephone. Shout, and your arena overpowers the arenas of others nearby.

To appreciate the concept of an acoustic horizon fully, take a moment to become aware of sonic events within your current acoustic horizon, and then notice how they change as time progresses. Writing this chapter in my backyard, I am located in the acoustic arena of the quieter sonic events of life: chattering squirrels, passing cars, and people engaged in their daily lives. But when the gardeners arrive with their power equipment, their invasive noise puts me outside even the acoustic arena of my laptop's clicking keys. When I move to my office with its closed doors and windows, my acoustic horizon is now determined by the physical boundaries of that room, which isolates me from the living sonic events outside my office.

Personally, I prefer the acoustic arenas I encounter when sitting in my backyard; I am then part of a social and natural world. Acoustic arenas can be private or social spaces. Some of us prefer to live in isolated arenas; others prefer to embed ourselves in life's multiple acoustic arenas. The concept of acoustic arena is limited when we assume that spatial designers have exclusive control over the aural properties of a space.

It becomes more powerful, however, when we think of the arena's occupants as aural architects who shape their arena, rather than as passive occupants who simply use a space designed by an architect.

Space as an Acoustic Arena

An acoustic arena is an intermediary between acoustic science and social science. Architects, acousticians, and engineers make decisions about spatial geometry, construction materials, and building technique, all of which influence the size, shape, and aural attributes of various acoustic arenas. Sociologists, anthropologists, and psychologists analyze how those who occupy these arenas react to them in terms of mood and behavior as well as a sense of privacy or social cohesion. Aural architecture bridges these two disciplines. An acoustic arena has both social and physical properties, serving as a shared concept for both disciplines.

Most descriptions of spatial boundaries arise from visual appearances and social markers, cultural signals that delineate a transition not just in social function, but in political rights. Acoustic arenas do not respect those transitions. When the windows of a private house are open during a summer afternoon, the acoustic arena of activities in the public street extends well into the private spaces of the house, and to a lesser extent vice versa. Yet ownership and social rights associated with both the house and street remain independent of the state of the windows. If you are the owner of a private space, you control who can enter and what they can do, but when you open the windows, you relinquish your control over the access of sonic events. The sounds of public life freely enter a private space, and an animated family discussion becomes part of the public arena, heard by any passerby. An open window fuses visually and socially distinct spaces into a single arena.

The social consequence of an acoustic arena is an *acoustic community*, a group of individuals who are able to hear the same sonic events. Within such a community, an individual who *broadcasts* some signal or information makes a sonic connection to everyone within the arena. The broadcaster can change membership in the acoustic community only by changing the size of the arena. We whisper to make an acoustic arena small and private, and we shout to make it large and public, thereby determining who is inside and who is outside. Using an inverted definition of a private acoustic arena, Leo Beranek (1960) describes it as a space where excluded conversations are inaudible. In the strongest manifestation of a private acoustic arena, acoustic privacy is bilateral: outsiders cannot hear broadcasts emanating from within, and insiders cannot hear broadcasts emanating from outside. Given the importance of acoustic arenas, the following discussion explores the social consequence of public and private arenas.

The concept of an acoustic arena applies equally to environments of all sizes and types: small private rooms, concert halls, large townships, and natural soundscapes. We expand our understanding of aural architecture by considering not only buildings and auditoriums that were designed according to a specific criterion, but also natural and accidental environments occupied by people and other mammals.

Human beings are only one of many species that evolved a sense of territory based on the size of their acoustic arena. Marc D. Hauser (1997a), in his analysis of animal communications among numerous species, described the complexity and importance of vocal signaling in a shared acoustic environment. Broadcasting vocal signals in a complex environment, such as a forest, is one of the most effective means of communicating because the acoustic horizon can be far larger than the visual or olfactory horizon. Many species therefore evolved specialized auditory biology and social systems, adapting to their specific acoustic environment, to their acoustic geography—nature's aural architecture.

Early humans first adapted to nature's acoustic geography: open savannas and mountain ranges. Modern humans adapt, in a weaker way, to the acoustic architecture of urban centers and of enclosed dwellings and gathering places. Both natural and fabricated environments are relatively constant and difficult to change, but by changing their vocalization behavior, those who occupy them adapt, whether as individuals, groups, or species. Every acoustic arena is an application of the principle that social groups create or select an environment, which in turn, determines the resources of their acoustic arena. The vocal behavior of a social group creates an acoustic arena as a geographic region that supports an acoustic community. Large arenas allow for larger acoustic groups spread over a larger area.

No single acoustic arena illustrates, or even manifests, all possible uses of a space. Use depends on the prevailing cultural values. At a basic level, acoustic arenas can be sorted into three categories—natural, private, and public. Natural acoustic spaces, at least historically, were shared by competing species. Use of private acoustic spaces, because of controlled design and limited access, is often the prerogative of those with resources and power, both financial and political. Public acoustic spaces, with sonically porous boundaries that connect several physical spaces into a single acoustic arena, are influenced by a multiplicity of occupants, designers, and owners. Whether in natural, or private, public acoustic arenas, occupants adapt their behavior to the properties of the arenas available to them.

In our technological society, mechanical and electronic interventions have largely obviated the need for social cooperation in regulating the public arena. The earlier social rules for creating and controlling sonic events become less relevant when everyone exists within his or her own isolation chamber. Simply put, there is less need to regulate sounds outside of encapsulated spaces. Technology has produced high-quality private acoustic arenas, making public acoustic arenas less relevant, whether these are quiet or noisy. Simultaneously, the function of the public acoustic arena has been replaced by other means for achieving social cohesion, mostly in the form of electronic communications. But, even though we have far greater control over our electronic

than our acoustic connections with others, these do not have the intimacy and immediacy of an acoustic community in a public arena.

Acousticians such as William J. Cavanaugh and Joseph A. Wilkes (1999) have sufficient knowledge to create nearly complete sound barriers, from recording studios and
home theaters to mansions for the rich and famous, and partial sound barriers even for
public environments, such as highways and airports. Over the years, new materials, installation techniques, and manufacturing methods have resulted in incremental
advances in the technology of sound isolation. Any competent acoustician can design
a sonic isolation barrier that approaches the theoretical limits of physics. The decision
to produce an arena of one type or another is only a matter of economic and cultural
choice. In some cultures, physical boundaries are sonically porous, and acoustic arenas
depend on social agreement. For example, in Japan, paper screens serve as walls; in
tropical islands, windows and doors are always open to allow the air to circulate. In
the United States, doors are often hollow and have intentional gaps at the bottom.
This contrasts with some Germanic countries where habitable spaces have rubber gas-
kets on solid doors, tight seals on windows, and thick concrete walls.

Although, to a lesser extent than physical boundaries, sound absorption can also
subdivide a space into multiple acoustic arenas by creating virtual partitions. A concert
hall, using a minimal amount of sound absorption, remains a single acoustic arena. In
such a space, a noisy individual disturbs everyone. In contrast, a large living room with
deep-pile rugs and well-upholstered furniture supports many independent conversations in separate arenas. They are private because sound absorption suppresses reflections that would fuse with the direct sound to make it louder and propagate farther.
In such spaces, conversation is possible only when the speaker is facing the listener.
Sound-absorbing surfaces allow the occupants to dynamically partition a space into
separate arenas, whereas sound-isolation barriers prepartition a space without the
active involvement of the occupants.

Somewhat paradoxically, a high level of background noise also partitions a space
into many small acoustic arenas, creating a matrix of tiny virtual cubicles. For example,
in an industrial factory, workers communicate with each other in private acoustic arenas that may have an acoustic horizon of a few inches from mouth of speaker to ear of
listener. At all greater distances, factory noise masks the conversation. There can be
hundreds of small acoustic arenas in such a space, and each of them is as private as an
arena created with acoustic isolation barriers. There is, however, a major difference in
comfort between a small acoustic arena created with high background noise and one
created with sound isolation or absorption. Given the choice, few would choose a
high noise level as the preferred means of creating a small acoustic arena. Using sound
isolation or absorption is more expensive, but the arenas are more pleasant.

The history of human societies can be viewed through the prism of their acoustic
arenas and acoustic communities. Like air, water, and land, acoustic arenas are

resources to be shared, divided, exploited, regulated, and even polluted, by those with political and social power. Because allocation of acoustic arena resources mirrors the culture's values, examining them reveals the social dynamics of acoustic communities. The assumption that small private acoustic arenas are desirable is a value in modern society, an ethnocentric bias resulting, in part, from advances in technology and changes in social structure, not just from elevated concepts of personal freedom. To illustrate how modern culture devalues natural and public acoustic arenas, we first need to explore the inverse case: historical cultures in which large public arenas were preferred. The contrast between small private acoustic arenas and large public ones demonstrates what has been gained and lost.

Scholars who have studied the soundscapes of older townships have noted that particular sonic events—*soundmarks*—were the auditory counterparts of landmarks (Truax, 2001). Soundmarks are sounds that are unique and high status, often with important social, historical, symbolic, and practical value. The sounds of church bells, foghorns, railroad signals, factory whistles, fire sirens are examples. Every soundmark has its acoustic arena. In many towns, only those individuals who lived within the arena of the most important soundmarks were considered citizens of the town. Indeed, the size of a township was effectively determined by its acoustic geography—terrain features having noticeable acoustic effects, such as flat plains, dense forests, gentle hills, deep valleys, craggy mountain peaks—and by the vagaries of the local climate. These features determined the radiation pattern of soundmarks, and the resulting acoustic arena marked the boundaries of towns and their citizens.

The chiming clock is one of the best examples of a soundmark that enlarged and determined the community. In tracing the history of time keeping, Daniel J. Boorstin (1983) describes how sundials and hourglasses were superseded by clocks that chimed the hours, using a synchronized hammer to strike a bell, and that thereby replaced a small visual arena with a much larger acoustic arena. For more than a century after chiming clocks were invented, they did not have faces or hands, which would have required literacy, proximity, and illumination. Time no longer flowed; it was broadcast to the community at punctuated intervals. Audible time functioned day and night over an acoustic arena that depended on the intensity and height of the bell. The technology of bells therefore became central to sustaining a large township; bell construction and its supporting metallurgy acquired the status of a valued craft, a peacetime equivalent of building cannons.

Only the most prestigious and powerful institutions, such as monasteries and civil governments, invested in bells. Bell towers built to announce the beginning of religious services acquired civic responsibility as broadcasters of public announcements. Bells warned of imminent danger from nature, called men to arms in defense of the community, honored the loss of great leaders, signaled the beginning of public ceremonies, and celebrated victory in battle. Centuries later, bells would be replaced by the factory whistles to signal the start or finish of a work shift. Towns were organized around these soundmarks, and no one outside its arena enjoyed social cohesion with the community. Public acoustic arenas were valued for their ability to integrate individuals into the social fabric of their community.

In his extensive study of bells in the nineteenth-century French countryside, Alain Corbin (1998) showed that self-esteem, emotional well-being, civic pride, and territorial identity all depended on hearing the town bells. When citizens heard the chiming of the bells, they felt rooted within a cultural geography that could easily be walked. Soundmarks provided local cohesion, a contrast to the modern concept of citizenship in a sovereign nation composed of millions of individuals spread over millions of square miles. Competition among towns and communes occasionally resulted in stealing one another's bells, and legal confrontation over the right to ring the bells resulted in riots. Corbin (1998) summarizes their attitudes with the well-known platitude "A town without bells is like a blind man without a stick."

Because the arena for a soundmark determined the scope of the town, those geological formations that would support sound propagation determined which regions could be absorbed into the township. Sound propagates farthest in valleys, which act like sonic conduits, and least over mountains, which cast acoustic shadows. As aural architecture on a grand scale, sonic geography controlled the social fabric of early rural communities. In the early twentieth century, when urban growth polluted the natural soundscape with noise, trolley lines rather than nature's sonic conduits defined social cohesion and its community boundaries. Transportation arenas replaced acoustic arenas. The public acoustic arena survived, but on a reduced and less personal scale.

Historically, for the average person without servants to act as messengers, living in a private acoustic arena meant social isolation. In contrast, a large public acoustic arena provided social inclusion. Schafer (1978) quotes a resident of a small town who remembers from the early twentieth century the importance of a large acoustic horizon, and the value of identifying horses by the sound of their steps: "The iceman had a couple of very heavy cobs...the coalman had a pair of substantial Percherons that always walked...the dry-goods store had a lightweight horses...and the Chinese vegetable men had very lazy horses." In a town with acoustically porous living spaces, you could hear the fishing boats returning to harbor, the children walking home from school, the rattling of leaves in the wind, and the dog fighting with the cat. You would know that it was time to visit your neighbors when you heard their wagon returning from shopping. Sitting at home, and without moving from your chair, you were intimately connected to the activities on your street.

As part of acoustic ecology, this is but one example of how a sonic environment creates a connection and cohesion among people. In her review of Steven Feld's documentary soundscape series, *The Time of the Bells*, Rachel Lears (2005) broadens the concept of soundmarks by mentioning the role of bells on bicycles, in carnivals, ceremonies, churches, government, and sheep farming. But she ignores the role of acoustic geography, even though it determines the scope of the arenas for those sounds. Thus the size of the acoustic arena for sheep bells, which effectively determines the protected grazing area for the herd, varies with the terrain, being larger in valleys and smaller on hillsides.

Modern society has a mixed attitude toward the size of public and private acoustic arenas. Radio, television, newspapers, e-mail, and telephone have replaced the public acoustic arena as ways to maintain social connections on a large scale. Cities are so noisy that residents treasure private acoustic arenas, often at the cost of feeling isolated, lonely, and anonymous. In contrast, within a modern household, a family arena exists when all family members sit together or keep their doors open. Our household, like many others, has no doors for any of the common rooms. When sitting in my office with the doors and windows closed, I am fully isolated from the public acoustic arena, whereas when I move to the backyard, I am fully immersed in the activities of my neighbors, the local squirrels, and the neighbor's cat. Some companies place workers in a single large acoustic arena, with only managers having private acoustic arenas. Similarly, later discussions on musical spaces illustrate a cultural progression from the shared acoustic arenas of public performances in churches and concert halls to the private arenas of sound reproduction in homes and automobiles. Headphones produce the most private of all acoustic arenas.

To summarize: the principles of acoustic arenas apply directly to the aural architecture of all spaces. To create an attractive space, be it a courthouse, school, civic center, family residence, or house of worship, an aural architect must also incorporate contemporary attitudes toward acoustic arenas and acoustic communities.

Social Spheres and Acoustic Arenas

A scarcity of acoustic arenas, as with all limited resources, provokes competition among the groups of people using those arenas. The social dynamics of human groups determine the outcome of that competition. Although stronger groups capture a larger percentage of available acoustic arenas than weaker groups, laws and social conventions provide cooperative mechanisms for regulating particular kinds of arenas. For example, concert halls have strict rules that give musicians the exclusive right to create sounds, whereas taverns have weaker rules that give any patron enjoying food and drink the right to sing. Airport agencies specify where and when airplanes can fly. Households have rules about the volume of television sound. In contrast, a self-indulgent motorcyclist riding through a neighborhood usurps the right to make noise, interrupting hundreds of conversations in less than an hour. Injecting noise of whatever kind into an acoustic arena is nothing more than the exercise of sonic power: social or political, autocratic or democratic, supportive or destructive.

When they design a space, traditional architects exercise power ultimately as potent as that of social or political agents in determining, however unwittingly, the size, use, and attributes of the acoustic arena of that space. For their part, those who occupy or live within a space have a dynamic, bilateral, and continuing relationship to space within their own aural architecture of created acoustic arenas. In contrast, the relationship of traditional architects to their spatial creations is severed at the completion of the project.

Whereas a traditional architect creates an acoustic arena in a space by erecting boundaries that are sonically impermeable, the occupants of that space create equivalent arenas by asserting their social or political right to generate sonic events. From the perspective of the acoustic community, voluntary silence and physical barriers produce equivalent arenas. Especially in the previous century, many societies passed laws to control nuisance noise, such as that made by vendors, barking dogs, radios, carpet beaters, and street musicians. In some cities, they attempted to enforce quiet on Sunday to emphasize the solemnity of a day devoted to religion. We have all been surprised at the large size of public acoustic arenas on a quiet Sunday morning. Like the airways, public acoustic arenas are common resources owned and thus to be regulated by the people.

For both architects and occupants, silence reveals more about the social and cultural aspects of acoustic arenas than sounds. Silence is far more than the absence of sound, a definition that considers only the physical properties of sonic vibrations. Rather, silence may be understood as an active choice by the creators of acoustic arenas: the occupants and the architects. The absence of sonic events—silence—is important because it leaves the acoustic arena available for low-level sonic events that add nuances to communications. Silence creates large acoustic arenas as a common resource, whereas loud sound consumes that resource. Only the highest-quality acoustic arenas, with very low background noise, communicate silence.

A few examples illustrate the social and psychological complexity of silence. It can signal: a cessation of both social and natural activity, a state of psychological tranquillity, a powerful emotion that transcends speech, a cooperative agreement to respect the public soundscape, a silent prayer communicating with a deity, a preoccupation with inner thought, a punitive response to social or political transgressions, or an acceptance of the right to be left in peace. Such nuances of communication are severely degraded when the aural environment falls victim to intruding noise.

The level of background noise determines the quality of an acoustic arena and the reliability of its auditory channels. A silent environment creates the best auditory channel; a noisy environment the worst. The sonic properties of the channel determine what messages can be transmitted. Communicating with ringing bells from a bell tower is more reliable than communicating a public announcement by voice, which is more reliable than communicating emotional intimacy by subtle tonal inflections. A noisy acoustic arena only allows for basic communications, such as a bell sound, because noise degrades the subtler aspects of vocal communications and social cohesion. For example, even at the most fundamental level, oral communication becomes more stressful when noise masks the short silent interval that distinguishes a voiceless consonant from its voiced counterpart, such as *t* from an *s*. Similarly, noise prevents signaling with a hesistant pause, which may signal the speaker's lack of confidence, or with a sudden cessation of speech, which may be intended to coerce an unwilling response from the listener. Such signaling requires a silent background.[6]

Unlike practitioners of vocal religions, Quakers value silent prayer as a way to distinguish that activity from the profanity of ordinary speech. They regulate silence using strong rules that forbid transgressing on the religious commons (Bauman, 1983). Group silence is the ultimate manifestation of social cohesiveness because silence can exist only if all members cease from speaking—total deference to the group's values. When silence dominates, vocalized prayer takes on special meaning: voices framed by the boundaries of silence rather than lost in an ocean of sound. Silence is the central component in many religions and rituals (Szuchewycz, 1997).

Teachers, judges, priests, and tyrants all have the power to silence others. To be silent in the face of authority can show either deference or defiance. The asymmetric relationship between those who give orders and those who must obey is always demonstrated by who controls access to the soundscape. The common command "Silence!" demonstrates political power because it defines who is allowed to express a point of view. Adam Jaworski (1993) called these interactions "the politics of silence and the silence of politics," and Wreford Miller (1993) stated that silence, or the lack of it, has been politicized in modern society to the point where the sounds themselves matter little.

Acoustic arenas are commercial as well as political. In exploring the history of background music over the last half century, Hildegard Westerkamp (1988) observes the unchallenged right of commercial organizations to exercise control over individuals in their acoustic community. For them, an acoustic arena is private property to be leased by the highest bidder. Marketing literature from companies that sell music services to commercial enterprises is explicitly blunt. With training in behavioral psychology and human engineering, the founder of Muzak claims that you will "see the difference in customers," and injected music will "teach your cash register to sing with the foreground music from AEI" (Westerkamp, 1988). Airport lounges, even as semipublic spaces, saturate their occupants with television advertising. Waiting passengers may avoid attending to the visual component of that space, but they cannot block its aural counterpart. Television sound creates a sufficiently large acoustic arena for its message that other acoustic arenas are reduced in size. Just as sponsors or owners may commission traditional architects to design acoustic arenas by manipulating acoustic parameters, they may also design these arenas themselves by injecting background and foreground sounds or by enforcing rules about who else can also inject sound.

More commonly, ownership rules of an arena are created informally when two or more individuals congregate for a social interchange. Territorial bubbles appear as if by magic around a group of individuals if they begin to interact, and the group quickly acquires rights to the arena. When encountering such a social bubble with its implied acoustic arena, outsiders are reluctant to intervene or to create sonic events (Lindskold et al., 1976). The strength of ownership rights to an acoustic arena depends on the distance between individuals, their perceived status, and the nature of their interactions. Cultures assign implicit rights to acoustic arenas, and there are complex unwritten rules governing the size of an arena being claimed.

Understanding the social rules for acoustic arenas requires the concept of social distance, as embodied in the term *social sphere*, which then becomes the means for evaluating arenas. The sounding of a foghorn is a public broadcast intended for everyone with the expectation that its acoustic arena will be large, whereas a whispered comment is a private communication intended only for an intimate companion with the expectation that its acoustic arena will be small. Social expectations determine the properties, especially size, of an acoustic arena, and social behavior then adapts to available arenas. For example, if an acoustic arena were large enough to signal emotional nuances over a great distance, its large size would conflict with expectations of privacy. Similarly, a small public acoustic arena conflicts with the need to broadcast public information to a large population. To be socially useful, acoustic arenas and their properties must match the cultural norms governing social spheres.

Whereas physical distance is measured in meters or feet, social distance depends on the social context. The social anthropologist Edward T. Hall (1966) divided social distance into four spheres: (1) the *intimate sphere*, which ends at about half a meter (1–2 feet) and is reserved for intimate friends and relatives; (2) the *personal sphere*, which ends at about 1 meter (3 feet) and is reserved for acquaintances; the *conversational sphere*, which ends at about 4 meters (12 feet) and is reserved for oral interchanges with strangers; and the *public sphere*, which is determined by the acoustic horizon and is impersonal and anonymous. How we experience a person, object, or sound depends on these distances, which Hall called "proxemics," the experiential manifestation of anthropological distance, which varies from culture to culture.

For each of these four spheres, a culture provides implicit ownership rules for the corresponding acoustic arena. Rules for the intimate sphere are rigid—lovers do not permit outsiders to enter. Strangers encountering an intimate sphere are likely to fall silent or speak softly. Rules for the public sphere are malleable—the social consequences of transcending sonic norms are minimal. Other spheres are intermediate cases between intimate and public spheres. Aural architecture fails when there are conflicts between social spheres and acoustic arenas. For example, individuals in a conversational sphere spanning a distance of 4 meters (12 feet) cannot coalesce if the arena diameter is only 3 meters (9 feet). Similarly, with an acoustic horizon of only 4 meters, sonic events in the public sphere are inaccessible.

To illustrate an application of social spheres, let us use proxemics to evaluate a chamber music concert. Musicians are located on stage in their conversational sphere, whereas listeners are located in their audience seats in the public sphere. Even if the management provides audience seating on the stage, some listeners are uncomfortable in a socially inappropriate sphere. The performers own their conversational sphere. But in the nineteenth century, performers and listeners often sat together in a small chamber, comfortably sharing a common conversational sphere. Today, if you put on binaural headphones by using spatial synthesizers, an audio engineer can place a virtual musician two inches from your left or right ear, well within your intimacy sphere. When such technology creates additional freedoms to move the location of a sound source, conflicts between the social and artistic expectations of the appropriate sphere may suddenly appear.

Proxemic distances are useful for evaluating the relationship between social spheres and acoustic arenas. If society does not provide the appropriate acoustic arena, then the corresponding social sphere is unavailable, and the corresponding social activities are not possible. Availability of an appropriate acoustic arena, in turn, depends on the aural architecture, which itself is a combination of acoustic design and the social rules for regulating sonic events. Aural architecture is not only the physical design of a space, but also part of a complete social system. We can only appreciate the importance of aural architecture when we recognize the interwoven relationship between spatial awareness, social behavior, and the design or selection of a physical space.

Navigating Space by Listening

Only listeners with motivation, dedication, and aptitude become expert at transforming the acoustic attributes of objects and geometries into a useful three-dimensional internal image of an external space. As with training sonar operators to identify underwater objects by how they modify incident sound, acquiring expertise of any form of auditory spatial awareness requires hundreds of hours of practice. Why would someone invest so much effort to acquire this proficiency?

Some listeners obviously benefit by having this ability. Musicians and composers include spatial attributes as a component of their art; acousticians depend on spatial awareness for designing concert halls; and audio engineers create spatial illusions with synthesizers. Listeners who must move around in places without light are likely to acquire the some basic abilities to recognize open doors, nearby walls, and local obstacles. But of the many groups of listeners who use auditory spatial awareness in their personal and professional lives, those with a visual deficit have the strongest motivation: hearing is a way to orient and navigate space, and their reward in developing their spatial awareness is the possibility of leading a normal and fulfilling life.

By itself, blindness never improves hearing. The auditory acuity of blind people as a population is average, spanning the same range of abilities to be found in the general population. On the other hand, some blind individuals are indeed motivated to enhance their spatial abilities far beyond the average. Practice is the most important predictor for achieving a high level of proficiency. With sufficient practice, some become expert, often displaying skills that are so extraordinary as to border on the magical. Such individuals illustrate what our species, in the limit, is capable of achieving. We are how we live—there is no generic human being.

There is evidence that those who practice a sensory or motor skill for thousands of hours change their brain wiring. Neurological studies, discussed in chapter 8, show that the cortical regions that process specific auditory cues are larger in conductors, musicians, and those with visual handicaps than in other people. Enhanced auditory spatial acuity is entirely a property of specialized sections of the brain that have been *trained* to interpret relevant audible cues. Listeners strengthen their neurological structure by repeated auditory exercise, just as athletes strengthen their muscles by physical exercise. Although the superb physiques of Olympic swimmers are plain to see, we cannot see the correspondingly superb "physiques" of "auditory athletes," except by observing their behavior while engaging in life's activities.

Cognitive strategies for decoding spatial attributes use such cues as the difference in time, amplitude, and spectrum between the sounds arriving at the two ears, as well as detection of changes in the expected spectral and temporal attributes of familiar sounds. Although some acoustic cues are specific to interpreting spatial attributes, most cues are unrelated to spatial acoustics. The cues that distinguish a *p* from a *b*, or a violin from an organ, are unrelated to space. Learning to hear space is mostly a matter of inventing a cognitive strategy that can decode the specific cues that arise from the acoustical behavior of objects and geometries in the world. From a physiological perspective, we all hear the sonic attributes of objects, but, absent training, we neither attend to their aural cues nor invent cognitive strategies for interpreting them. Although placing your hand a few inches from one ear illustrates that the hand's presence is audible, to translate audibility into a conscious sense of a hand, with its corresponding size, location, and skin surface, you must adopt a unique cognitive strategy. Far more difficult to detect, a traffic sign at a distance of a few meters also produces a set of cues that allows a skilled listener to detect the sign's existence and shape. At best, even when highly developed, auditory "seeing" of space (echolocation) is comparable to extreme visual nearsightedness, identifying physical objects that are relatively nearby or comparatively large. Small or remote objects simply do not produce aural cues that can be interpreted by any human being.

Echolocation is directly relevant to aural architecture because it conclusively demonstrates that our species has the neurological endowment to make judgments about objects and spatial geometries just by listening. Yet most aural architects, both amateurs and professionals, are unfamiliar with the native ability of human beings to hear space. A listener using a cognitive strategy to transform auditory cues into an image of a space, by sensing the doorway to the bathroom late at night, for example, is experiencing the *navigational spatiality* of aural architecture.

Experts at Hearing Objects and Geometries

Although history is replete with anecdotal accounts of blind persons "seeing" space, it was only in the mid-twentieth century that this ability came to be understood as an auditory skill. Curiously, the auditory ability of bats and dolphins to navigate without vision or smell was also discovered at about the same time, and it is now known that other species have a residual ability to sense their environment in the dark. Rather than thinking of this ability as a curiosity, such as sensing magnetic fields or infrared light, scientists now recognize that hearing space is more common than first imagined. Even though most animals and people with adequate vision and available light have little need to enhance their residual ability to hear space, it remains a viable alternative for supplementing vision.

The scholarly language to describe orienting and navigating in a space through hearing is ambiguous and confused. For example, the literature incorrectly uses the term *echolocation* (locating by means of self-generated echoes) for all forms of spatial awareness. This name originated from studies of bats and dolphins, which have a synchronized means for vocalizing and then decoding the responding echoes. Currently, the term *echolocation* applies to sensing spatial attributes with any kind of sounds, not just with self-made ones, whether by vocalizing, clicking fingers, or tapping canes. Background noise, for example, may provide sufficient sonic illumination to "see" aspects of a space. Moreover, the concept of echolocation, as now understood, also applies to acoustic cues other than echoes. Terminological confusion arose because the phenomenon of spatial awareness was recognized long before its physical and perceptual basis were understood.

One of the earliest written records of *face vision*, the early name for echolocation, was recorded by Denis Diderot (1749), who described the amazing ability of some blind individuals to perceive objects and their distances. Two centuries later, as part of his work at the Perkins School for the Blind, Samuel P. Hayes (1935) reviewed and cataloged the evidence for echolocation from a scientific rather than philosophical perspective. In his review, Hayes notes that scientists began to study echolocation only after sufficient anecdotal evidence and personal testimonies demonstrated that it was a real phenomenon. He describes a particularly impressive example of blind navigation he himself witnessed:

Martin was a native of New York City and had been blind nine years. He was of a fearless and im-
petuous disposition, and went about the city without a guide. He passed up, down and across
great thoroughfares frequently and only a few times colliding with a bicycle, which vehicle he
detested. I was with him on occasions when I marveled at the perfect freedom with which he
walked along crowded streets, showing not the slightest timidity, and requiring no aid whatever
from me. . . .

I was amazed to see him cross Broadway at 14th Street with perfect ease, and imagine my aston-
ishment when he shied around timbers that had been set up across the sidewalk to prop the wall
of a building undergoing repairs. He got on and off street cars without a blunder and made his way
across narrow streets without betraying his blindness. He used no cane nor did he feel his way
with his hands. Had I not known that he was actually blind I would have thought that he was
feigning.

I asked him how he knew his way and avoided collisions, and he invariably told me that he did
not know. He seemed to be guided by what I shall term a miraculous instinct superimposed by a
subconscious mental condition. I am inclined to the belief, in the absence of a better theory, that
he was directed by what Hudson terms "the subjective mind"! (Hayes, 1935)

The historical literature contains many such testimonies from many periods and cultures. Accepting the introspective comments of those who are adept at echolocation
provides the kind of insight that is not yet available from scientific studies, which reveal little about the underlying cognitive strategy for sensing space. These testimonies
emphasize several important aspects of echolocation. First, the skill is not conscious,
and even those who have a highly developed skill cannot describe how they do what
they do. Second, the exclusive use of echolocation for navigation requires great cour-
age. Third, using hearing for navigation, at least at this high level of performance, is
unusual; more frequently, blind persons depend on touch with their cane, using echolocation only as a supplement to their tactile sense of space.

How blind persons acquire a cognitive strategy for echolocation is still somewhat of
a mystery. Ved Mehta (1957), blind from childhood, described his experience of navigational space. Wanting to live a normal life in Calcutta, he learned to jump from banister to banister, from roof to roof, and rode his bicycle through unfamiliar places.
When he later attended the Arkansas School for the Blind, he participated in their
echolocation program, which was based on motivating students to avoid the pain of
colliding with suspended objects. Teachers simply believed that echolocation could be
learned by anyone, and their task was to provide motivation to invest in such learning.
Mehta described the environment, not the process of learning the skill.

One day in early spring, all the totally blind students were herded into a gymnasium and asked to
run though an obstacle course. Plastic and wooden slabs of all sizes and weights were suspended
from the ceiling around the gymnasium. Some of them hung as low as the waist; others barely
came down to the forehead. These slabs were rotated at varying speeds, and the blind were asked
to walk though the labyrinth at as great a speed as possible without bumping into the obstacles.
The purpose of keeping the slabs moving was to prevent the student from getting accustomed to

their position and to force them to strain every perceptual ability to sense the presence of obstacles against the skin—a pressure felt by a myriad of pores above, below, and next to our ears. Some of the slabs were of an even fainter mass than the slimmest solitary lamppost on a street corner. This obstacle course helped gauge how well an individual could distinguish one shadow-mass from another and, having located the one closest to him, circumvent it without running into yet another. . . . The gymnasium was kept so quiet that the blind people could hear obstacles, although I could not help feeling that I could have run through the labyrinth with a jet buzzing overhead. . . . For me, going through this obstacle course was child's play. (Mehta, 1957)

Although the details of learning echolocation vary, there is common attitude shared by those who are determined to "see" with their ears. Ved Mehta was not unique. The world-famous jazz musician Ray Charles eloquently describes a similar approach to living as a blind child (Charles and Ritz, 1978): "Being blind wasn't gonna stop me from enjoying the bike. . . . Somehow in the back of my mind I knew I wasn't going to hurt myself. Sure, I rode pretty fast, but my hearing was good and my instinct was sharp. . . . On another day Momma asked me to chop wood. . . . I was treated like I was normal. I acted like I was normal. And I wound up doing exactly the same things as normal people do." A few years later, he went to a special school for the blind, but his attitude toward echolocation was already solidified. "There were three things I never wanted to own when I was a kid: a dog, a cane, and a guitar. In my brain, they each meant blindness and helplessness." Being sensitive to the nuances of sound in general, he taught himself music and echolocation by listening carefully to the world of sound. Ray Charles never used a cane for navigating a space.

During the ensuing half century, modern methods have evolved for teaching echolocation, but the assumption that it can be taught is still controversial. Many, if not most, schools for the blind have abandoned teaching it. What explains the current lack of interest? In reviewing the literature, I noted that, with the exception of Kish (1995) and a few others, those who teach echolocation are themselves fully sighted, as such, they are very unlikely to develop sophisticated echolocation abilities. In contrast, Kish was blind from childhood, and taught himself echolocation by an intuitive sense of how to acquire that skill. He is now a licensed teacher for orientation and mobility, having created his own teaching methods (Kish and Bleier, 2000). Along with a colleague, Kish founded TeamBat, a program that guides blind teenagers into the mountains on bicycle trips, shown in figure 2.3. The answer to the earlier question is, in part, that echolocation is more a commitment to learn than a teachable skill.

Those blind individuals who use echolocation belong to a unique sensory subculture that has transformed a latent ability to hear navigational space into a high art form. Although there is no question that most listeners possess only the most rudimentary ability to detect spatial objects and geometries by listening, the difference between experts and beginners is only a matter of degree because the underlying cognitive and personal issues are the same.

Figure 2.3
Blind teenage bicyclists in TeamBat. Courtesy Cal State L.A Today; photographer Stan Carstensen.

Like ear training for musicians (Ottman, 1991) and for audio engineers (Moulton, 1993), learning echolocation also involves attending to the subtlest auditory cues. Unlike such training, however, echolocation involves an additional step—using a cognitive strategy to convert *binaural* cues into spatial images. Those cues originate from a multiplicity of transient sound sources interacting with a range of moving objects and surfaces. Consider the number of sounds and surfaces on an urban street. The cognitive strategy for echolocation must process all of them. Acquiring this ability therefore requires an individual to practice in a real sound field in a real space. For this reason, echolocation is best learned as part of daily life in a real-life environment, unlike other forms of ear training, which can take place in a studio or classroom. It is difficult, if not impossible, to artificially create or record teaching examples that faithfully replicate realistic sonic environments.

The ability to create an internal picture of external objects and geometry is greatly enhanced when strong motivation, greater than average skills, and an extended opportunity to practice are present. For blind individuals, enhanced echolocation ability correlates with several key factors. Engaging in echolocation, if begun in childhood when brain substrates are evolving, can readily adapt neural structures to become optimized for different purposes. A child without any residual vision is simply more likely to discover hearing as an alternative means for navigating a space if permitted to do so. Because practicing echolocation includes the risk of injury, the child needs to be comfortable taking risks, and the child's parents must avoid excessive protectiveness. In fact, participating in activities that normally assume adequate vision is the best predictor of acquiring auditory spatial awareness for navigating, as attested by the personal examples of Daniel Kish (1995, 2001), who categorically rejected the guidance of those who urged him to learn to use the cane, and Ved Mehta (1957), who moved about the streets of Calcutta without supervision. Investing in auditory spatial awareness is always a free choice that any of us can make, but few do.

Even though there are numerous examples of individuals who learned echolocation, the rehabilitation literature is, at best, ambivalent about using hearing rather than the tactile sense for navigating space. When large numbers of soldiers returned from World War II with visual disabilities, formal training programs became a priority, and echolocation was an obvious technique (Bledsoe, 1980). After prolonged controversy and passionate debates, rehabilitation workers involved in helping blind soldiers eventually concluded that tactile navigation—using a cane—was simply easier to teach. Many soldiers could not, or would not, learn to sense subtle auditory cues and invent cognitive strategies. Some schools for the blind explicitly taught auditory spatial awareness, which is fundamentally different from navigational skills (Campbell, 1992), although such teaching proved problematic because most rehabilitation professionals were themselves sighted and could not teach from personal experience. Scientific studies of blind persons using echolocation do not reveal the underlying cognitive processes. As a generalization, cognitive strategies are learnable but not necessarily teachable; for those who cannot echolocate, such strategies have little, if any, practical value in daily life.

The literature on echolocation actually illustrates a larger principle: sensory skills are acquired, rather than innate; they are based on personal utility and lifestyle. Blind persons with the ability to echolocate are an obvious example of a sensory subculture that has the ability to use a specific cognitive strategy to interpret spatial cues arising from one aspect of aural architecture: navigational spatiality. In contrast, professionals who are actively engaged with other aspects of aural architecture, such as designers of concert halls or composers of music, become very adept at other cognitive strategies for interpreting other spatial cues.

Hearing Specific Spatial Attributes

Insights into the sensory and cognitive aspects of echolocation contribute to our understanding of aural architecture. And for this reason, it is worth shifting the discussion from anecdotes to research. By the mid-twentieth century, explaining the intractable phenomenon of echolocation became a scientific challenge. As with many perceptual phenomena that are complex, researchers broke echolocation down into many small, simplified questions and special cases. Theories about how we hear the distance to an isolated wall or how we judge the size of a door opening are examples of special cases. At the current state of knowledge, the cognitive and perceptual sciences are more collections of disconnected theories and experiments than unified wholes. On the other hand, when a blind person rides a bicycle in a city, that person is merging a great number of special cases into a holistic strategy. Navigating real spaces involves hearing walls, openings, passive acoustic objects, and extracting their relationship to the location and properties of sound sources. The whole is far larger than the sum of the parts. Space is experienced as an unconscious unity rather than as a collection of recognizably separable processes.

To appreciate the acoustic complexity of an urban street, consider that the environment is composed of multiple objects and numerous sound sources, some stationary, some mobile. Each traffic sign, parked automobile, or telephone pole has a surface that produces both sonic reflections when the sound source is in front of it and acoustic shadows when it is behind. A reflection will be heard as an echo if the sound is impulselike and the surface is more than 10 meters (33 feet) away, or as tonal coloration if the source is continuous and the surface is nearby. A sonic shadow may be diffuse and blurred for low frequencies, or sharp and clear for high frequencies. Sonic illumination is the aural equivalent of a space illuminated with multiple lights: some bright, some dim, some colored, some blinking, and some moving. In a real-life environment, the sound field is indeed complex.

Now consider that, because you have two ears separated by the width of your head, each ear senses sound at a slightly different location in space. By moving or rotating your head, you reposition your two ears at another location. The physical sound field actually varies in three dimensions: left-right, front-back, and up-down. Obviously, if we had more ears and if our heads were larger, the auditory cortex would acquire far more information about the spatial distribution of sound. But even with our limited abilities to sense a three-dimensional sound field, the sounds arriving at the two ears are often sufficient for the auditory cortex to build a perceptual model of the objects and geometries that *could* have produced those particular sounds. Perception is an unconscious inferential process that synthesizes a hypothetical collection of objects and geometries. This process is the result of having learned the subtle, ambiguous, and inexact relationship between auditory cues and spatial attributes. Those who have developed echolocation skills cannot describe how the spatial image suddenly appears in their consciousness.

Scientists are still probing for important clues and theories to explain echolocation. Since the phenomenon of echolocation was first recognized by Michael Supa, Milton

Cotzin, and Karl M. Dallenbach (1944) at Cornell, explanations of its mysteries have been of periodic interest to small groups of researchers. The science of echolocation is far from the mainstream of auditory research, being supported mostly by those with an interest in rehabilitation of people with visual deficits.

Before reviewing what science has learned about echolocation, we need to explain the tentativeness of research conclusions. Scientists are wrestling with a confounding methodological problem: individual listeners are remarkably inconsistent in their abilities to hear space. Auditory spatial awareness ranges from raw sensation to unbelievably high levels, corresponding to an equally wide variability in sensitivity to acoustic cues and effective cognitive strategies. Is a scientist actually studying a general phenomenon, or the unique ability of specific individuals on specific tasks? In practice, scientists ignore this question when they use randomly selected subjects. Even within the sorted population of blind subjects, there is a wide range of abilities.

Human echolocation is actually a collection of independent abilities to perform a variety of tasks, from hearing spectral changes produced by a nearby wall, to hearing the acoustic shadow produced by a telephone pole, to hearing the reverberation arising from two coupled spaces. A given listener might be very good at one task but mediocre at another. Experiments are designed to focuses on a single task under controlled conditions. For example, blindfolded subjects might be asked to walk along a long hallway with a single continuous noise source located at the end. Because there is only one sound source and a very simple geometry, acoustic shadowing, diffraction, and reflections cannot exist. In this restricted case, the experimental paradigm is evaluating the degree to which a subject's cognitive strategy incorporates only auditory cues explicitly included in the experiment. Although good scientific studies produce modestly consistent results, it is unclear how or when such insights apply to real life.

Even with these limitations, scientific results explain certain aspects of echolocation. Daniel H. Ashmead and colleagues (1998) showed that blindfolded subjects walking through a hallway without colliding with the wall detected low-frequency tonal coloration near walls. The ear closer to the wall surface senses background coloration different from what the farther ear senses. In the center of the hall, the coloration is the same in both ears. Differential coloration corresponds to distance to the wall. The same mechanism allows subjects to detect when they are passing an open door, which is equivalent to a missing wall.

In addition to hearing an open door as the absence of a wall, the door's frame creates acoustic shadows of sounds originating from within the room. When presented with an open doorway of unknown width and height, subjects can estimate its dimensions relative to their own body size with remarkable accuracy (Gordon and Rosenblum, 2000). Walking past an open door into another room therefore involves at least two cues: the absence of coloration from the missing wall segment, and the sonic shadows produced by sounds emanating from the room. The relative contribution of each type of cue depends on the sonic illumination in each space. If one space has stronger sonic illumination and the other has none, only one of the two cues would be available for sensing the doorway. Moreover, if the door were partially open, the door surface would itself become a source of reflections, which would then become yet another set of cues. The door is an additional object, separate from, but related to, the open doorway and the doorframe. In this simplified example, a trained listener uses a cognitive strategy that melds three sets of cues into a single image of the space: a partially open door in a doorframe leading to another room.

Listeners can sense not only doors and walls, which are relatively large, but also small objects and small differences in larger objects when they are relatively nearby. Charles E. Rice (1967) showed that listeners can detect a difference of 1 centimeter ($\frac{3}{8}$ inch) in a 9-centimeter ($3\frac{1}{2}$-inch) disk at a distance of 60 centimeters (2 feet). Winthrop N. Kellogg (1962) showed an even higher level of discrimination: listeners detected an area difference of 5 square centimeters ($\frac{3}{4}$ square inch) on a square of 60 square centimeters (9 square inches) at a distance of 2 meters (7 feet). One blind subject could reliably detect a 1-inch disk located at a distance of three feet (Rice, 1969, 1970). Even more remarkably, Steven Hausfeld and colleagues (1982) demonstrated that listeners could distinguish square, circular, and triangular objects. One blind subject was able to recognize a stop sign by its octagonal shape. Kellogg (1962) found that on the most difficult discrimination tasks blind individuals performed significantly better than sighted subjects who were blindfolded.

Although only a few studies have been designed to explore why some individuals performed better than others, Connie Carlson-Smith and William R. Wiener (1996) showed that two specific aspects of auditory acuity were partial predictors of echolocation ability. Those subjects who performed best at detecting spatial attributes were also better at sensing small changes in the amplitude and the frequency of continuous sounds. When a sound field is not uniform, moving through it converts spatial differences into time differences. As listeners move through the space, they hear spatial differences as temporal changes. Although the ability to detect soft or high-frequency sounds at threshold is not related to echolocation, the ability to hear and interpret small changes in sound is.

Apart from genetic endowment, learning is the dominant component of acquiring echolocation skills. We are not, however, speaking of 20 hours of practice but of thousands of hours. Say you are a 20-year-old adult. You have already spent well over 100,000 hours listening to the physical world of spaces. If, during that time, you had also engaged in self-directed practice exercises, as would a blind person moving through life's spaces, you would likely have much improved your perceptual acuity to aural cues, and have become highly proficient both at inventing cognitive strategies and applying them to convert those cues into spatial perception. Like athletes who love sports, those who want to become more proficient in echolocation engage in complex sensory activities that simultaneously exercise a wide range of skills and methods. They invent methods to teach themselves how to become proficient—customized pedagogy. Formal training managed by a (usually sighted) teacher in a classroom is far more limited than a lifetime of training managed by the individual listeners themselves.

Sensory practice changes the brain. When examining blind subjects who had engaged in extensive practice, Brigitte A. Röder and colleagues (1999) found that their neurological responses to sounds in the peripheral field were significantly better than those of normal subjects. With enough practice, the improved ability of the blind subjects is observable in the neurological response of the relevant cortex. Similarly, Christo Pantev and colleagues (2001b) found that the brains of pianists who began their careers as children responded more intensely to piano notes than those who began later. Because immature brains have greater plasticity in their neurological wiring, practice produces larger brain changes during early developmental periods.

Learning is far more specific to the task being practiced than you might expect, and acquired skills do not readily transfer from one task to another. Just as exercising one muscle group does not strengthen other muscles, exercising one sensory skill does not enhance other skills: each sensory skill involves specific brain substrates. An audio engineer who has acquired enhanced acuity to tonal coloration in reverberation is unlikely to transfer that skill to navigating a corridor without vision. Although the concept of task-specific learning is well understood, only a few isolated experiments confirm the phenomenon. A curious experiment on pitch discrimination dramatically illustrates the extreme specificity of auditory learning. Laurent Demany and Catherine Semal (2002) trained subjects over the course of 11,000 sessions to discriminate the pitch of a 3,000 Hz tone from tones at slightly different frequencies, a very specific task indeed. Subjects improved by a factor of 3, and would likely have improved further had training continued. Not only is it surprising that intensive practice produces improvement on such a basic psychophysical task; it is even more surprising that improvement at this one frequency did not transfer to other frequencies. Pitch discrimination at 8,000 Hz remained unchanged. Subjects were not learning generic pitch discrimination; they were learning pitch discrimination of 3,000 Hz tones. Although I believe that this result applies to a large number of other phenomena, scientific studies have not yet revealed the extent to which spatial cues can be learned with extensive practice.

These somewhat speculative conclusions have broad implications. First, extensive practice produces dramatic changes in perceptual ability, and those changes are observable using neurological imaging techniques. Brains reflect how individual listeners live their lives. Second, a culture that motivates and rewards listeners to learn auditory spatial awareness is likely to have a population that can better appreciate aural architecture. And conversely, without such a population, aural architecture is likely to be irrelevant to the culture. Third, auditory spatial awareness is a collection of independent sensitivities. Some listeners may be acutely aware of reverberation and the enclosed volume of a space, whereas others may be aware of local objects and geometries in a navigational space. Finally, any discussion about aural architecture must include an understanding of various aural subcultures, each of which has its own idiosyncratic investment in the ability to detect and appreciate attributes of spaces.

Cognitive Maps as a Spatial Framework

Although our internal representation of space usually originates from an external reality, internal and external representations are not as tightly linked as you might expect. To use a misleading analogy, we often speak of an internal image as if it were a neurological "photograph" created by the brain. But internal images are not replicas of the external world. How does an external space become an internal space, and in what ways are these two spatial concepts related? The answer to this question involves cognition as well as perception and lifestyle as well as biology.

Although our knowledge of how the brain creates its internal representation of an external reality is, at best, rudimentary, a diverse collection of fragmentary insights reveals a consistent picture. Evidence shows that cognitive processing of spatial attributes is plastic, flexible, adaptive, and dependent on the way individual listeners conduct their lives. Evidence also shows that auditory spatial awareness merges with visual spatial awareness, together creating a holistic spatial awareness—a high-level cognitive process.

An internal spatial image is a cognitive map of space—a private construction that includes a mental response to sensory stimuli modified by personal experience. Roger Downs and David Stea (1973) provided a basic definition of cognitive mapping as a "process composed of a series of psychological transformations by which an individual acquires, stores, recalls, and decodes information about the relative locations and attributes of the phenomena of everyday life." A cognitive map of a space is a combination of the rules of geometry as well as knowledge about the physical world. It is this extra environmental knowledge that allows us to perceive a ball as moving away from us rather than as simply shrinking. This knowledge associates reverberation with enclosed space, echoes with remote surfaces, and high frequencies with hard objects. These associations are learned. Because this knowledge is acquired in childhood and continually modified in our experience as adults, we are not conscious of its existence. When sensing a spatial environment, an individual builds a cognitive map of space using a combination of sensory information and experiences accumulated over a lifetime. The cognitive map of space in our consciousness is subjective, distorted, and personalized—an active and synthetic creation—rather than a passive reaction to stimuli.

Individuals have choices about which sensory inputs they use to create their cognitive maps of space. Blind individuals who navigate a space by listening are choosing auditory cues to build their maps, but when navigating with a cane, they are choosing tactile cues. When light is sufficient, sighted listeners usually ignore auditory and tactile cues when navigating a space, but may use them when light is inadequate. When listening to live symphonic music, such listeners may merge both auditory and visual inputs in forming a sense of the concert hall. More generally, individuals have personal biases toward their senses, as for example, favoring vision over hearing, or vice versa.

Although, normally, each of us can fuse any combination of aural, visual, tactile, and olfactory inputs into a cognitive map, it is only a single mental map because there is only one single external reality. For example, when touching, hearing, and seeing a violin, there is still only one violin, not separate visual, aural, and tactile violins. The same principle applies to space: different senses provide access to different aspects of a single space. Vision is better for sensing an object's distance; hearing is better for sensing the volume of an enclosed space; and touch is better for sensing surface texture. We are able to see the "rough" texture of a surface because we have experience touching rough objects. The olfactory sensation of volatile hydrocarbons allows us to see (interpret) a shimmering surface (visual) as wet paint (tactile). We combine sensory cues and then interpret them using our memory of previous experiences to create a compelling internal sense of an external world.

To further emphasize that cognitive maps are not biological photographs, sketches of familiar spaces drawn from memory deviate from realistic maps. Frequently, important spatial attributes are larger than reality, and unimportant attributes are smaller or missing. The nature of a distortion also depends on which sense dominated the construct of the map (Jacobson, 1998) because each sensory system is better at some attributes than others. Errors and distortions in what you perceive are a complex mixture of your sensory system, your sense of what is important, and your memory of historical experiences. In his study of how Parisians represented the geography of their city, Stanley Milgram (1976), demonstrated the lack of consistency in their mental maps. His subjects could not preserve complex spatial details and relationships, instead using personal symbols, omitting unused regions, and expanding personally important areas.

There is increasing evidence that cognitive maps of space have dedicated neurological substrates that combine visual and auditory input. These substrates contain a fused representation of spatial attributes independent of the sources of sensory information. John O'Keefe and Lynn Nadel (1978) initially suggested that such maps reside in the hippocampus, but recent neuroscience studies have identified specific neural substrates that respond when objects are spatially aligned in both vision and hearing (King and Schnupp, 2000). When multisensory inputs are aligned, we experience a single object with aural and visual properties; you do not experience an aural object and a visual object. But when sensory attributes are not aligned, you experience two objects, one with visual and the other with aural attributes. As neuroscience uncovers details about specific brain substrates, we find that some intellectual abstractions, such as cognitive maps, have an observable manifestation in the brain.

In attempting to solidify the vast collection of experimental data on sensory fusion, the neuroscientist Alvaro Pascual-Leone (2000) took the concept one step further. He argued for a metamodel of the brain where neural substrates act as "operators" to implement a given functionality regardless of the sensory modality. In his conceptualization, there would be a *spatial operator* in a brain substrate that operated on aural and visual cues to create an internal representation of space. Similarly, there would be an emotional operator that created an affective response to that same space. As a rule, an operator appears to be dominated by a particular sense modality. Thus, for a sighted individual, a spatial operator might be dominated by visual inputs, and for a blind individual, by auditory cues. For a deaf individual, a speech operator might be dominated by visual or tactile cues. Dominance is far from universal or complete, and operators incorporate inputs from multiple senses without explicit awareness. For example, Beatrice de Gelder and colleagues (1999) showed that the emotional responses to hearing a voice and viewing a face influenced each other when the emotional content of the two modalities was not in agreement. We might expect that the emotional responses to hearing space and seeing space influence each other as well.

The separation of a cognitive map from its sensory inputs is illustrated by how individuals imagine an object when they have no visual input. Oliver Sacks (2003) observed that some blind individuals experience "deep blindness," an inability to imagine the shape of an object without tracing it, whereas other individuals experience a "hallucinatory visual world," rich and full with real and imagined objects. In one case, the visual cortex had atrophied, whereas in the other case it remained active using a combination of inputs from internal memory and the aural and tactile senses. Some part of the visual cortex may actually serve as a spatial operator. Sacks (2003) commented, "studies on the effects of blindness on the human cortex have shown that functional changes [in brain substrates] may start to occur in a few days, and can become profound as the days stretch into months and years." Even after being blindfolded for only a few hours, sighted subjects begin to experience changes in spatial and object images. These changes reflect a rewiring of the spatial operators, thereby compensating for the lack of visual input. An internal representation (cognitive map) of space depends on the way you teach your brain to use *all* your senses. For Sacks, the visual cortex is only the "inner eye," a concept that has nothing to do with sight itself. Auditory and tactile information also contribute to the functioning of this inner eye. Because we use a visual vocabulary to describe spatial experiences rather than a sensory-neutral language, we assume that spatial experiences are visual both in origin and in representation. In common discourse, the word *map* itself means a visual picture of an environment. In fact, the "inner eye" is, not visual, but multisensory, "seeing" the present combined with the past.

Consider that perceived size and distance are not just a visual measure of a physical reality but also involve subjective and personalized concepts derived from multisensory data. The experience of large distances is also an indirect consequence of experiencing time, as exemplified by the time to walk from one place to another, or by the time for an echo to return from a distant surface. The vastness of an enclosed space is revealed by decaying reverberation. In contrast to distances that can be experienced as the passing of time, small distances can be measured in terms of the length of an arm. You experience the size of a doorway opening, not in terms of a ruler measurement, but in terms of its ability to accommodate the width of your body when walking through the opening. In an earlier discussion, we explored the concept of the acoustic horizon, which is also a measure of distance, using social spheres as the metric. The aural, visual, and tactile experiences of space contain different *perceptual* units for size, which are then fused into a single spatial map. Conversely, a single map can be converted into different units of sensory size: the object is at arm's length, it takes ten strides to reach, or it returns an echo in 100 milliseconds. We should think of spatial cognition as the process of fusing and reconciling overlapping contributions from all sensory modalities.

Having established that size and distance are multisensory abstractions that are fused into a single cognitive map of space, we now turn to the issue of spatial relationships among objects and the perceiver's relationship to those objects. This, too, is an abstraction that depends on a given reference point or viewpoint. A cognitive map of space implies a spatial framework. At the most basic level, saying that a boy is standing in front of the tree implies a specific location for the viewer, but saying that the boy is standing north of the tree implies an abstract spatial reference independent of the viewer. Where is the boy? In the first case, the relative location of the boy changes if the observer changes location. In the second, the boy's relative location changes only if the environment, including the observer, is rotated relative to the reference frame.

In all spatial experiences, there are two perspectives: *allocentric*, from which objects are perceived relative to a fixed external framework; and *egocentric*, from which objects are perceived relative to the perceiver. Rotate a concert hall and, depending on which perspective you adopt, the relative location of the orchestra either changes or remains the same. Although mathematically equivalent, in that one reference frame can be converted into the other, each perspective is experienced differently. For example, musicians are at the front of the concert hall (allocentric), but the person with a large hat is sitting in front of you (egocentric). Cognitive maps of space contain aspects of both perspectives, but emphases vary from culture to culture.

Because an allocentric framework situates you within a fixed external environment, philosophically, it implies that reality exists apart from your self. In contrast, an egocentric framework situates your self at the center of an experiential universe, where everything is interpreted relative to you. A cognitive map of space can be egocentric, allocentric, or some combination of both. The choice of framework modifies the experience of space.

There is evidence that the brain contains substrates for encoding space in a multiplicity of allocentric and egocentric perspectives (Behrmann, 2000). Although neural substrates exist to support both perspectives, cultural values and personality biases usually emphasize one over the other. One culture's language and religion may focus on egocentric representations of space; another's may focus on allocentric representations. It is easy, but presumptuous, to expect cognitive maps of space to be consistent across cultures, or even across individuals from the same culture.

Because a cognitive map is, by definition, entirely private, we have access to it only by observing behavioral differences among cultures, such as difference in the language of space, or in the ability to perceive spatial attributes. Benjamin Lee Whorf (1956) first advanced the thesis that language influences how we experience life, and vice versa. Although still controversial, his thesis remains a major component of cognitive theories (Lucy, 1997).

A manifestation of differences in cognitive maps of space can be observed by analyzing a culture's language, and by testing individuals on behavioral tasks. As Stephen C. Levinson (1999) notes, some languages do not employ the spatial notion of left-right-front-back but rely on north-south-east-west. These differences are more than merely linguistic. They are fundamentally different ways of viewing the world and placing oneself into the world. The type of cognitive map of space changes one's behavior on spatial tasks. For example, on various tests, Dutch subjects consistently performed better at encoding and referencing relative locations, which is characteristic of modern cultures, whereas Tenejapan Mayan subjects performed better at encoding absolute locations, which is better for navigation and orientation in natural spaces. Similarly, Levinson (1999) observed that modern European languages favor using self-referencing body parts to identify building sections, such as the head, wings, back, or face of a structure. Other languages refer to component parts using absolute references, such as seaward or northerly.

The discussion on cognitive maps of space demonstrates that we cannot consider the navigational spatiality of aural architecture in isolation. And just as aural architecture is an inseparable component of sensory architecture, so aural spatial imaging is inseparable from spatial awareness, which is a high-level cognitive process separate from specific sensory modalities. The creation of a navigational space depends on the cognitive map of the aural architect, just as auditory spatial awareness depends on the cognitive map of the listener. Both designer and listener have acquired their maps through experiences. Unfortunately, cognitive maps of space are difficult to observe, even though they are central to spatial experience. Although the ability to use auditory spatial awareness for navigating space is present in all human beings with adequate hearing, the degree to which that awareness contributes to cognitive maps of space is specific to individuals and their cultures.

Aural Enrichments in Architecture

Architecture is more than the design of a utilitarian space; architecture is also an expressive art form that communicates. Using the broadest definition of architecture, we also include decorations, ornaments, adornments, and embellishments as important elements of spatial design. These elements are aesthetic supplements to the utility of the spaces we occupy or live within. Although they are traditionally considered part of interior design, they are as relevant to the experience of a space as the structural framework that encloses a space. Every picture, statue, tapestry, archway, mirror, dome, textured surface, and ceiling molding, to name but a few, is an architectural embellishment. There are embellishments that produce or admit light, such as candles, chandeliers, or frosted-glass panels, and there are embellishments that absorb light, such as dark tapestries or black walnut panels. There is no functionality in the aesthetic aspects of these adornments—flat white walls illuminated by industrial lamps are adequate for ordinary living—yet such embellishments enhance aesthetics by creating a pleasant or reflective mood. They may also convey symbolic meaning, such as wealth, political power, social status, or historic legitimacy.

Architecture includes aural embellishments in the same way that it includes visual embellishments. For example, a space we encounter might contain water spouting from a fountain, birds singing in a cage, or wind chimes ringing in a summer breeze—active sound sources functioning as active aural embellishments for that space. Producing aural rather than visual illumination, these are the aural analogues of decorative candles and lamps. In contrast, passive aural embellishments, such as interleaved reflecting and absorbing panels that produce spatial aural texture, curved surfaces that focus sounds, or resonant alcoves that emphasize some frequencies over others, create distinct and unusual acoustics by passively influencing incident sounds. Passive aural embellishments are the aural analogues of pictures, tapestries, mirrors, arches, and statues.

For both visual and aural embellishments, there are two independent oppositions: active versus passive and local versus global. A water fountain and a resonant alcove are both aural embellishments, but the first serves as an active source of sounds whereas the second passively filters them. Similarly, a candle and a mirror are both visual embellishments, but the first actively generates light whereas the second passively reflects it. Affecting only an area of a larger space, fountain, alcove, candle, and mirror alike are *local* embellishments. We experience them only when we are relatively close. In contrast, affecting the entire larger space, both reverberation and diffuse lighting are *global* embellishments. We experience them throughout the space. Parallels between visual and aural embellishments are not generally recognized because visual objects are most often local, whereas acoustic objects are most often global.

Almost every visual embellishment has some acoustic influence. Thus a mirror, a statue, or a tapestry changes the acoustics of the space around it. If these changed acoustics are unintended, their role as aural embellishments may not be recognized or appreciated. Nevertheless, they are relevant to our experience of aural space. A large mirrored wall reflecting light also functions as a perfect reflector of sound. An elegant tapestry absorbs sound and a marble statue diffuses it. Conversely, a sonic diffraction grating designed as an aural embellishment might also be considered as a modern visual sculpture. Depending on the sensibilities of the designer or the perceiver, every embellishment can be either visual, aural, or both at the same time.

Aural embellishments give a space an aural personality. Without them, every space, be it bathroom, concert hall, military barracks, or other space, would sound like every other space of similar size and shape. In addition, without local aural embellishments, every area of a space would be aurally indistinguishable from every other area of that space. When you move into a new house, you add personal touches—visual embellishments—by your selection of art and furniture, thus making the space of the house visually unique. By analogy, and for quite the same reason, you also add aural embellishments, whether intentionally or not. The antique rug that contributes visual elegance also adds aural warmth. Customizing a space to give it a unique and personal feel, perhaps to make it a symbol of yourself (Cooper, 1974), operates both aurally and visually.

We are now ready to define *aural embellishment*. It is an acoustical object or geometry, whether local or global, that produces aesthetically recognizable acoustic attributes, adding aural richness and texture to the space. An alcove in a cathedral is a local embellishment, providing aural privacy. Extensive carpets and thick drapes, by removing high frequencies from reverberation, are global embellishments that create an aural sense of warmth. As a generalization, aural embellishments produce acoustic attributes that are not related to the functional aspects of an acoustic arena, spatial navigation, or musical aesthetics.

Because of the extensive interest and research in the architecture of musical spaces, many assumptions that apply to those spaces have been implicitly carried over to other applications of aural architecture. In the design of a concert hall, aural embellishments are considered to produce unwelcome acoustic effects and should be avoided whenever possible. According to our musical norms, the aural experience of a concert hall should ideally be uniform throughout the space. The acoustic shadows produced by a balcony, for example, are tolerated, but unwelcome. Similarly, specific global aural embellishments are unwelcome because the acoustics of a musical space, as extensions of the musical instruments, should match the musical repertoire. In contrast, aural embellishments are welcome in a social or religious space, providing aural variety, symbolic meaning, and spatial texture.

Just as Japanese Noh drama and ancient Chinese opera convey little to an inexperienced audience without extensive exposure and knowledge, so aural embellishments may convey little to inexperienced listeners. All three are art forms and serve as evolving vehicles for expressing our relationship to ourselves, the world, and the cosmos. Understanding their message requires experience with the cultural symbols they use to convey it.

Spatial Distortions in Aural Geometry

An aural architect can design a space such that the acoustics at selected areas magnify the aurally perceived size, mass, and intensity of a speaker. Unlike optical magnification, however, acoustic enlargement is inconspicuous, arising from the shape of the enclosing surfaces. Strong sonic reflections arriving shortly after the direct sound increase the apparent aural size of the sound source. Even when the total sound energy remains constant, shifting energy to the early sonic reflections enlarges the perceived size. In contrast, late sonic reflections are perceived as echoes or reverberation, degrading intelligibility. Concentrating sound in time and space is one means of creating local acoustics in aural architecture.

The same phenomenon is well recognized in musical spaces. When early sonic reflections from the sidewalls and ceiling reflectors are appropriately combined, musical instruments on the stage of a concert hall sound closer—aurally larger—than they would otherwise. The musicians playing on stage are, by their special location, like a judge sitting on a dais, a politician or lecturer standing at a podium, or a minister preaching from a pulpit. These individuals are deemed to have socially dominant status; their special locations should have acoustics consistent with their dominant status, their relative social prestige. Thus, to symbolize the social relationship, the acoustics of the podium area in a lecture hall should raise the aural status of the speaker, whereas those of the auditorium should lower the aural status of the listeners.

The same natural amplification that increases the apparent size of the speaker also increases the size of the acoustic arena. In addition to sounding larger, the voice of the dominant speaker covers a wider acoustic arena, and is heard by a larger audience. A socially dominant location thus has two acoustic attributes: larger aural size and a larger acoustic arena. From extensive research on concert hall design, the knowledge required to create such local acoustics is well known and readily transfers to social and religious spaces. Architectural design therefore includes, intentionally or incidentally, the aural symbolism of dominance.

Just as a visual architect specifies the shape of the physical space, an aural architect specifies the shape both of the acoustic arenas and of the areas where aural magnification occurs. Whereas physical boundaries clearly delineate a visual shape for the space, through a more complex process, those same boundaries determine the shape of acoustic arenas. Shaped acoustic arenas thus become tools of aural architects. While most of the following examples are, unfortunately, unrelated to the more conventional social spaces, they do illustrate the wide range of freedoms available. An analysis of these examples shows that they are the acoustical analogues of placing numerous lenses and curved mirrors about the space.

When a space has curved surfaces, its acoustics can readily change the aurally perceived geometry of that space. Like the side mirror of an automobile warning that (visual) objects are closer (larger) than they appear, curved surfaces also change the apparent location of aural objects. Particular curved surfaces can focus sound such that the source appears aurally closer or farther, larger or smaller. We can think of these curved surfaces as distortions of a circular acoustic arena. Curved surfaces can also produce acoustic dead zones such that a source is inaudible, as if it were in a acoustically isolated arena. Aural privacy does not require walls. In contrast, some curved surfaces can give you the aural impression that a speaker is sitting on your right or left shoulder. Science museums often demonstrate how a parabolic sound reflector displaces a speaker 30 meters (100 feet) away to an aurally perceived distance of 3 cm (1 inch)—a thousandfold shift in location.

The concept of shaping an acoustic arena for aural effect is not new. In the early part of the last century, Wallace Clement Sabine (1922) described numerous examples of "whispering galleries," large enclosed spaces where a listener could hear the whisper of a speaker at remote distances. The more famous ones in Sabine's time included the Dome of Saint Paul's Cathedral in London, Statuary Hall in the Capitol at Washington, D.C., Saint John Lateran in Rome, and the Ear of Dionysius at Syracuse in Sicily. Most, if not all, of these whispering galleries are architectural accidents resulting from curved surfaces presumably designed for their visual impact. The time delay for the sound to return from the ceiling, combined with its focused direction, gives the visitor standing in the center of such a gallery the "effect of an invisible and mocking presence." This is not an echo. If you are the visitor, the sound of the distant speaker's voice is focused directly at you, as if the speaker were right next to you. The experience is unforgettable.

Similar effects are found with elliptical enclosures, such as the Mormon Tabernacle in Salt Lake City. In such spaces, if you stand at one of the two foci, you can readily converse with someone at the other. Using our spatial language, we can describe the situation as two physically separate regions of space that are joined by a bilateral auditory channel into a single acoustic arena. Even widely separated or oddly shaped physical spaces can be acoustically joined. In an example described by Sabine, the Cathedral of Girgenti in Sicily, by an unlucky coincidence, one focus is located at the confessional; secrets of the most intimate nature are broadcast to a remote location in the church. There is a story, assuredly apocryphal, that John Quincy Adams eavesdropped on visiting government dignitaries by placing them at one focus of the Capitol's Statuary Hall and himself at the other. Besides creating an acoustically joined pair of foci, an elliptically curved surface can also create a sonic conduit, as sound bounces along its periphery hugging the wall. Geometrically complex spaces have complex acoustic arenas.

As these examples illustrate, visual and acoustic arenas of the same physical space can differ, sometimes surprisingly. But even though specific geometric designs can create acoustic arenas and aural distances that differ dramatically from their visual counterparts, the aural experience of distance between a speaker and a listener can change when they enter even a simple enclosure. In more complex spaces, visual proximity can correspond to aural remoteness, just the opposite experience to that in whispering galleries.

In summary, aural architecture determines the aurally perceived size and location and the acoustic arena of a speaker in each area of a physical space. Although these acoustic properties have a social meaning, however unintended, the most impressive historical examples of aural architecture are famous chiefly as spatial curiosities. Indeed, there is little evidence that the architects intentionally designed the acoustic arenas of these spaces based on the social, navigational, or aesthetic needs of those who were to use them.

Illusions of Expanded Spaces

Windows, mirrors, and pictures belong to a specific class of architectural embellishments: visual space manipulators. A window expands visual space by establishing a visual connection between the observer and an additional physical space; a mirror expands space by connecting the observer to a replica of the existing space; and a picture expands space by inserting the image of another environment. The size of the window, mirror, or picture determines the degree of coupling between two or more spaces.

When physical constraints force a traditional architect to work within a limited space, the art of visual illusion becomes important in making a space seem larger than it is. Mirrors, in particular, are visual space expanders: they create the visual illusion of added space. Small rooms with many mirrors give the impression of being far larger than their actual size. A mirror is a window into a virtual copy of the same room, located on the other side of the wall. The experience of the enclosing surfaces then disappears. With mirrors on multiple surfaces, as in dance studios, replicated virtual spaces grow exponentially as if the visual space were infinite.

Having drawn analogies between seeing and hearing, we can search for aural parallels to these visual space expanders. Although, from the physical perspective, light and sound waves closely parallel each other, from the experiential perspective, they diverge widely.

To understand this divergence between physics and experience, consider a crackling (noisy) candle, emitting both light and sound energy in a room with a mirror surface that reflects both forms of energy. The light and sound energy waves radiate spherically, following the same trajectories and producing the same reflections from the mirror. An observer sees the candle and its replicated image in the mirror. The image is equivalent to a virtual candle located in a virtual space. Similarly, a listener hears the direct sound from this noisy candle and hears the reflected sound from the mirror, which is *physically* equivalent to the sound that would have radiated from a virtual crackling candle in a virtual space. The optical and acoustical phenomena parallel each other closely. Indeed, parallels between light and sound in enclosed spaces are found in most elementary textbooks on spatial acoustics.

Because of physiological differences between hearing and seeing, however, the *experience* of reflected light and that of reflected sound diverge. Whereas multiple sonic reflections are generally perceived as a single fused sonic event even when sound arrives from different directions and at different times, multiple visual reflections always remain distinct. Under normal circumstances, aurally, we would perceive only a single noisy candle in our example, along with the reflecting wall; visually, we would perceive two candles, an actual and a virtual one.

A sonic reflection creates the illusion, not of a new virtual candle in a new virtual space, but rather of a louder (aurally larger) noisy candle—and it induces the aural perception of a solid wall. If the delay between the direct sound and its reflection is large enough to produce a distinct echo, and if we experience the echo as *un*bound from the direct sound, then, and only then, the sonic reflection creates the aural illusion of a separate virtual candle. But normally, we experience a distinct echo as bound to the original sound. A sound-reflecting surface is the aural equivalent of an opaque wall—a spatial boundary, a spatial reducer.

What kinds of acoustic objects and designs create the aural illusion of a larger space? What are the aural analogues to mirrors, pictures, and windows in creating this illusion? Unfortunately, such analogues remain in the hypothetical realm: they have not yet been realized in physical spaces.

To create the aural illusion of an expanded space, we must simulate the sound field at a virtual window, that is, we must replicate the sound field that would have been present if an additional space were actually present.

Sound absorption is an aural space expander. *Complete* sound absorption would simulate a virtual window into an infinite, unbounded space, a space without the ability to respond to sonic illumination, and with no sound sources of its own. Thus a thick panel of dense, completely sound-absorbing materials, one that could absorb *all* sound waves that arrive, would aurally replicate a window into an absolutely open space. Sound arriving at the panel would completely disappear, as if it had actually encountered an open window into an absolutely open space—an infinite void.

But if the virtual space is to be the equivalent of an actual room, rather than an infinite void, the appended space must have its own sound-reflecting surfaces, sound absorption, and sound sources. The virtual space would reverberate sound entering from the real space through the virtual window. To experience the appended space as an actual environment, we would need to reproduce the appropriate sound field at the window. Hypothetically, we might create this illusion in a sequence of stages.

First, to create the sound field, we might embed an array of small loudspeakers driven by a spatial synthesizer in a sound-absorbing panel. These loudspeakers would then duplicate at the surface of the panel the sound field of a space as it would appear at the virtual window. We might simulate the sounds of a bird sanctuary with chirping birds and babbling brooks together with its acoustics, including reverberation and sonic reflections. Our simulation would need to replicate the sound field only at the virtual window since listeners could not actually enter the virtual space. Walking near the panel with their eyes closed, they would have the impression of a window opening onto a bird sanctuary. The aural experience would be analogous to a visual picture of a bird sanctuary. In fact, if we had the panel also contain a visual display of the sanctuary, we would have a multisensory space expander.

Second, to refine our virtual window onto a virtual space, we would need the virtual space to respond to sound originating from the actual room. If the bird sanctuary were a real space, listeners could shout through the window into it and then hear the reverberation of their voices. Hypothetically, this is also possible. We might embed an array of microphones into the panel such that the sound waves arriving at that surface would feed a spatial synthesizer that created the virtual reverberation, which the loudspeakers would then reproduce.

Third, to make our virtual space simulate an extension of our actual room, we might expand the area of the sound-absorbing panel to cover an entire wall such that sound arriving from the actual room would be completely absorbed. This would effectively remove the aural perception of a wall. Sound would impinge on the absorbing surface and disappear. We might then have the spatial synthesizer create the sound field at the surface that would have been there had the actual room been larger. We might, for example, have the synthesizer add a delay of 10 milliseconds to the sound that arrived at the wall. The sound field would then be the same as that of an actual room 3 meters (10 feet) wider, with a wall 3 meters farther away. Or, using the same approach, we might simulate still larger and more complex spaces to create the illusions of larger and more complex actual rooms.

Our scenario is compelling if we assume the synthesized sound field could be made identical to its natural counterpart, paralleling an optical hologram, which re-creates the light field of an actual object at the surface of the holographic image. Primitive versions of an artificial acoustic wall have been demonstrated in the laboratory, but the technology has not yet sufficiently evolved to make the dream practical. I have no doubt that such hypothetical scenarios will eventually become reality if technology continues to advance at its current rate. Primitive versions are currently used to make musical spaces feel larger and more reverberant. Many concert halls now incorporate active acoustics with arrays of microphones and loudspeakers. Eventually, perhaps within a decade, aural architects will be able to use "acoustic holography" as an additional tool to create virtual space expanders.

Local Anomalies as Aural Texture

Nobody remembers a visual space that is without unique features. A rectangular room with blank walls and minimal furnishing acquires a unique visual personality only when embellishments such as pictures, wallpaper, colored surfaces, and mirrors are added. Likewise, a prosaic aural space acquires an aural personality only when aural embellishments are included. Openings such as windows and alcoves add aural personality; by absorbing sounds, thick drapes, large tapestries, and upholstered furniture create aural texture, as do statues, pillars, and complex geometries, which diffuse sounds. Such aural embellishments create local acoustic attributes, supplementing global ones such as reverberation. The aural personality of a space is especially apparent to blind persons, who experience embellishments chiefly by listening.

Although the concept of a local aural embellishment is not yet recognized as such, we can easily demonstrate its role in creating a personality for spaces. As children, many of us first experienced an aural embellishment when we placed a conch shell to our ear and listened to the sounds emanating from inside it. Because of the shell's complex inner hollows and passageways, its interior creates resonances that filter background noise to produce a sound that resembles that of the ocean. The region of space near the opening of the shell creates an acoustic anomaly—a spatial filter that changes the spectrum of the background sound. The conch shell is a miniaturized version of a cave or alcove, which is also a hollow that can be experienced at its opening.

There are examples of acoustic hollows other than caves and conch shells. Objects in the shape of a large vase with a narrow neck, called "Helmholtz resonators," change the background sound at their openings. Depending on their construction, they can amplify or suppress particular frequencies. Archaeological and written evidence from ancient Greece and into the Middle Ages indicates that theaters and churches once had acoustic vases scattered about their auditoriums. Although scholars still argue about how effective such vases may have been in enhancing voices, those sitting or standing close enough would have likely heard them as some sort of aural embellishments. The acoustic vase is the man-made equivalent of a conch shell, but with different resonant properties.

To appreciate the extent to which acoustic objects can create aural texture, first consider the visual analogy. Wallpaper produces visual texture because of nonuniformities in its visual pattern. At a distance, the details of the pattern may not be visible but they still create a texture that is quite different from a painted surface. When all surfaces of a space have the same hue, intensity, saturation, and reflectivity, the environment is visually sterile, in contrast to the effect of elegantly decorated and richly textured wallpaper. As the aural analogue of wallpaper, consider a wall that had a pattern of conch shells embedded in it, thus creating a pattern of resonances at different frequencies—like variations in aural color. Such a wall would have aural texture. By standing at the optimum distance, you would hear that texture. This example illustrates how small objects, each of which cannot be perceived individually, can be multiplied and extended to produce aural texture. We can take the idea further. An aural pattern might include small regions of absorbing mats, planar reflectors, dispersing wedges, and diffraction gratings. The art of aural wallpaper is as unlimited as that of its visual counterpart.

Besides creating a large acoustic surface from an array of small acoustic elements, we might also design larger acoustic objects that have a recognizable aural personality—the aural version of modern sculpture. After a search of the architecture literature, however, I failed to find any examples of acoustic objects characterized as aural embellishments. Yet many artistic and religious objects have acoustic properties that match our definition of aural embellishments, even though they were never intended to be aural art or acoustic sculptures. Although creative artists can design such objects for their explicit impact on listeners, there are also vast repositories of historical artifacts that have unusual acoustics. Combining mastery of both archaeology and acoustics, acoustic archaeologists have discovered ample physical evidence in ancient sites that older cultures valued objects and structures for their acoustics. Leaving extensive discussion of the cultural relevance and symbolic meaning of these objects and structures to chapter 3, let us briefly consider three examples of unintentional aural embellishments.

Our first is from the Mayan culture. Acoustic consultant David Lubman (1998) discovered that, when illuminated by the sound of clapping hands at a particular location, the staircases at the Pyramid of Kukulkán at Chichén Itzá produce chirplike echoes that bear an uncanny resemblance to the call of the Mayans' sacred bird, the resplendent Quetzal. This readily perceived resemblance most likely invested the staircases with special religious meaning.

Our second example, also a religious one, is the medieval shrine to Saint Werburgh in Chester, England. As described by Lubman (2004), the shrine's six recesses, where kneeling pilgrims would insert their heads while pleading their petitions, serve both as amplifiers and as filters, giving the petitioners' voices dramatic and emotional emphasis with only modest vocal effort. The shrine's recesses thus create uniquely private acoustic arenas that exclude external sounds without walls. (Their modern social counterparts might be alcoves designed for the aural intimacy of lovers.)

Our third example is a sculpture by the respected twentieth-century Spanish minimalist artist Eusebio Sempere. Composed of a three-dimensional array of polished

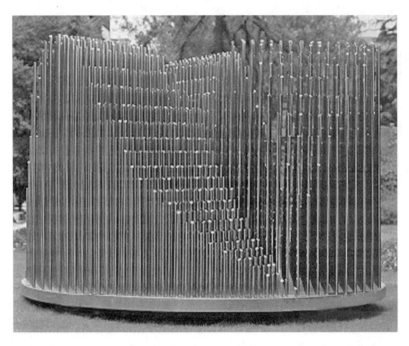

Figure 2.4
Eusebio Sempere's stainless-steel tube sculpture. Courtesy of Collection of Fundación Juan March, Madrid.

stainless-steel tubes, the sculpture rotates at its base, as shown in figure 2.4. The moving surfaces serve not only to dramatically reflect the sunlight but also to selectively filter transmission of particular frequencies of sound. Listeners on one side hear a tonal modification of sounds coming from the other side—the moving surfaces acting like the aural equivalent of colored glass prisms. Although scholars took several decades to recognize the sculpture's acoustic properties (Mártinez-Sala et al., 1995; Sánchez-Pérez et al., 1998), there can be little doubt that, by changing the sounds that propagate through it, Sempere's work serves as an aural embellishment. If the artist had had a background in acoustics, we would assume he had intended to design a multisensory sculpture.

A search of the literature revealed that the phrase "aural sculpture" applies almost exclusively to experimental art based on active sound sources, often interacting with the listeners and often prerecorded. Artists are sculpting the sound field by manipulating sources and their location. I did not find any reference, even using alternative search phrases, to any form of aural sculpture experienced by illuminating an object with the natural sounds of a living environment. Most likely, the aural effect is too subtle for a population more familiar with high-impact computer-generated sounds that do not occur in nature.

The previous examples also illustrate a plausible process by which aural embellishments come into existence. Without any formal knowledge of acoustics and aural perception, an artist creates an aural embellishment as an unintentional artifact of another design process. We, the listeners, are then left to sort and evaluate objects for their aural aesthetics. But with the appropriate knowledge, an artist can also explicitly create aural sculpture. And if that art, however created, is then included in a space, it becomes an embellishment of aural architecture. Nevertheless, the aesthetic value and symbolic meaning of these aural embellishments still depend on the attitudes of those who listen to them. Aural adornments can be overlooked, barely noticed, or even dismissed, appreciated or even revered.

Although aesthetic space, like social, navigational and musical space, is always a reflection of the prevailing culture, even when not recognized by auditory experts and professional architects, aesthetically pleasing aural spaces and their aural embellishment may still arise. They are there to be discovered. And they may be consciously experienced by those who have developed a refined sense of aural spatial awareness.

The Affect of Enveloping Reverberation

Aesthetically pleasing at an appropriate level in musical spaces and the label for millions of sonic reflections, reverberation can be mixed blessing in ordinary living or gathering spaces. Excessive reverberation degrades the intelligibility of spoken communication, raises the background noise level, and makes a living or gathering space aurally unpleasant, whereas inadequate reverberation makes a space seem aurally dead, unresponsive, and uninviting. Energy in the late-arriving sonic reflections reduces the size of the arena by creating corrosive noise, whereas energy in the early-arriving sonic reflections increases the size of the arena by amplifying and focusing a speaker's voice. In an unenclosed space with no sonic reflections, oral communication between speaker and listener is difficult unless they are close to and facing each other.

Each specific area of a space may have its own reverberation profile. An alcove with deep-pile rugs and a low ceiling has less reverberant energy than the large open space to which it is connected. The acoustic properties of a space are locally distinct when the profiles of early and late sonic reflections are not uniform throughout the space. Indeed, reverberation is uniform only when the space is large, open, and acoustically uniform. This is desirable for performance spaces, such as concert halls, but not necessarily appropriate for other kinds of spaces.

The physical properties of reverberation tell us little about their experiential meaning. From a social perspective, reverberation does not intrinsically produce a specific affect; rather, the affect is indirectly determined by the listeners' aural expectations. Spaces that match the listeners' aural expectations are pleasing to them; spaces that do not are not. Listeners have expectations about the way that reverberation should respond to sonic events (responsiveness), and about the way that reverberation should create acoustic arenas of particular sizes (social spheres). We cannot specify what the listeners' affective response to reverberation will be—whether stress, anxiety, comfort, or well-being—without examining the social context; this aspect of reverberation in aural architecture is culturally relative.

Aside from its influence on acoustic arena size and listeners' spatial responsiveness, reverberation is unlike all other sounds. Because enveloping reverberation cannot be localized as a sound originating from a particular place, we refer to it as "enveloping aural ambience." Just as we experience water visually, tactilely, and aurally as an enveloping environment when scuba diving, so we experience reverberation aurally as an enveloping environment when we find ourselves within it. The difference in affect between being underwater and on dry land parallels the difference between a cathedral and an anechoic chamber. For this reason, reverberation has an affective component apart from its associations with social expectations. How then is enveloping reverberation experienced, what properties should it have, and what role should it have in aural architecture?

The ability to determine the location and direction of a sound has undeniable survival value. When you hear a stampeding herd of animals, knowing which way to run can be a matter of life or death. If the natural acoustics of forest or savanna destroyed their ability to locate the direction of an approaching herd or predator, our ancestors would very likely never have survived. Fortunately, natural environments typically produce low-level sonic reflections, not enveloping reverberation. Over millions of years, our auditory cortex evolved the means to determine the location and direction of a sound source by using the direct sound, which is reliable, while disregarding sonic reflections from a multiplicity of surfaces. That process fails, however, when enveloping reverberation from enclosed spaces competely overwhelms a weak direct sound. Evolution could not adapt to the reverberation of enclosed spaces because they were the exception rather than the norm.

In a modern context, the ability to aurally localize the blaring siren of a fire truck in an acoustically complex metropolis is central to a city driver's safety. Moreover, the driver's feelings of anxiety upon hearing the siren are instantaneous and automatic. Unlocalized sounds are associated with potential danger; danger triggers either anxiety or a heightened state of arousal, which is a biological state of enhanced alertness.

Reverberation gives rise to an interactive experience, with the space entering into an acoustic dialogue with its occupants. It is difficult to enter a reverberant space surreptitiously because the sound of your footsteps produces an acoustic reaction for all to hear. Metaphorically, the reverberated sound of footsteps is the reactive voice of the space; the spatial acoustics of a reverberant space announce the presence of active life by responding with an audible hello, as either a whisper or a shout. The acoustics are like the voice of a receptionist, with aural architects determining how that voice should greet entering visitors. Aesthetically pleasing reverberation produces a dialogue that is neither unresponsive nor domineering—a pleasant voice. (As we will see in chapter 3, the idea that a space has a voice provides a plausible explanation of how pre-scientific cultures experienced spatial acoustics.)

Or, to use the metaphor of dining out, enter a space, and it responds to your footsteps with a serving of reverberation. But unlike dining in a restaurant, you cannot choose the taste of reverberation from a menu. But if you could select its taste, what would you choose? The clearest distinction among the choices involves the frequency content: at every frequency, reverberation fades away slower or faster—has a different decay time. Frequencies that last longest dominate tonal color because the other frequencies have already decayed to inaudibility. Ideally, you would choose tonal color to match your mood and aesthetic taste.

From our ordinary experience as Western listeners, we acquire associations to tonal color. We associate low frequencies with objects that are soft or malleable, and high frequencies with objects that are hard or brittle (Freed, 1990). In the language of experience, the two categories are often called "warm" and "cold," respectively, even though they are unrelated to temperature. Although connections between physical objects and tonal color are, no doubt, learned, they are consistent across large populations for one simple reason. Objects that are soft and malleable, such as wood or fiber, produce weaker high frequencies when bent, hammered, or otherwise manipulated. Hard materials, such as glass, steel, or porcelain, produce stronger high frequencies. The two categories of objects absorb sound in the same way that they create sound. A room with a deep-pile rug is heard as warm and soft; a barren room with hard plaster walls is heard as cold and hard. Interior decorating, which is part of aural architecture, determines the tonal color of reverberation. To the extent that enveloping reverberation is analogous to being underwater, tonal color can be thought of as the water temperature.

Although smaller spaces still produce reverberation, as a listening visitor, you experience it as changing the tonal color of the direct sound, not as enveloping you. The acoustic dialogue between you and the space changes, but it remains a dialogue nevertheless. The spatial acoustics of a shower stall may induce you to sing because a small space has numerous discrete resonances. When the pitch and overtones of your voice coincide with these resonances, its loudness is greatly enhanced; when they shift away from the resonances, the intensity of your voice decreases dramatically. Rather than remaining neutral, the space reacts to the presence of some frequencies and not to others. Spaces may thus be said to have tonal preferences. A singer is an aural detective exploring an environment the way a child explores a toy.

Even though a space reacts to all sonic events with its own characteristic response, nobody from our modern cultures imagines that an enclosed space is actually alive.

Using a similar concept, but without realizing that it still applies today, acoustic archaeologists speculate that ancient shamans heard cave acoustics as the voice of a cave's spirit. In ancient cultures, objects were animate, containing living spirits. Although, in modern terms, spatial acoustics have replaced animating spirits in describing the aural personality of a space, nevertheless, I prefer to believe that, however subliminally, some sense of spirits animating spaces resides within us even now.

Application of Spatiality Principles

Having explored some of the experiential attributes of auditory spatial awareness, we are now in a position to examine their relevance to aural architecture. Depending on which cognitive strategy they adopt, those who occupy or live within a space can experience it in any of four distinct modes: social, as an arena for community cohesion; navigational, as local objects and geometries that combine into a spatial image; aesthetic, as an enhanced aesthetic texture; and musical as an artistic extension of instruments. The four modes exist simultaneously for all listeners even if some listeners are aware of only one or two of them. Both the aural architect and the occupants or inhabitants of a space decide on the relevance of each mode, whether consciously or unconsciously. We experience a concert hall, for example, primarily as a musical space, but should the lights fail, we almost certainly would experience it as a navigational space as we tried to find an exit. When small tables and chairs replace the audience seats during Boston Pops performances, we experience a concert hall as a social space. And when attending to the local acoustics produced by statues and alcoves, we experience the hall as an aesthetic space.

Although the aural architect focuses on particular aspects of the aural design of any space, those who use the space control the nature of their aural experience. As a listener, you may be aware of the large spatial volume created by a high-domed ceiling at a given moment, but using those same cues at another moment, you may experience reverberation only as the blending of individual sounds. You aurally sense the location of nearby stairs, doors, walls, and low-hanging chandeliers; and when talking to your partner, you respond to an acoustic arena that is mismatched to the social sphere. Furthermore, those who use space also determine, consciously or unconsciously, its sonic illumination, which in turn influences their experience. A musical space requires music, a social space requires people having conversations, and a navigational space requires transient and continuous background noise. The inhabitants then are the final aural architects of a space.

When a space is being designed, the aural architect must balance how the range of physical properties specified by the acoustic engineer influences various aspects of experiential space: social, navigational, aesthetic, and musical. In many cases, spatial attributes produce conflicting experiences. Large, open spaces are weak on acoustic attributes that enhance navigational cues and local acoustic embellishments. Aural privacy in a multiplicity of small acoustic arenas conflicts with having a single public acoustic arena. A space with a socially dominant region that magnifies a speaker's aural size conflicts with an egalitarian space having uniform acoustics throughout. Conflicting requirements call for choices. For the aural architect, these choices depend on the values of the sponsors, as well as on the expected use of the space.

Of all attributes, throughout the history of architecture, the size of an enclosing space is, perhaps, *the* major source of conflict. Motivated by theology, economics, or politics, the need for large audiences dominates the architecture of public spaces. An intimate space for chamber music with an audience of 6,000 is impossible. For the same reason, the Protestant Reformation shifted to smaller churches, in part, as the means to elevate the importance of the spoken liturgy, which would have been unintelligible in the acoustics of a large cathedral.

Aural and visual architecture converge insofar as every object and every geometric shape has both visual and aural attributes. Because, however, we experience many architectural elements with more than one of our senses, not all of which can be best served at the same time, architects must make sensory trade-offs, which vary from culture to culture. For example, an open window couples one space to another by allowing the passage of light and air. But that same opening also provides a path for extraneous noise, and the opening functions as a perfect sound absorber with no reflected energy. Windows are thus multisensory acoustic structures. Similarly, statues are aesthetically pleasing to the eye as sculpture, but they also diffuse sound and may therefore affect the acoustics of a musical space. Panels suspended from the ceiling may produce welcome amplification through early sonic reflections, but may also produce an unwelcome visual sense of confinement. Where diffusion of sound is desirable, using an acoustic diffraction grating may simply be too visually unaesthetic to include in a space.

The aural and visual architecture of a space may diverge in other ways. Visual illumination is determined by the way that architects place lamps and windows; light sources are mostly static and built into the spatial design. In contrast, sonic illumination is mostly a consequence of some human activity. As a rule, then, aural architects have less influence than visual architects do over illuminating energy. As with any rule, however, there are clear exceptions: visual architects sometimes give control of visual illumination to the users of a space and aural architects sometimes assume control of sonic illumination.

Aural architecture can influence, both directly and indirectly, the mood and emotions of those who occupy or live within a space. Such influence can be the direct consequence of how the space changes sounds: amplifying background noise to an uncomfortable level, creating enveloping reverberation, destroying aural localization cues, or pleasantly blending a sequence of musical notes. In these cases, listeners are responding to sounds modified by the aural architecture. And it can also be the indirect consequence of spatial acoustics: acoustic arenas that are too small to include the companion of a listener within the social sphere, a listener's personal associations to familiar aural embellishments, or a listener's comfort at navigating a space in the dark using strong aural cues. Listeners' responses to a space thus depend on the direct and indirect manifestations of spatial acoustics, as well as on culture and context and the listeners' individual biases, histories, and personalities.

In controlling the sonic illumination as part of the design process, an aural architect becomes a soundscape architect. This is seldom possible, however, because the dynamic and ephemeral activities of those who use a space are the dominant source of sound. Yet in certain art forms, the artist is also allowed to control sound. Japanese garden design, an ancient art form that stylizes and miniaturizes natural environments by creating the illusion of larger ones, includes the aural experience of space. Not only are objects and plants arranged for their visual pattern, but also for their ability to shadow and reflect sound from active sources. David A. Slawson (1987) mentions how muffling the sound of a waterfall makes it seem farther away, thereby enlarging the perceived size of the garden. By including the aural experience in its design, a Japanese garden becomes the artistic union of a landscape and a soundscape, and its designer a truly multisensory architect.

3   Aural Spaces from Prehistory to the Present

The historical sense involves a perception, not only of pastness of the past but of its presence.
—T. S. Eliot, 1975

Investigating the relationship between culture and aural architecture would be easy if we could study a wide variety of cultures with dissimilar attitudes toward auditory spatial awareness. There are two problems with such a study. First, compared with their historical counterparts, modern societies are remarkably similar to one another. International professional societies, which transcend cultural boundaries, further standardize the attitudes of those who build spaces. Second, societies distinctly different from Western culture, potentially with contrasting examples of aural architecture, have not contributed to Western scholarly literature. Although there are likely to be important examples of aural architecture in, for example, Asian, African, and Middle Eastern cultures, I could find no published information about them, despite a thorough search of the literature and consultation with numerous experts. Hence, however regrettable it may be, this book has an unavoidable Western bias.

As an alternative to examining extant cultures for variations in aural architecture, we can examine historical cultures, with their varied politics, music, religion, science, and social structures. In exploring the relationship between aural architecture and social values, the history of aural architecture, far from simply cataloging historical spaces and their associated acoustics, has found clear evidence that older cultures employed alternate cognitive frameworks for experiencing sound and space. Our modern approach to aural architecture is only one of many possibilities.

Because spaces and buildings endure, often for centuries or longer, the two-way relationship between culture and aural architecture is passed along to many successive generations. Once constructed, the aural architecture of a space memorializes the values of those who built it. Later generations, in turn, develop their own cognitive frameworks from experiences with those inherited spaces; newly constructed spaces are then created from those cognitive frameworks. Just as we may trace Western attitudes toward politics to ancient Greek culture, so we can trace our attitudes toward aural architecture to earlier cultures as well. Indeed, residing under an overlay of science, technology, and rationality, ancient cognitive frameworks are still part of our sensory legacy; we can understand our current aural architecture only by relating it to that of our ancestors.

Historical evidence suggests that aural architecture resulted from unplanned and inadvertent acoustic accidents, which were then passed through the cultural filter of social and religious values. By examining a variety of societies and showing how they incorporated aural space into their culture, this chapter supports the hypothesis that aural properties of spaces were *not* the result of conscious design.

The Aural Experience of Space as a Cultural Filter

Although modern scholars can examine ancient structures that have survived, and although a modern audio engineer can synthesize their acoustics from archaeological artifacts and written records, it is impossible to re-create the aural experience of the original listeners who used or lived in the structures, just as it is impossible to re-create their music or ceremonies. We will still hear acoustic environments, however accurately simulated or reconstructed by engineers to replicate what the ancients heard, from the perspective of modern listeners. The spatial experience of our ancestors is forever buried with them. Nevertheless, we can at least partially reconstruct their cultural frameworks to show the degree to which aural architecture and the experience of sound depend on culture.

Chapter 2 described the sound of church bells in nineteenth-century rural France as a soundmark whose acoustic arena delineated the membership space of a town. Although the bells may have the same intensity, pitch, and timbre now as then, the social and cultural meaning of their sound is now dramatically different: hearing them, modern listeners do not have the same experience of social inclusion. Similarly, modern listeners experience the aural architecture of a twelfth-century cathedral without the religious feelings, faith, and worldview of listeners of that epoch. Even though the acoustics of the cathedral have not changed in the intervening nine centuries, modern listeners are unlikely, upon entering the edifice, to feel transported to heaven on earth, as many, if not most, medieval listeners very likely did. An even starker contrast becomes apparent when we consider how differently a visiting twenty-first-century geologist and the prehistoric celebrants might experience the acoustics of a ceremonial cave used 10,000 years ago by a tribe of hunter-gatherers. Focused on the cave's geology, our modern scientist would find its unusual acoustics of passing interest—certainly without the deep symbolism and emotional associations we can safely suppose they had for members of the prehistoric tribe.

Although scholars have made widespread use of cultural relativism in interpreting the artistic and sacred objects of older cultures, they have seldom applied it to their aural architecture. Thus scholars appreciate how a visual symbol, such as a circle, a cross, or a star, can have profoundly different symbolic meanings across cultures and time. Yet they tend to ignore the culturally various interpretations of aural architecture, in part, because sound is ethereal. For modern intellectuals with a scientific and rational bias, itself a cognitive framework, understanding the symbolic attitudes of preliterate cultures is difficult, and interpreting the aural experiences of historical spaces almost impossible. Just as these experiences have changed over the millennia, so, too, have the corresponding cognitive frameworks. The social interpretation of spatial attributes has, in fact, undergone greater change than the spaces themselves.

Issues in Using Historical Evidence

To interpret historical records and archaeological artifacts that pertain to the aural architecture of a given culture, we need to reconstruct that culture's attitudes toward auditory spatial awareness. Unfortunately, acoustic archaeology is a highly speculative field, supplementing sparse evidence with culturally linked inferences that necessarily include a modern perspective. Whereas sound is today understood mostly as information, entertainment, or background noise, older cultures were more sensitive to their rich soundscape, which implicitly included aural architecture. However thoroughly a modern guidebook may describe the history and architecture of a town, it rarely mentions the town's soundscape, except perhaps as a curiosity. And even if the structures and spaces of a town have been faithfully preserved from ancient times, its soundscape is distinctly modern: ancient Greek philosophers and subsistence farmers never heard a car or truck drive by, much less an airplane fly over. For all these reasons, the history of aural architecture is even more fragmented and speculative than most cultural histories.

In older cultures, the written history of the use and design of spaces was the exclusive domain of an educated elite who possessed the means for keeping written records. Meanwhile, the craftsmen and laborers who actually created aural architecture, though trained, were unschooled and illiterate. Because of the social divide between these two groups, knowledge about acoustics, whether sophisticated or otherwise, was not necessarily incorporated into aural architecture. The traditions of craft guilds and laborers almost certainly played a strong role in creating spaces, but neither group left any written records.

Another problem with the written history of aural experiences is readily illustrated by the etymology of aural words. Ancient cultures described their auditory experiences in terms of observable external events and objects. Thus, for example, the word *reverberate* is derived from the Latin root *verberare*, which meant "to flog or beat with a leafy branch [*verbena*]" (Souter et al., 1968). *Perceive* is derived from the Latin *percipere*, meaning to seize. Such definitions are actually referencing external events not aural experience.

Still older cultures were more comfortable explaining their aural experiences in symbolic, religious, mystical, or spiritual terms. Evidence from preliterate peoples suggests earliest humans heard the voices of their gods in the acoustics of sacred spaces. Without modern technology and without the need to use an objective language, these older cultures were free to experience aural attributes more directly; for them, sounds had a mystical quality.

Aural experience, whether of sonic events or acoustic spaces, is itself a combination of auditory perception (neurobiology and learning), which is relatively consistent over the millennia, and cognitive interpretation (use and meaning), which is subject to the viscissitudes of intellectual, religious, and emotional translation. Cultural values convert physical phenomena into experiential phenomena. The history of aural experience is therefore the progressive change in the nature of this conversion—from the mystical religion of prehistoric tribal shamans to the rational explanations of modern acoustic and perceptual scientists.

There is no evidence that an aural architect can *internally* auralize—aurally visualize—a novel acoustic space.[1] Converting a mental image of a prospective spatial design into its acoustic properties and aural architecture is still too complex for all but the most sophisticated acousticians. Without the ability to aurally visualize an imagined space, aural architects cannot create a novel acoustic space purely as a mental activity. This contrasts with visual architecture, where architects can readily visualize novel visual spaces. History provides an almost limitless portfolio of visual sketches of spaces, but no corresponding portfolio of aural records. The radical asymmetry between our capacity to visualize the visual aspects of a space, and our incapacity to *aurally* visualize its acoustic aspects dramatically affected the history of visual and aural architecture.

Even today, most aural designs are arrived at by trial and error, with designers selecting an interesting variant out of hundreds generated by scientific analysis and spatial simulators. That done, the variant can, in most instances, be constructed without undue delay. Lacking modern technology, older cultures could evaluate only the spaces that existed and, having selected one, set about laboriously to build another like it. But older cultures were also more patient. Whereas modern concert halls take only a year or two to complete, and virtual spaces, only an hour or two, cathedrals took centuries.

The history of aural architecture is thus the study of spaces that were selected, valued, preserved, and extended, regardless of how they originally came about. Spatial attributes were selected using criteria based on religion, mythology, social utility, or aesthetic pleasure. If a space with novel properties was found to be socially useful, the culture incorporated it; otherwise, it was discarded. In this respect, older cultures had much the same aural goal in mind that technically sophisticated modern ones have—creating spaces whose size and acoustic properties serve the needs of their people, whether through scientific application of electroacoustic technology or the intuitive selection of a natural space with nearby hills.

Mystical Voices from Heavenly Spaces

Before examining how earlier cultures might have used auditory spatial awareness, we need first to explore how sound connects listeners to an external event. What would hearing have meant to our distant ancestors? A speculative answer to this question provides clues about how we as modern humans experience space without depending on scientific explanations of acoustics.

To begin, imagine that, as an infant, you had been adopted into a community of early humans 50,000 years ago, and that you grew up in a world without clocks, calendars, electricity, telephones, or even paper and pencil, with no scientific knowledge of the physical properties of sound or acoustics. Without even the basics of high school biology, physics, or psychology to interpret the events of life and nature, you would have had no idea that your head contains a brain, or that the physiology of your eyes and ears converts physical energy into neurological signals. In such a culture, you would hear sound immediately, directly, without intermediate analysis. The absence of factual knowledge of physics and biology would not, however, reduce the role of intelligence, experience, curiosity, or perception.

Suppose now, as an adult searching for food, you came upon an opening into a mountain that led into a vast cavern. Standing at the entrance, you would have heard it speak to you in the same way that a conch shell speaks. Sound entering a cavern is changed sufficiently that, when it reradiates back through the opening, it seems as though it is coming from within. The cavern would not be quiet: as you passed by its opening, you would have heard the cavern speak to you. The *voice* of a resonant cave is more than a literary metaphor. You would have felt the cave was *alive* when it acknowledged your presence by responding to your footsteps with a voice of its own. From an experiential perspective, a cave is something that has a voice and sounds alive. Only from a modern, scientific perspective is it simply a natural hollow with sonic reflections and resonances.

Lacking a scientific explanatory framework preliterate cultures used religion, along with its associated art and myths, to explain a wide range of otherwise inexplicable events, to invest the unknowable with some kind of causal meaning. Such explanations provide a memorable, enduring, and external description of experience, one that can be passed along to future generations.

To the extent that we, as modern humans, can suspend our educated minds when listening, we become like our distant ancestors. Banging two sticks together makes a distinctive sound, but how would our ancestors have explained that sound? Were the sticks "talking" to each other? Without even a rudimentary understanding of atmospheric physics, how would they have explained the sound of thunder? If we suspend our modern understanding of voice, hearing can be understood as listening to the voice of the spirit of a thing or space. The reverberating sound of a cavern then becomes the voice of the cave *spirit*. Voice becomes the means by which a spirit, whether near or far, talks to us, gets into our heads. For our ancestors, voices included those of powerful spirits, the sounds from large acoustic structures and objects like caves and mountains, from thunder, and from wind itself.

Aural architecture traces its origins to the voice of the space spirit. Believing the power of the voice measured the power of the spirit, early humans ignored the whispers of ordinary spaces and focused on large caves with commanding voices.

Anthropological studies of the few remaining preliterate tribes in the early twentieth century, before technology and globalization had intruded on them, provide some clues about their alternative cognitive frameworks. An ethnopsychiatrist working for the World Health Organization, J. C. Carothers (1953) described the education of preliterate peoples as verbal and dramatic, with no demands placed on logic, reasoning, or inference. By the time children were 12 years old, they were comfortable with and accepting of all that was unknown. These peoples made no clear distinction between subject and object—experience and the source of experience were unified. In such a monoideic psychology, external reality, internal needs, and memories of the past are one, and sound—personal, emotional, and immediate—is a major contributor to that unified one.

Walter J. Ong (1982), a scholar of anthropologic psychiatry, asserted that early humans would have been aware of sound as revealing interior events, as opposed to visual appearance, which reveals only the surface of objects. The bellowing of the elephant revealed the animal's interior state of being. In the *aural consciousness* of our early ancestors, sounds would have originated from the interior and magically appeared inside the self.

Thus, then as now, sound acquires its power by its experiential immediacy, its direct connection of source with listener. Because they connect the interior of one person to the interior of another, voice and music are some of the most powerful sounds.

There are many older cultures that revere the power of sound. Among the Eskimo, all sculpture speaks; silent idols are unknown, and deities are masked dancers who speak and sing (Carpenter and McLuhan, 1960). In his study of Hinduism as a sonic theology with roots into the ancient past, Guy L. Beck (1993) has argued that Hindus experienced the Divine by listening to its voice, as embedded in Vedic music, through which they attained peace, release, and liberation. To this day, Judaism and Christianity alike make repeated references to dialogues with God: "God spoke to him," "God heard his cry," "God listens to His people."

Because sound was so powerful, spiritual leaders were needed to control that power. Social groups followed spiritual leaders who possessed special abilities to create ceremonies and to interpret experience using the power of sound. Many ancient religions had shamans, whose extraordinary powers, it was believed, could heal the sick, change the weather, foretell the future, communicate with spirits, and induce altered states of consciousness. The shaman translated ethereal voices into holy messages. Sharing this viewpoint, Jean Clottes and David Lewis-Williams (1996) noted that "at all times and in all places people have entered ecstatic or frenzied altered states of consciousness and experienced hallucinations. Indeed the potential to shift, voluntarily or involuntarily, between different states of consciousness is a function of the universal human nervous system. All people have to cope with different states of consciousness in one way or another. Some people—by no means all—became shamans." Our modern society also has its "sonic shamans"—priests, musicians, and politicians—who are revered for their ability to evoke powerful emotional responses with the sound of their voices or instruments.

Modern cultures have an ambivalent attitude toward manipulating the mental state of people, often reserving the phrase "altered state of consciousness" for socially un-acceptable experiences, such as the mental distress induced by complete sensory depri-vation (Cohen et al., 1965) or hallucinations induced by psychotropic drugs. But under the right conditions, loud music, protracted exercise, and deep meditation can also in-duce altered, if not trancelike, states. Mary Florentine and colleagues (1998) describe maladaptive behavioral patterns, typical of substance abusers, among those who nor-mally listen to excessively loud music. The term *altered state* is misleading because it presumes the existence of a normal state as a reference point. In fact, every sensory ex-perience, not only unusual or extreme experiences, has the possibility of changing emotions, which are also altered states. Priests, musicians, and politicians are merely experts at using aural stimulation as social and religious tools to affect their listeners. When the aural experience of an acoustic space is sufficiently strong, its voice contrib-utes, however slightly, to creating an altered state of consciousness in listeners, even in modern listeners. By extension, the aural architect who designs a space is also an aural manipulator—a modern-day version of an ancient shaman.

Acoustic Archaeologists Interpret Ancient Spaces

Archaeologists and anthropologists convincingly argue that sacred caves, as sites of mystical experience where ancient rituals were performed, constituted a special sha-manic cosmos.

The way in which each individual cave was structured and decorated was a unique result of the interactions of four elements: the topography of the cave, its passages, and chambers; the univer-sal functioning of the human nervous system and, in particular, how it behaves in altered states; the social conditions, cosmologies, and religious beliefs of the different times at which a cave was used; and lastly, the catalyst—the ways in which individual people and groups of people ex-ploited and manipulated all of these elements for their own purposes. (Clottes and Lewis-Williams, 1996)

The cave wall paintings at Altamira, dating from the Upper Paleolithic period some 20,000 years ago and discovered at the end of the nineteenth century, shifted our conception of paleolithic humans, who would no longer be labeled as "primitive." Indeed, the Altamira cave has been called the "Sistine Chapel of Quaternary art" (Beltrán, 1998). With their limited tools and materials, Stone Age artists created works of art that modern painters can only envy. Picasso himself has been quoted as saying that "not one of us could paint like that." Later, in the twentieth century, several thousand sites with wall art were discovered in hundreds of countries; some in eastern Germany have been dated as far back as 300,000 years ago. David Coulson and Alec Campbell (2001) estimate that there are perhaps a million cave art images in southern Africa alone. Although, as evidenced by the written reports of the Chinese philosopher Han-fei-tzu, 2,300 years ago, cave art is not a recent discovery (Bahn, 1998), only recently has it become a prominent contributor to our understanding of human nature and the origins of civilization. Clearly, prehistoric humans were artistically sophisticated, sensitive to their world, and strongly invested in giving meaning to their experiences.

Protected from destructive atmospheric influences, cave wall images are tangible, enduring manifestations of the visual art of early humans. In contrast, their auditory art has no enduring manifestation, nor of course could it have for any pretechnical peoples. Although certain primitive objects, mostly made from bones, have been interpreted as flutelike musical instruments, and although there are images of a man playing a musical bow (father to the harp) in the Caves of the Three Brothers at Ariège in the French Pyrenees, available data are too sparse to draw strong conclusions.

We should not, however, conclude that these distant ancestors lacked aural sophistication. Researchers are only now beginning to speculate on the auditory arts of prehistoric peoples. It is unlikely that humans with the sophistication to produce complex images on cave walls would not have also discovered auditory art as a parallel to their visual art, especially since all preliterate cultures appear to have displayed a strong interest in sound. Just as the grand caves and caverns presented early humans with unusual visual environments, so they also presented them with the unusual acoustics of large enclosed spaces. Caverns were nature's bequest of concert hall acoustics to peoples who would otherwise have experienced only open-air acoustics.

Steven J. Waller (1993), a pioneer of acoustic archaeology, suggested that the paleolithic art found in the caves of Lascaux and Font-de-Gaume was influenced by the acoustic character of the chambers in which it was created. Pictures of bulls, bison, and deer were more likely to be found in chambers with strong echoes, spaces whose acoustics created percussive sounds similar to the hoofbeats of a stampeding herd. A typical example of such cave art is shown in figure 3.1. In contrast, acoustically silent chambers are more likely to contain drawings of felines. Cave art may well have incorporated echoes as a supernatural phenomenon that brought life into visual images. Waller and others speculate that multisensory art was part of the hunters' rituals to

Figure 3.1
Two bison in the cave at Lascaux, Dordogne, France. Courtesy of Nacq Partners, Ltd., at www.nacq.com.

summon game. Extensive observations of prehistoric sites support the notion that the subjects of cave wall pictures and the acoustics of their locations were deliberately related. After having personally studied over 150 sites around the world, Waller (2002) observed that pictures of animals whose movements generated loud sounds were frequently placed in spaces having enhanced echoes, resonances, and reverberation. When such spaces are excited by sound, the animal portraits seem to come aurally alive.

The concept of a cave wall surface as a veil that separates the spirit world from that of ordinary mortals is evident in South African rock painting (Lewis-Williams and Dowson, 1990). In this regard, borrowing from Lewis Carroll's *Through the Looking Glass* (1871), Waller (2001) has advanced a compelling theory about the aural perception of echoes in caves. Just as Alice, when viewing a reflection in a mirror, felt she was seeing another world beyond the mirror's surface, so early humans, when exposed to echoes (sound reflections) in a cave, would have felt they were hearing the sounds or even voices of spirits from a world beyond the cave wall.

Aside from echoes, numerous other acoustic attributes would have been experienced within caves. Sonic "hot spots," regions of resonances where certain frequencies are amplified, also have also been correlated with cave wall images. Similarly, Michel Dauvois and Xavier Boutillon (1990) found a relationship between cave art and lithophones, natural stalactites and stalagmites that produce marimba-like sounds. Indeed, any object or space with a strong resonance had the ability to acquire spiritual meaning.

Acoustic archaeologists, unlike classical archaeologists and anthropologists without special training in sound, believe that early humans readily discovered the "unusual" acoustic properties of their spaces. Paul Devereux and Robert G. Jahn (1996) studied six Bronze Age structures in Great Britain, dating from around 3500 B.C. Chambers of varying geometries and configurations were thought to have been used as sites for ritual burial and other ceremonies. Regardless of the builders' intent, all six configurations manifested sustained, discernable acoustic resonances in the vicinity of 110 Hz, well within the vocal range of a male adult. Devereux and Jahn (1996) suggested that such spaces would have enhanced male chanting in ritual ceremonies. Moreover, stones were configured to produce dominant standing waves along the radial or longitudinal axis, and like the cave art, images correlated with these spatial resonances. It is tempting to think that these man-made spaces were a conscious extension of naturally occurring acoustics.

Extending this hypothesis to more complex prehistoric monuments in Scotland, Aaron Watson and David Keating (1999) observed a wide range of acoustic attributes in the chambers of these monuments that could have had social and religious meaning. At different frequencies, sounds appeared to originate from different locations, and in some cases, they seemed to come from inside the heads of the listeners. Small head motions changed the perceived pitch and intensity. Listeners within the chamber could detect the approach of others from the acoustic disturbance their bodies made as they moved through long passageways. At certain locations inside a chamber, listeners would hear unexpected tremolos, periodic changes in the intensity of sounds; their speech would acquire an unusual quality, often becoming unintelligible. Watson and Keating explored the acoustical properties of these spaces using musical instruments that might have existed at the time, in effect, re-creating their speculative concept of early soundscapes. Curiously, the long passageways combined with a large enclosed volume produced a Helmholtz resonator, an acoustic structure that amplifies narrow bands of frequencies, in this case at about 4 Hz, well below the lower limit of audibility. Rhythmic drumming would have excited a Helmholtz resonance at this frequency strong enough to be felt. Such infrasounds have been associated with otherworldly experiences and, if sufficiently intense, produce discomfort, disorientation, and sensory distortion.

In their speculation on the purpose of these megalithic structures, Graeme Lawson and colleagues (1998) noted that these spaces were acoustically isolated from the sounds of the environment and, more significant, that, when coupled, they would have become reverberation chambers—voice changers. It was well within the capacity of paleolithic humans to have created devices that could produce sounds matching the unusual acoustics of such spaces. Standing in a focal point, during a ritual ceremony, the presiding priest would have been able to raise the intensity of his voice and any noisemaker he might use far above the sounds of the other celebrants. In short, these spaces would have created an auditory experience uniquely disconnected from the familiar earthly soundscapes of open plains and dense forests.

Animism, which holds that all things, animate and inanimate, contain a soul or spirit, is one of our oldest belief systems. Bushman folklore, among others, is openly animist: "O beast of prey! Thou art the one who hearest the place behind, it is resonant with sound" (Bleek and Lloyd, 1911). The experience of echoes appears in the legends of the South Pacific islanders (Jobes, 1961) and the Paiute Indians (Gill and Sullivan, 1992). Moreover, Waller (1999) described an indigenous tribe in India that still chose the caves for its rock art by the quality of their echoes. Ancient Greece, as a transitional culture between prehistoric tribes and modern society memorialized the myth of Echo in plays, stories, and myths. In the myth of Narcissus, Echo was the nymph who simply repeated what she heard (Bulfinch, 1964). Rather than interpreting an echo as a delayed sound reflection from a surface, the ancient Greeks heard it as a distinct voice with symbolic meaning.

Evidence from other sources parallels the analysis of prehistoric rock art: sounds represent the presence of spirits. Greek myths, using concepts from earlier societies, made extensive use of oracular sites, special places where the gods spoke to ordinary mortals through oracles, priestesses who could communicate with the spirit world. The acoustic attributes of these sites reinforced and even created the notion of a dialogue with the spirit world. Devereux (2001) suggests that the voice of Trophonius at Lebadea in Boeotia, Greece, was actually a roaring underground stream, and that the Colossi of Memnon at Luxor, Egypt, spoke to their human audience, perhaps creating sound as a result of the thermal stresses induced by the hot sun.

The history of aural architecture in preliterate cultures reveals an aspect of experiencing space that is somewhat alien to modern listeners. The aural experiences of early cultures, which did not have science to explain sensory perception, were almost entirely subjective, emotional, and affective. This is seldom the case for modern listeners, whose aural experiences add the objective and the scientific to the subjective, emotional, and affective: indeed, we are as apt to think about and analyze the sounds of spaces as we are to experience them.

With that in mind, some modern composers and sound engineers have created musical sounds and spaces whose aural impact on modern ears approximates that of sacred spaces on prehistoric ears. Spatial experiences are again being decoupled from physical reality; virtual spaces are pure experience with disembodied sonic voices appearing suddenly from any location in an imaginary space that is fluid, dynamic, and otherworldly. Because contemporary electroacoustic sounds are unrelated to natural acoustic spaces or sounds, modern listeners are less likely to evaluate their listening experience and more likely to simply feel it.

Beneath the overlay of our scientific sophistication and factual knowledge, primitive cognitive constructs persist, however unaware we may be of them. Over the millennia, and even with advances in audio technology, scientific insights, and rationalized language, our aural *experience* of space has remained relatively consistent.

Aural Experience without Scientific Knowledge

Modern aural architects, with scientific knowledge about physical acoustics and perceptual psychology, probably assume that their historical counterparts also incorporated a rudimentary version of these disciplines when designing spaces. But the evidence does not support this assumption. During the last two millennia, the science of sound progressed at a much slower rate than the corresponding sciences of light and mechanics. And when scientists acquired primitive insights about sound, they were seldom incorporated into spatial designs. Rather than using empirical and theoretical science, builders and designers used whatever rules came to hand. Lacking in generality, stability, consistency, and reliability, these rules were, at best, folk science and, at worst, cultural myths. A short history of acoustics illustrates this thesis.

Michael Forsyth (1985) noted the proliferation of dedicated musical spaces in the eighteenth and nineteenth centuries. Examples of such grand spaces, whose designs were mostly derived from tradition, crude experiments, visual aesthetics, and dogmatic beliefs in imaginary science, and whose acoustics ranged from magnificent to disastrous, include the Royal Albert Hall in London, the National Opera House in Budapest, the Neues Gewandhaus in Leipzig, and the Grosser Musikvereinsaal in Vienna. Scientific analysis of the acoustics of old concert halls was not undertaken, however, until the twentieth century.

We begin our discussion of modern architectural acoustics with its widely acknowledged father, Wallace Clement Sabine (1922), whose pioneering work incorporated mathematical physics and theoretical acoustics into the design of buildings. At the very end of the nineteenth century, as an assistant professor of physics at Harvard, he was asked to improve the acoustics of a lecture hall at Harvard's Fogg Museum. He formulated both an experimental paradigm and a mathematical framework whose predictions were consistent with empirical data.

Having validated his scientific methodology, Sabine then used it to *replicate* the specific acoustic parameters of the Neues Gewandhaus in Leipzig in the new Boston Symphony Hall (figure 3.2), with its greater seating capacity. Although replicating acoustic parameters would remain a tradition well into late twentieth century, Sabine was not expected, encouraged, or even allowed to be a true aural architect by imposing his ideas of aural properties on the new space. Indeed, he lacked sufficient musical training to have had an informed opinion about the desired acoustics. As a reference model, the Neues Gewandhaus, which replaced the Alte Gewandhaus constructed in 1781, had acoustic properties that were historical accidents. Many of these historical accidents

Figure 3.2
Classical shoebox shape of Boston Symphony Hall. Courtesy Boston Symphony Orchestra Archives.

were experienced as musically pleasing. When composers wrote music to be performed in a specific space, they were memorializing and proliferating the attributes of these spaces. Sabine's design of the Boston Symphony Hall was therefore an extension of musical traditions that predated modern acoustics. Musical repertoires and performance spaces would be forever linked.

The aural success of Boston Symphony Hall was, in part, the result of three centuries of accumulated knowledge about sound as a physical phenomenon. At the beginning of the twentieth century, using this knowledge, Sabine transformed the basis of acoustics—from philosophy to science and engineering. A short review of the history of acoustics reveals two patterns: acoustic knowledge evolved relatively slowly, and that knowledge was not integrated into the building arts.

Aristotle (350 B.C.) was perhaps the first to speculate on the nature of sound, which, he contended, is the result of the impact of two solids against each other. The vibrations of the impact "thrust forward in like manner the adjoining air." An echo occurs when the enveloped air set in motion rebounds like a ball from a wall. As an astute observer, Aristotle also recognized that background noise is reduced when the orchestra floor of a theater is spread with straw.

The earliest writings on the acoustics of performance spaces, however, are those of the Roman architect and engineer Vitruvius (30 B.C.), considered the father of spatial acoustics as an observational discipline. In what might well be the first handbook for designing theater spaces, he presents an extended discussion of acoustic principles. Summarized in Book V, these advised, for example, the use of ray tracing to avoid acoustic shadows when objects block sound; they recognized both the destructive influence of echoes produced by large surfaces and the role resonances play in enhancing or degrading intelligibility. "Whoever uses these rules," Vitruvius concludes, "will be successful in building theaters." Although some of his insights would be confirmed by modern science, others would prove to be nonsense.

Not until the seventeenth century, at the start of the classical period in acoustics, did natural philosophers establish the mechanism by which sound moves from vibrating object to listener. Marin Mersenne (1644) and Robert Boyle (1662) demonstrated that air is the medium for sound transmission, and Isaac Newton (1686) proposed a mathematical framework for computing the velocity of sound in a fluid. The scientific advances of the seventeenth century were a logical extension of the much earlier speculations by Aristotle (350 B.C.).

In the nineteenth century, the basic mathematical foundations were laid for the science of acoustics, and auditory psychology became an accepted field of inquiry. John W. S. Rayleigh (1877) established that sound was a radiating wave. George Green (1838) formulated the concepts of reflected and refracted sound. And by exploring the psychophysics and physiology of hearing, Hermann von Helmholtz (1863) separated the perceptual experience of sound from its physical nature. Other great thinkers of the period, such as Poisson, Laplace, Wheatstone, Faraday, Stokes, and Ohm, to name but a few, also contributed to the scientific infrastructure used by Sabine and his colleagues in the twentieth century.

We can summarize the history of spatial acoustics as having three milestones. Aristotle introduced the concept that sound had a physical rather than a mystical basis. Vitruvius established observation as the means for creating pragmatic rules that could be used by spatial designers. And Sabine merged theoretical physics with empirical measurement. During the two millennia spanned by these three milestones, aural architecture gradually acquired a theoretical framework, but one with limited predictive reliability. The experiential aspects of spatial acoustics were not recognized until the twentieth century, when perceptual psychology and neurophysiology established a relationship between physical sound and aural experience. And as mentioned earlier, "face vision" was discovered to be auditory spatial awareness only in the middle of that century.

Yet even with increasingly scientific explanations for sound, spatial acoustics remained intellectually isolated. Because the ways sound is transmitted are more abstract than the ways it is generated, and because its transmission has a less direct influence on listeners, spatial acoustics are elusive, difficult to understand, let alone change for the better. Thus manipulating the spatial acoustics of an immense open-air amphitheater constructed of stone blocks embedded in a hillside was arduous at best, with no assurance of success. In contrast, it was easy to manipulate the length, mass, and tension of vibrating strings, and the perceptual consequences were immediately apparent in terms of consonant and dissonant combinations of sounds. Although the physics of musical instruments and of small spaces are similar, natural philosophers contributed only to the former. Early Chinese craftsmen discovered how to visualize the resonant standing-wave patterns on the vibrating surface of large drums (Kuttner, 1990), but it would be another two thousand years before the corresponding discovery was made for spaces. Sound generation has always been the more prominent manifestation of sonic phenomena, both scientifically and experientially.

Natural philosophers were preoccupied with rules that predicted pitch from vibrating strings for two related reasons. First, creating pleasing tones was directly relevant to the interpretation of music. Second, predicting the pitch based on string lengths was central to discovering universal mathematical ratios of integer numbers, the key to unlocking the secrets of the cosmos. The Greeks treated music, mathematics, geometry, and astronomy as aspects of a single, unified philosophical framework, one that would survive for almost two millennia, until the Renaissance. The study of spatial acoustics, however, was not part of that framework.

Historically, explanations about aural experiences were more mystical than rational. Charles Burnett (1991) states that the "dominant impression that one gets from reading the medieval philosopher's account of sound is their fascination with the illusiveness of the entity." Optical phenomena appear to have been more tractable, which is consistent with the observation that "in Western scholarship, visual perception has been studied much more than aural perception." Too abstract to be readily understood, sound was therefore seen as the voice of gods, people, and resonant objects or spaces.

Even with the modest advances of acoustic science during the two millennia from Aristotle to Sabine, the insights of physical scientists and natural philosophers did not spread throughout the general culture. Intellectuals and craftsmen (including architects), sharing neither a common education nor a common social class, were not tightly coupled until the beginning of the twentieth century. As a result, scientific explanations paralleled but did not contribute to the evolution of aural architecture. Centuries old traditions dominate modern spaces: a twenty-first-century concert hall supports nineteenth-century music, which was composed for seventeenth-century spaces, which themselves were modeled on spaces inherited from yet earlier centuries.

Auditory Awareness as an Extension of Religion

Older cultures were not only aware that some of their revered objects and spaces had unusual acoustics; they also wove sound into the fabric of their religion. They made no distinction between objects actively producing sound, such as bells, and objects passively modifying those sounds, such as a caves. Aural experience resulted from a composite of all contributory elements. And that composite was an integral part of all experience, which included mythology, religion, and philosophy. Life was holistic, not segmented.

In remote sites around the world, scientists and nonscientists alike have observed unusual acoustics in the structures created by ancient and prehistoric cultures. Acousticians, archaeologists, and amateurs with an unfettered curiosity about human history have formally studied some of these sites. However, due to the highly speculative nature of acoustic evidence, support for this kind of research is limited. This has not, however, prevented those with a passion for the auditory sense from pursuing acoustic archaeology. Eventually, some speculations will no doubt be confirmed by scholars; others will be discarded as baseless. Nevertheless, the overall patterns in the aural architecture of these structures are too compelling to dismiss as accidental.

Aural Icons and Acoustic Spaces in Early Cultures

Existing in almost every culture, whether as pictures, as statues, or as small, unremarkable objects, icons have special meanings linked to particular ideas, people, events, or other objects. Icons are experienced through one or more of the senses—vision, hearing, touch, and smell—but the experience expands beyond immediate sensation by including cultural, religious, or collective memories and associations. As perceptible manifestations of abstractions, icons are especially prevalent in religions: they strengthen the relationship between believers and their beliefs. In our technological culture, computer icons, representing complex data and actions, link users to operating systems.

The aural analogue of a visual icon, an *earcon* is a sonic event that contains special symbolic meaning not present in the sound wave. The concept has recently appeared in specialized vocabularies but has not yet spread into ordinary lexicons. In a computer environment, special sonic signals such as spectrally chirped tones can represent user success, failure, or acknowledgment. These are earcons. In earlier cultures, earconic sounds merged religious and philosophic views of the cosmos with life on earth. Sound in general, and earcons in particular, connect the *here* with the *there*, be it spiritual leaders with their followers or heavenly spirits with earthly beings.

Earcons acquire symbolic meanings by repeated exposure to a particular event in a corresponding context, which then creates an associating linkage between the sound and its context. Subsequently, such sounds, even without the original context, trigger the thoughts, emotions, and memory associated with that context. Consider the family dinner bell. Its sound stimulates the appetite just because it is associated with the eating event in the family dining room. The symbolism is not deep, but it can still be considered an earcon. In this case, the linkage to food is acquired through family dining. And that particular bell sound might have no such symbolic meaning in another family. Similarly, the extended reverberation of a cathedral becomes an earcon to those who frequently attend religious services in that particular space.

We can hypothesize how the earcons of aural architecture come into existence. An architect designs a structure or a space for its visual or utilitarian properties, while being generally oblivious to its acoustic attributes or aural personality. Then, over time, with increasing exposure and familiarity, the aural attributes become associated with the visual attributes in the minds of those who use that structure or space, and, together, these attributes share a common symbolic meaning. An earcon has come into being.

To test our hypothesis that aural architecture is invested with symbolism, we would need to describe and analyze a wide variety of cultures whose objects and spaces manifest themselves as earcons. Ancient and prehistoric cultures, because of their mystical attitude toward sound in general, are more likely to have had earcons. Yet, whereas icons can survive for centuries, earcons disappear in a moment. Nevertheless, speculative evidence supports the concept of earconic sounds and earconic spaces.

For many ancient civilizations, sacred objects produced sacred sounds. When the tribes in pre-Columbian west Mexico discovered metallurgy, they treated their crafted objects as an extension of their aural religion (Hosler, 1994). Religious leaders used metallic bells as a novel replacement for early materials that did not produce pure resonances. The sounds of small bells were central to their rituals and served to celebrate human and agricultural fertility, to protect warriors from injury, and to create the garden of paradise. Bells and rattles figured prominently in ritual and ceremonies throughout indigenous American societies because of their special aural powers.

The early Greeks are credited with having invented the Aeolian harp, named after Aeolus, their god of wind, and often called a "wind harp." Although constructed like a harp, the Aeolian harp does not function as a musical instrument because its sounds are unpredictable, ethereal, and not under human control. As the wind passes over its taut strings, they vibrate in an oscillating vortex, with a series of overtones determined by the wind velocity. Its melodies and harmonies, if one could call them such, *sonify*—make audible—an otherwise inaudible natural phenomenon, in this case, the wind. Yet, for Aristotle, the sounds of this instrument were the spirit of the wind carrying the heavenly Muses to the earth, where they sang to their earthly children. Aeolian harps produced the music of the spheres.

The modern counterparts to Aeolian harps are wind chimes; some artists have constructed giant versions as acoustic sculptures. When large, the harplike wind chimes become aural architecture, creating an acoustic arena, with its own aural personality, within which people can move; when small, they are simply an "instrument," although earconic symbolism does not depend on size.

The religious structures of ancient Greece, like the Aeolian harp, also made connections between earth and heaven. To supplement their study of the few that have survived intact, S. L. Vassilantopoulos and John M. Mourjopoulos (2001) used historical records to explain how acoustically complex spaces were transformed into aural expressions of religion. The Acheron Necromancy, which served as a temple around the eighth century B.C., was associated with a ceremony where the soul of a deceased person was separated from the body and led via chasms and caves to an underground world populated by the spirits of the dead. The temple, with its many acoustically coupled rooms, was situated over a cave that had been modified into an underground chamber. The exterior walls were 3 meters (10 feet) thick, thereby ensuring both structural integrity and acoustic isolation from the outside world. The rooms and chambers had minimal reverberation, which, when combined with the extreme acoustic isolation, let listeners hear even the softest whisper from a priest located in acoustically coupled but visually remote chambers. When the space was dark, listeners experienced the priest's disorienting voice as coming from a remote and unknown chamber, as if from another world. By separating the image of the priest from his aural manifestation, the aural architecture of the temple aurally separated the priest's spirit from its physical body. The temple's "spirit voices" would have been clear, enveloping, and intimate, yet invisible. Many mythical figures, such as Ulysses, Hercules, Theseus, and Orpheus, are said to have participated in such rituals.

Comparable auditory phenomena were also found in cultures that were unrelated to our Western tradition. The Mayan culture in Mexico and Central America had a relatively sophisticated grasp of mathematics, astronomy, agriculture, and social organization. Perhaps as early as 1500 B.C., the Mayans settled in small villages, which eventually grew into large cities containing ceremonial centers, temples, pyramids, palaces, courts for public games, and plazas; at its peak, the Mayan population exceeded 2 million. Unlike the Greek tradition, however, written evidence describing the meaning, purpose, or recognition of the acoustic attributes of structures at the Mayan sites is conspicuously missing.

Nevertheless, archaeological evidence supports the thesis that the Mayans had a heightened auditory spatial awareness. The Great Ball Court, a large gathering place of some 10,000 square meters (350,000 square feet) surrounded by sloped and vertical stone walls, contains a raised temple at one end. It hosted a combined sporting and religious event, where losers were sacrificed to the gods. Many visitors have noted the unexpected pleasant acoustics: a whisper at one end can be heard clearly at the other end, making it an ideal place for a ceremonial leader to guide the audience. By adding reflections and resonances, the Ballcourt augmented the perceived mass and size of the leader's voice, raising his stature and perceived power.

As the de facto, self-appointed chairman of amateur acoustic archaeologists collecting reports from tourists, Wayne Van Kirk (2002) compiled stories and testimonials about archaeological sites with notable acoustics. He described a series of stones shaped like artillery shells in a Mayan ruin upon which you could play a tune by tapping them with a wooden mallet. Other visitors observed distinct howls and whistles from stone structures when the wind arrived from a particular direction, as if it were giving early warning of imminent storms and hurricanes. The Mayans also had configured three pyramids such that you could conduct a three-way conversation when you and two others stood at their tops. In general, many Mayan structures and spaces possessed both religious meaning and distinctive acoustic properties. It seems highly likely that the aural experience of these structures and spaces also had religious interpretations.

David Lubman (1998), an acoustic consultant and a scientist with zeal for examining the acoustics of historical spaces, studied the chirplike echoes produced by the staircases of the Pyramid of Kukulkán at Chichén Itzá, shown in figure 3.3. Although the echoes can be physically explained as a series of periodic sonic reflections from the small tread steps, their sound bears an uncanny resemblance to the call of the Mayans' sacred bird, the resplendent Quetzal. Both the call of this bird and the echoes from the staircases show the same decreasing frequency of the dominant formants. Chirplike echoes appear in many other Mayan sites where a long stone staircase faces an open plaza. The Quetzal, now near extinction, symbolized the spirit of the Mayans; however

Figure 3.3
Pyramid of Kukulkán at Chichén Itzá. Courtesy of Wayne Van Kirk.

they came to be, the echoes could well have been heard by the Mayans as the call of their sacred bird immortalized in stone.

More than other acoustic artifacts, the Mayan staircases at Chichén Itzá have given rise to heated debate among scientists. Are the chirplike echoes of the staircases merely an accident with no religious meaning, or were they deliberately created as an earcon of the Quetzal? Those who have heard both the echoes and the call of the Quetzal concede that they do, indeed, sound alike. The following discussion presents a plausible justification for treating the sound of the staircases as an earcon of the Quetzal.

Consider the properties of the staircases from a scientific perspective. Wave interferences from evenly spaced regular geometric shapes, be they jars, tubes, or stair treads, create a frequency-dependent response in sound reflections or sound transmission. Nevertheless, we must also acknowledge that geometric regularity has aesthetic appeal in itself, quite apart from its acoustic consequences.

Acoustics and aesthetics aside, why might the Mayans have constructed staircases with such physical properties? Historians speculate that the shallow stair treads arose from the decision to construct four staircases of 91 steps each, together totaling 364, which, with the platform on top as a final step, equaled the number of days in the Mayan year. There is ample evidence that the Mayans possessed considerable knowledge about astronomy and mathematics. The Mayan staircases, it would seem, are linked to astronomy.

With continuing interest in this ancient Mayan pyramid, Niko F. Declercq and colleagues (2004) explained why observers seated on the lowest steps of the pyramid hear the sound of raindrops falling in a water bucket instead of the footstep of people climbing the stairs. Somewhat later, it was observed that a mask of the Mayan rain god Chac was located at the top of the pyramid. We know that patterns can always be discovered when one looks for them, even if they are accidents, and yet we also know that older cultures often had a refined aural sensitivity to their environment. The Mayans could certainly have recognized the sound of raindrops, even if it resulted from an architectural accident, but once having perceived the sounds as that of raindrops, they might have then intentionally placed an image of their rain god at the top.

The frequency effects of both the Mayan staircases and Eusebio Sempere's stainless-steel tubes, described in chapter 2, required only a constrained relationship among the numerous physical parameters. Objects intended for a nonauditory purpose may still have interesting aural properties, and once recognized, they become earcons or acoustic sculptures with symbolic or artistic meaning. The sound of Sempere's tubes could just as easily have become a religious earcon if the sculpture had been embedded into a religious context, and if listeners had then become aware of its aural properties.

Lubman (2004) describes a perfect example of an aural embellishment arising from a religious context. As shown in figure 3.4, the shrine to Saint Werburgh, a seventh-

Figure 3.4
Shrine to Saint Werburgh, Chester Cathedral. Courtesy of Nicholas Fry.

century Saxon princess, located in England's Chester Cathedral contains six recesses where kneeling pilgrims inserted their heads while petitioning their saint. The geometry of these recesses, with their strong resonances and powerful amplification, created the feeling of an intimate encounter with the saint, whose spirit was visually and aurally accessible. Lubman's recorded demonstration of the acoustics of these recesses is dramatic. Resonances contribute to the sense of being in another world; amplification contributes to intimacy; visual isolation contributes to privacy. Absent any evidence that it was deliberately chosen, we assume that the earconic aspect of this embellishment was an incidental consequence of its religious context.

After the acoustic properties of a structure or space were recognized and integrated into the culture, the rules for replication would have been obvious: make copies or variants from the original reference. Architectural rules then become their own traditions—like religious rules. With a high tolerance for trial and error, even a scientifically unsophisticated culture can duplicate value-laden acoustic designs that originated from unrelated forces and ideas.

Following the accidental creation of a structure with unusual aural properties, it may then acquire earconic meaning, which induces the society to replicate the structure for its symbolic meaning. This conclusion requires only that the culture have an elevated sense of the aural experience, a well-documented characteristic of many early cultures. Knowledge of physical acoustics or recognition of aural architecture per se is unnecessary.

Religion and Philosophy Dominate Architecture

The ancient Greeks and Romans, who created the foundations of Western culture, believed in a unified cosmos; in their world philosophy, ideas and concepts fit together in glorified, harmonious unity. How did their harmonious view of the cosmos influence their aural architecture, and their legacies to modern society?

The ancient Greek belief in a universal and harmonious natural order was all-pervasive. "Works of art or of society were seen less as a contrivance of man than as a reflection of nature . . . and the natural world was viewed in terms of life and mind." With this comment, Edward A. Lippman (1964) captured the essence of this belief—everything was a reflection of a single unified and integrated cosmos—a grand, well-oiled assemblage of all elements of the universe. Theater, politics, music, religion, architecture, and government, along with the twelve gods of Mount Olympus, were all part of the integrated harmony.

From earliest Greek times, music was integral to religious ceremonies and closely connected to astronomy, both because of a shared mathematical foundation, and because of links to the harmony of the cosmos. As exemplified by Pythagoras's "music of the spheres" (James, 1993), early intellectuals held the ratios of integer numbers to be fundamental to understanding experience. Indeed, the numbers and integer ratios found in music and astronomy joined the harmony of the cosmos to that of human experience. Rather than being simply utilitarian means for creating physical comforts, arithmetic, geometry, and science were also vehicles for representing that harmony. Likewise, rather than being distinct pursuits in the own right, religion, the arts, music, dance, theater, and poetry were all natural parts of the harmonious universal whole. Much of ancient Greek art, science, philosophy, and technology served as the foundation for Christian culture. More than a thousand years later, the great Gothic cathedrals would still be designed on the basis of integer ratios (Wittkower, 1971).

In ancient Greece and in the succeeding Roman and Christian cultures, using abstract thinking as the means for attaining truth, philosophers influenced spatial concepts by elevating the importance of their cosmic rules; indeed, spatial concepts originated from religious philosophy.

This raises interesting questions. How did philosophy influence the design of the large basilicas and cathedrals? Were their acoustics, with long reverberation time and enveloping sound, an *intentional* imitation of God's house on earth? Why and how did early Christian spaces, which initially were small and clandestine, and which supported a spoken liturgy, evolve into grand cathedrals?

Although cathedrals may have been the first man-made structures of such great volume, large enclosed spaces also occurred in nature. Cathedrals were the acoustic equivalent of the largest natural caverns found in many countries. Some, such as the Kateřinská Jeskyně cavern in the Czech Republic, approaching 50,000 cubic meters (1,750,000 cubic feet), are comparable in volume to a cathedral. Caverns and cathedrals alike are large enclosed spaces with irregular geometries, randomly shaped surfaces, minimal acoustic absorption, and uniform diffusion of sound arriving from all directions.

Long before cathedrals, prehistoric cultures are known to have used large enclosed spaces. The underground Oracle Chamber at the Hal Salfieni Hypogeum on Malta is but one example of a natural space that had become a sacred space four millennia before the flowering of Greek culture. Archaeologists discovered a series of temples, large man-made caverns, carved into the solid limestone before the advent of metalworking tools. After settling on Malta, perhaps around 5000 B.C., this isolated agrarian civilization evolved skilled builders and engineers who constructed numerous megalithic temples, both below and above ground. The underground oracular and ritual burial chambers at the Hal Saflieni Hypogeum were the acoustic equivalents of Gothic cathedrals.[2]

Were the Gothic cathedrals an *intentional* extension of earlier cultures that used the acoustics of large enclosed spaces for their religious meaning? As a working hypothesis, the notion that religious spaces were designed or selected for long reverberation time is attractive. But at least in the case of Christianity, reliable evidence contradicts this intuitive conclusion.

Beginning in the fourth century, when Rome converted to Christianity, and continuing to the fifteenth century, when secular forces redefined music and space alike, the acoustic properties of cathedrals and churches were actually an *unintentional* consequence of religious, philosophic, and social forces. When Emperor Constantine proclaimed Christianity the official religion of the Roman Empire in the early fourth century, various basilicas, which had been designed using a style of Greek architecture of some two centuries earlier, and which had served as courthouses and as meeting places for commerce, were converted to churches. William Smith (1875), in describing the rectangular shape and two rows of columns of these early basilicas, noted that they had open sides. When Roman society became wealthier and more refined, outer walls were added to create an enclosed space, and the supporting columns became part of the interior. Although walls dramatically changed the acoustics, they were added for other reasons: to protect against the weather, to demonstrate political power, and to visually separate the space they enclosed from the external environment.

Thus basilicas were of utilitarian design, initially unrelated to art or religion. As the traditional architecture of the period, this style also became the model for early church buildings. Rapid construction of new religious spaces, many of which followed the shape, form, and size of public buildings, dispersed the basilica style of church architecture throughout the Roman Empire (Platner, 1929). The Christian view of architecture was not rigid, but adapted to local regions; with the rapid growth in new converts, church spaces grew large enough to hold thousands of congregants (Krautheimer, 1965). Thus, even though size was and is the dominant parameter of aural architecture, large church size actually reflected the desire to have large congregations share religious events and was unrelated to the resulting acoustics.

With their enormous floor area, dramatically high ceilings, and stone surfaces, cathedrals have a reverberation time at the theoretical limit of an enclosed space, often approaching 10 seconds for middle frequencies (1000– Hz), where air absorbtion begins to dominate acoustics. The three key components of such spaces, area, height, and surfaces materials, each arose from independent social forces. Whereas increasing floor area allowed churches to contain large congregations, increasing height, the other dimension of volume, was of no practical value, serving instead to express the grandeur of God's home. Thus, for example, the ceiling of the Basilica of Constantine is 34 meters (110 feet) from ground level. Stone surfaces, which replaced wood, resulted from advances in building technology. Stone buildings were durable, strong, and immune to the ravages of fire, a common event at a time when illumination was provided by torches and other flammable material. Hard, acoustically reflective surfaces were therefore incidental, a by-product of using stone as a building technology to avoid fire, and to support large heavy ceilings over large floor areas.

A thousand years after the founding of Christianity, architectural design reached its pinnacle with the construction of the majestic Gothic cathedrals of Europe. Beginning in the twelfth century, when Abbott Suger sponsored the cathedral at Saint-Denis, just north of Paris, cathedrals were built with volumes well over 200,000 cubic meters (7,000,000 cubic feet), compared to the early churches of the fourth century with less than 5,000 cubic meters (175,000 cubic feet). Although closely linked to religious traditions and building technology, these enormous sizes created overpowering and enveloping acoustics. Ultimately, they determined the acoustic scope and nature of liturgical music.

Early Christians (like many pre-Christians) believed that particular places were associated with God's presence. Even today, the word *church* retains its dual meaning of religious faith and building for religious services. Other religions have separate words to distinguish the place from the religion: notably, mosque and Islam, temple and Judaism. In his study of cathedrals, Otto Georg von Simson (1989) portrayed a religious culture dominated by a single conceptualization that merged all aspects of life into a unified view: as the "symbol of the kingdom of God on earth, the cathedral gazed down upon the city and its population, transcending all other forms of life as it transcended all its physical dimensions." The architecture and sculpture of the medieval sanctuary were images of heaven, while the music therein was its sounds. Everything was symbolic—objects were neither illusions nor allusions, but a representation of religious truth. The creative legitimacy of architects, sculptors, and musicians was determined by their ability to represent this truth.

Biblical descriptions represent church structures, mystically and liturgically, as a vision of the Celestial City with its Heavenly Mansions. In their role as heavenly architects, the designers of religious structures called on the cosmic geometries passed down through the ages from Pythagoras. For example, all of the ribs under the vault of the Reims Cathedral circumscribe equilateral triangles, a fact few observers are likely to notice. Surviving documents report debates about the religious meaning of squares or triangles as the basic geometric shape for cathedral design. As cited by von Simson (1989), Thierry of Chartres argued that geometry and arthmetic provided divine inspiration.

Similarly with music, Saint Augustine (387), though not denying that music could be produced by instinct or practical skill as a vulgar art suitable for popular audiences, asserted in his treatise *De musica* that music becomes an important expression of universal truth only when based on the science of arithmetic—ratios of integers. Without numbers, music and space return to chaos. For medieval Christians, "auditory and visual harmonies are actually imitations of the ultimate harmony which the blessed will enjoy in the world to come" (von Simson, 1989). The attitudes toward music and architecture arose from this concept, which permeated all aspects of Christian Rome and arose from the earlier Greek concept of harmony.

As the Roman Empire was dissolving and Christianity was spreading, the church gradually became the only central authority governing politics, spirituality, and artistic expression, eventually merging music, religion, and administration into a unified social system. In the Western heritage, auditory arts enjoying church support were fully incorporated into the Christian religion; their dedicated spaces determined the nature of music for a thousand years.

When music was linked to a dedicated space, it became stylized and constrained in order to fit with the unique acoustics of those spaces. Christianity was a prolific builder of dedicated spaces and an enthusiastic sponsor of dedicated music. Such spaces were expensive, and the costs had to be borne by a civil or religious organization with power, resources, and architectural vision. Just as Greek and Roman theaters were dedicated spaces supported by the state and wealthy patrons, Christian basilicas and cathedrals were dedicated spaces supported by an increasingly powerful church hierarchy.

As the church became the only viable social structure, its evolving concepts of space dominated architecture. Looking at several examples of basilicas and cathedrals that have survived, we see a progression in their size, shape, and design. The Rotunda of Thessaloniki, a unique fourth-century monument of late Roman art, was typical of the kind of building that served early Christianity. As a modest-sized space with a volume of 15,000 cubic meters (530,000 cubic feet), it has a comfortable reverberation time at middle frequencies of about 2.5 seconds when fully occupied (Tzekakis, 1975). Its acoustics were acceptable for a wide range of uses. In contrast, Roman basilicas with volumes of over 100,000 cubic meters (3,500,000 cubic feet) have reverberation times that range from 5 to 10 seconds even when full (Raes and Sacerdote, 1953). Of the four Roman basilicas studied by Robert S. and H. K. Shankland (1971), Saint Peter's stands out in terms of sheer size; it is the largest church in the world with a length of 180 meters (600 feet) and a volume of 500,000 cubic meters (1,750,000 cubic feet). Because of its extensive interior surfaces, however, its reverberation time is only 7 seconds. Having fewer interior surfaces, the smaller San Paolo fuori le Mura, also in Rome, has a longer reverberation time.

Although most forms of music and vocalization do not work effectively in spaces with so much reverberation, the Gregorian chant does. A type of slow, monophonic, unison singing, it is believed to have originated as a dominant component of the Christian liturgy after a progressive series of simplifications of a more complex vocal tradition that had existed much earlier (Grout, 1960). However beautiful in themselves, Gregorian chants were a utilitarian adjunct to worship, as well as a functional music that defined the temporal prayer cycle of monastic life. Selecting such chants as the vocalization of choice was an inevitable consequence of the high reverberance of most cathedrals and monasteries (Lubman and Kiser, 2001). Only slow, simple singing would avoid the aural soupiness of reverberation that seemed to last forever. The more rapid and complex singing that had existed earlier (and would again later) would have been acoustically degraded to total unintelligibility by long reverberation. Even in the relatively small monastery at Santo Domingo de Silos, near Burgos, Spain, with a volume of only 5,000 cubic meters (175,000 cubic feet), reverberation was excessive (Lopez and Gonzales, 1987). The simpler chants are so tightly linked to their original spaces, that in 1994, when the Benedictine monks of Santo Domingo de Silos recorded *Chant*, one of the best-selling albums of Gregorian chant music, they selected the same space their brethren had used some thousand years before.

Within the vast literature of the church, extensive theological discussions describe the intent and goals of church architects and spiritual leaders over a 1,500-year span. But, even though the forms of church music and liturgy alike clearly responded to its presence, there is no mention of *intentionally* creating reverberation for its theological relevance. Instead, evidence suggests that long reverberation was simply an unintentional consequence of the spatial grandeur of God's earthly home. Nevertheless, for those who repeatedly attended services in these religious spaces, aural and visual symbolism became tightly linked. In this context, long reverberation indirectly acquired its meaning from the religion, with its liturgy, icons, and visual designs. And this link was further strengthened by religious music written for this highly reverberant space.

Christianity was certainly not the first religion to give reverberation a theological meaning. The Temple of Zeus, constructed about 460 B.C. in Olympia, was one of the largest, most prominent religious structures of ancient Greece. Owing to a lack of sound-absorbent materials, its reverberation time was a relatively long 3 seconds, which impaired speech clarity. With sound arriving from all directions, the space created a listening experience similar to that of Christian churches of comparable size, where reverberation enveloped the listener with the grandeur of God's voice.

Thus the acoustics of temples, cathedrals, monasteries, and churches acted as cultural filters, excluding those art forms that were aurally inappropriate. Moreover, because the Christian church was the locus of literacy and power, other subcultures would not have been in a position to support alternate forms of the auditory arts, and they certainly were not in a position to leave a written record of their activities. For a thousand years of Western civilization, aural architecture was, in effect, the Christian view of their music in their spaces. We might argue that the symbolic meaning of reverberation, even in reduced amounts, was partially a legacy of Christianity. When spaces became that large, they had limited utility for theater, spoken liturgy, musical detail, and any aural form that required clarity and intelligibility.

Social Forces Influence Aural Spaces

Having seen how Western religious ideas dominated European aural architecture up to the Renaissance, let us now consider the historical role of secularism, including artistic, political, social, and economic forces, all of which existed in parallel with the prevailing Christian theology. Historically, religion was so fused into the social fabric that clear distinctions between secular and religious spaces are nearly impossible. But where secularism existed separately from theology, it is possible to observe its unique role in the use and design of spaces. For example, artistic and political forces determined the aural architecture of spaces used for Shakespearean plays and Greek political orations.

Regardless of the motivation for creating or selecting a particular space, whether religious or secular, once a space was used for social activities, the occupants then gained a heightened awareness of its acoustic properties. Just as priests adapted to the acoustics of their churches, so actors adapted to the acoustics of their theaters, musicians to those of their performance spaces, and politicians to those of their meeting halls. In each case, auditory awareness led to a social response, a pattern that has existed since the start of recorded history and probably much earlier.

The Acoustics of Public Spaces in Ancient Greece

Given their strong interest in all forms of aural activities, including music, oration, rhetoric, and religion, the ancient Greeks were likely to have been aware of how these activities were influenced by spatial acoustics. Although acoustical attributes are rarely mentioned in surviving documents, sound was generally viewed as an important social and political resource. Benjamin Jowett (1964), in looking at Plato's dialogue *Laws*, commented that music was viewed as a pragmatic extension of political education and military training. Plato worried that music could make people lethargic, indolent, or irrational because it fed the waters of passion rather than drying them up (Bowman, 1998). Furthermore, in order to preserve the path to truth and goodness, Plato argued that music must be controlled by the state, and should not be left to personal preferences. Sound had power, and spatial acoustics, even if not so recognized, also had power.

The acoustic differences in the spaces of ancient Greece were not subtle. Although only a few buildings from ancient Greece have survived intact, historical records provide commentaries on their design, social use, and in some cases, on the aural experience of their spaces. Two examples will illustrate the role of small and very large acoustic arenas in ancient Greece.

Our first example, the Echo Hall, once stood in the ancient city of Olympia. A long structure measuring some 100 meters by 10 meters (350 feet by 35 feet), it had three enclosed sides and one open side with 44 Doric columns. The renowned traveling geographer Pausanias described how a voice in this hall would echo seven or more times. These strong echoes would have prevented communications over a wide area, creating multiple small acoustic arenas, whose aural privacy would have been ideal for any number of small groups wishing to discuss politics and commerce without fear of being overheard.

In our second example, the Greek amphitheater, poetry, drama, music, dance, and religion fused into a single type of aural experience in a very large public acoustic arena. Greek theater could tolerate neither the excessive reverberation time of large

Figure 3.5
Open-air theater in Epidauros. © 1993 Gebrüder Mann Verlag, Berlin.

enclosed spaces nor, with the political need to accommodate large audiences drawn from a democratic society, the limited audience size of smaller enclosures. The open-air amphitheater (figure 3.5) would remain the only means of combining a large audience with oratorical clarity until the advent of electronic broadcasting in the twentieth century, with its widely distributed audiences.

The Greek amphitheater was also the result of geographic and climatic accident. Many major Greek cities were located on rolling hills, which provided ideal acoustic settings for open-air theaters. (In contrast, flat plains, wide valleys, or steep mountains would not have provided good acoustics.) And Greece's mild climate made unsheltered public spaces feasible. Indeed, we might speculate that geography and climate contributed to the success not only of the amphitheaters but also of Greek democracy, which might not have flourished without the frequent, publicly shared experiences these theaters made possible.

Theater was an important part of ancient Greek culture, and is still studied in schools across the world. Because of its artistic, social, and political relevance, then and now, there is a large body of information about its theatrical content and its spatial acoustics in surviving historical texts and archaeological evidence. And because Greek theater has broad interest to modern scholars, they study the acoustics of the open-air amphitheaters with the tools of modern science.

Peter Walcot (1976) described the social context of Greek theater as the driving force that defined its art form. Massive audiences attended the Theater of Dionysus in Athens during the festival held in honor of the god to enjoy the plays of Aeschylus, Sophocles, and Euripides. The state sponsored these festivals as part of the annual calendar of religious ceremonies, and the audience was drawn from the general population rather than from a small elite group. As a democratic state, Athens provided theater venues that could seat more than 15,000. The ancient Athenians have been described as a critical and demanding audience, both emotionally and intellectually.

Because plays often had only a single performance, there was intense competition among the resident population of 150,000 for that, relatively speaking, limited seating, with occasional violent confrontations.

From the perspective of satisfying an artistic and political requirement—intelligibility and democracy—Benjamin Hunningher (1956) analyzed the implications of the acoustics on the acting style. Indeed, given the size of the audience, which was seated in a semicircular area set into the rolling hills, acoustics became the central issue. The size of these open-air theaters was immense, even by present-day standards of sporting events. The distance from the performers to the farthermost spectator in the fifty-second row was some 80 meters (260 feet). In comparison, a modern opera house, such as the Prinz Regententheater in Munich, has a distance from the curtain line to the farthest seat of less than 30 meters (100 feet). More important, open-air theaters do not add sonic energy from reflecting surfaces the way that enclosed spaces do. Without special design efforts, a large percentage of the audience would not have been able to hear the performance. Regardless of their location, spectators expected intelligibility throughout the seating area.

Applied spatial acoustics was born of the necessity to solve the problems of large-scale spaces. In his treatise on architecture, Vitruvius (30 B.C.) included an extensive discussion on rules for improving theater acoustics. Over the years, many scholars and researchers have tested these rules as well as theories about Greek theater to determine which artistic styles and architectural solutions would have solved the first major problem in spatial acoustics: amplification without electronics. Several ideas have emerged from these studies. First, the large front wall of the skene, positioned behind the performers, would have reflected sound to the audience in much the same way that the front wall of the stagehouse in many sixteenth-century theaters did (and does in some modern theaters as well). Second, increasing the angle of rise in the seating area would have placed the audience closer to the performers. (Amphitheaters with sharper angles of rise do indeed have better acoustics.) Third, the mouth openings of theatrical masks may have functioned as miniature megaphones. Fourth, through special training, performers learned to project their voices for maximum intelligibility. Finally, by singing, performers could project their voices still farther than by simply speaking—much farther, perhaps reaching the most distant seats.

When Robert S. Shankland (1973) correlated the variations in acoustic quality among Greek amphitheaters with their physical parameters, he observed that geometries that optimized sight lines also optimized acoustics because they both follow the same rules. This fortuitous relationship between seeing and hearing, rather than an understanding of acoustic architecture, may have produced the remarkable acoustic clarity of the best Greek amphitheaters, although, even when these theaters are refurbished for contemporary performances, their acoustics are by no means without major problems (Schubert and Tzekakis, 1999).

There is no doubt that Greek performers invented creative ways to compensate for the acoustics of their open-air theaters. J. Michael Walton (1984) argued that their playwrights used "a whole armoury of visual signs and devices to amplify and often to take over from the spoken word." Dance and exaggerated movements do not depend on sound. Even in a society where political influence depended on skilled rhetoric and fixed speech patterns, with appropriately dramatic gestures, the consequence of weak acoustics would not have been severe. This view is consistent with the notion that acoustic limitations forced the performing arts to be multisensory, thereby compensating for reduced sound quality. Greek theater provides the first concrete example of the way in which space controls performers, as well as the art form, and space itself is determined by social, political, and technical forces in the society.

Size matters: audience size determines aural architecture. The problem is not one of creating a good listening experience for a small number of people located in the "sweet spot," but rather of creating a satisfying experience for an entire audience. Larger audiences create larger problems. For the Greeks, the need for large size was a social consequence of their democracy. For the next two millennia, until technology created the means for listeners to move into a small family room, audience size would be central to aural architecture. But with modern radio broadcasting, millions of families in their parlors could share a listening experience, making the audience profoundly larger than that of a Greek amphitheater. In terms of social function, the aural architecture of Greek open-air theaters is analogous to the aural architecture of modern radio broadcasting. Size matters.

Discussions about Greek aural architecture yet again illustrate the complex interactions among the various social and acoustic issues. Integrating religious spaces into social activities in ancient Greece was similar to doing so in many other cultures, but connecting open-air theaters to social democracy was unique in history. Without the benefit of electroacoustics, Greek amphitheaters were among the world's largest manmade acoustic arenas. They were created to serve a specific need. Yet regardless of how their spaces were used, the relationship between acoustics and social values worked both ways: acoustics responded to social values, and society then responded to the influence of acoustics. Although the specifics vary, the principle remains consistent: culture both filters and reacts to the ways aural architecture is created in any given society.

Shakespearean Theater in Sixteenth-Century England

Whereas the ancient Greeks and Romans established the tradition of large, open-air theaters, the sixteenth-century English emphasized small interior spaces, with improved intelligibility and greater aural intimacy, in their theaters. Given the English climate, enclosing theater spaces was mandatory. And given that theaters served as entertainment, rather than as a political expression of open democracy, they did not need to be large. English theaters of this period were one of the earliest documented examples of customized aural architecture—acoustic spaces intended for a single use. Eventually, the idea of a reserved space for a specific aural art would be extended to such arts as operas and concerts.

In the late 1960s, working from fragmentary historical records, restorers reconstructed Shakespeare's Globe Theatre in its original form. The current interest in experiencing sixteenth-century theatrical works within their original aural and visual context gives us a unique opportunity to explore historical attitudes toward the aural architecture of theater. Documents give us at least a sense of how participants adapted their acoustic space to dramatic productions, and vice versa. Those who reconstructed the Globe Theatre realized that it was far more than just a place to seat spectators.

Even without any appreciation for acoustics, early theatrical producers overtly recognized that aural architecture was an important part of their arts. In his comprehensive analysis of soundscapes in early England, Bruce R. Smith (1999) argued that the acoustics of theater spaces were recognized as an extension of the human voice. Plays of this period were based largely on verbal rhetoric, a descendant of the Greek tradition of oratory as the highest aural art form. Rather than merely being a place for actors to perform and the audience to listen, the theater was an extension of the actors' mouths, producing, shaping, and propagating their voices. By modern theatrical standards, visual props were sparse and simple. Since oratory was the dominant form of emotional communication, theatrical voices required adequate clarity and loudness to achieve dramatic impact. As discussed in chapter 2, spatial acoustics can provide early sonic reflections, enlarging sonic mass, broadening the acoustic arena, and increasing intelligibility. For all these reasons, sixteenth-century theater spaces were sonic instruments that were extensions of the actors' mouths. Acoustics were an important aural prop, even more important than visual props.

In Shakespeare's time, theaters were generally portable and temporary, installed within such existing spaces as schools, courts, and inns. Moreover, such theaters, each with its own aural personality, were the prized assets of a theater company. Timbers were marked so that they could be dismantled and easily reassembled for the next season or another locale. Early theater spaces were less buildings than portable appurtenances of the company, not unlike stage props. Smith commented that after the theater company at Shoreditch, in Somerset County, had moved to new quarters, they returned two years later to retrieve the theater's wooden framework from their old facilities. Without the old framework, the new space simply did not have the same aural personality as their traditional space. Musicians transported their instruments; acting companies transported their portable theaters. Spatial acoustics had a theatrical personality.

As theater increased in popularity, its spaces became permanent. In 1599, the Globe Theatre was large, even by modern theater standards, with a volume of some 10,000 cubic meters (350,000 cubic feet), holding an audience of perhaps 3,000. It was shaped as a twenty-sided cylindrical polygon with a diameter of some 30 meters (100 feet), and had an open-air top. Having a projecting stage and three tiers of raked seating, the Globe was a true theater in the round. The walls and stage canopy provided many surfaces for sound reflection, yet without an enclosing ceiling, the reverberation remained at a modest 1.4 seconds, at least as measured by Russell Richardson and Bridget M. Shield (1999) in the reconstructed version. With ray tracing of sound paths, the researchers demonstrated that the dominant first reflected sounds would have appeared at an optimum delay of about 20 milliseconds, thus fusing with the direct (incident) sounds.

In their analysis of the reconstructed Globe Theatre, Richardson and Shield commented that certain aspects of Shakespeare's plays make more sense when considered in their original setting. Historians believe that sixteenth-century audiences would have been noisy and boisterous, bringing ambient street life into the theater environment. Being on the central stage with a rear canopy as a reflecting surface located actors where their acoustic arena was largest, thereby elevating their voices above this noise. Asides to the audience, which appear superfluous in a modern theater, become meaningful in this context. Although the theater would never be considered as having "great" acoustics, the acting company adapted to the theater space—both theatrically and acoustically—to maximize their impact.

It is tempting to think of these theater companies as being aural architects. In fact, the original design for a circular performance space with three levels of audience galleries was copied from existing bearbaiting and bullbaiting houses. Actors and playwrights then adapted to the acoustics of these replicated structures, and through empirical experimentation, made minor improvements to them. These spaces served their intended purpose: aurally, visually, and socially. Historical records suggest that theater companies acquired increasing awareness of spatial acoustics, rather than beginning with a proactive understanding of acoustic principles.

Builders copied bearbaiting houses when building the Globe Theatre, just as three centuries later, Sabine copied a European concert hall when designing Boston Symphony Hall. In this respect, little changed. In both cases, sponsors evaluated existing choices and then selected the best compromise among competing social, economic, and acoustic requirements. The practice of aural architecture was mainly a process of evaluating existing models and selecting ones for new spaces.

Scholars in sixteenth-century England wrote of a heightened awareness of sound. As the urban soundscape replaced the tranquillity of pastoral life, the utility of manipulating aural space became apparent to the intellectuals of the period, who extended the earlier work of Greek philosophers. As exemplified by Sir Francis Bacon (1626), intellectuals were already espousing the notion that what could be heard could excite passions far beyond what could be seen. Bacon envisioned the creation of new aural experiences. If nothing else, he recognized the importance of sounds of space being altered, decomposed, recombined, and then broadcast in altered forms, which included synthetic echoes bouncing from virtual surfaces. But with limited technology, his creative inspirations remained only thought experiments. Proactive manipulating of aural experiences—designing a space for a particular aural experience—would have to wait until the twentieth century.

The Advent of Public Performance Spaces

The Renaissance marked a major shift in the character and size of musical spaces, away from religious toward secular, and away from larger toward smaller. Beginning in the fifteenth century, elite institutions that supported cathedrals, palace theaters, and royal ballrooms were no longer the exclusive sponsors of aural architecture. After two thousand years during which Greek and Christian political and religious thought dominated concepts of music and space, the ascending middle class opened musical spaces to broader segments of society.

With their expanding social power and political influence, tavern owners, small shopkeepers, and wealthy merchants determined the acoustic properties of spaces. The Protestant Reformation shifted resources and institutional power away from religious organizations and toward secular ones. Such newly enfranchised groups had their own ideas about artistic sponsorship, and organized religion was no longer the only viable social structure. Princes, parliaments, municipalities, craft guilds, and enterprising merchants took over many of the functions that had been controlled by the church; economic resources and political powers were now shared. And that sharing allowed music and musical spaces to be more than religious.

These social changes accelerated the shift away from the acoustic extremes of high reverberance in cathedrals and low reverberance in open-air theaters. In varying degrees, spaces now had modest reverberance, and concomitantly, aural clarity and intimacy. As music chambers, concert halls, opera houses, and theaters proliferated; they became the dominant manifestation of artistic spaces. Not only artistic sensibility, but also social forces drove this transformation, thereby creating a new generation of aural architects.

The Reformation signaled the end of Roman Catholic dominance of the aural architecture of public gathering spaces. For the newly formed Protestant sects, churches were more a utilitarian place for sharing religion than a vision of a heavenly home. The service was led by a minister whose liturgy focused on words, ideas, and reasoning. In support of these theological changes, a new generation of church builders began to emphasize acoustic clarity and spatial intimacy through lower ceilings and smaller room size. Automatically, with smaller spatial volume and denser congregations, the new architecture produced shorter reverberation time. Unobstructed sight lines and increased intelligibility became a requirement for the new architecture, which better suited the spoken sermon.

The larger Dominican and Augustinian churches in Germany, for example, were renovated with the addition of galleries, tribunes, and private boxes, changes that increased sound absorption and decreased reverberation time (Bagenal, 1951). The original Thomaskirche in Leipzig, the space for which Bach wrote the Saint Matthew Passion and Easter Mass, was acoustically more like a small concert hall than a grand cathedral. Leo Beranek (1962) estimates that, when full, the Thomaskirche had a reverberation time of no more than 1.6 seconds, which supported the delicacy of stringed instruments, and a more rapid ebb and flow of musical tempos. Bach and other composers adapted to these acoustic changes by altering their phrasing and inventing new musical forms.

Just as the aural architecture of Reformation churches encouraged new styles in religious music, so small taverns in the century to follow would encourage secular music by hosting public concerts (Elkin, 1955). In a room specially set aside for performance, musicians would sit on a raised podium where they would entertain the drinking guests. Investing in the new musical spaces, proprietors would provide comfortable furnishings for patrons who paid admission fees or purchased food and drink. Listeners would give tips to musicians. The earliest historical record of such musical taverns dates from the mid-seventeenth century. Some music houses published schedules, similar to the marketing and advertising of a modern concert series. Music, as consumable entertainment, became a public business, and every business needed its space. Aural architecture was now an investment based on economic yield.

Music houses brought together individuals who shared an interest in music. Enterprising leaders created amateur music clubs, whose members were from all social levels, including the working class. Members had access to instruments, scores, and a place for informal performances. Handel, during his first visit to London, attended one such club, as did poets, painters, and nobility. The Music Club of the late seventeenth century, run by Thomas Britton, the proprietor of a small coal delivery business, was so famous that the B.B.C. recently did a special program of a concert performed in that space. To have a place to practice their art and earn a living, professional musicians built concert rooms in various fashionable residences. Newspapers routinely printed advertisements for locally organized concerts. Music developed a passionate following among amateurs and professionals alike. Sponsorship of music and performance spaces moved away from prevailing organized institutions, with their theological and philosophical rules. The aural properties of space were now controlled by other ideas.

The ingenuity, creativity, and resourcefulness of these seventeenth-century individuals pursuing music were an extension of a pattern that predates written history: music

as entertainment. Throughout history, troubadours, entertainers, and wandering min-
strels traveled from town to town and from festival to festival. Because they would play
their music wherever crowds gathered, their performance spaces had unpredictable
acoustics. Being sensitive to the effect of their music on the audience, these performers,
no doubt, selected their music, tempo, instruments, and playing style to match the immediate acoustics as best they could.

In contrast, many of the social changes in the seventeenth century raised the importance of dedicated spaces for musical performances. Audiences went to hear musicians
in these spaces, each with unique, stable, and predicable, which is not to say good,
acoustics. Ordinary living spaces were transformed, at least in name, into "music
rooms." A wealthy merchant might organize a music night in his elegant parlor.
Listeners and performers now had the opportunity of experiencing the effects of
space on music. Some spaces were more ideal for certain types of music, while others
degraded the listening experience. With this diversity of music and spaces, auditory
spatial awareness expanded.

By the eighteenth century, music rooms had grown in size, proliferating throughout
London, which was the musical capital of the world. The popularity of music, combined with the attractiveness of commercialized entertainment, created pressure to
seat larger audiences and orchestras. Music rooms became concert halls. This in turn
spurred the invention of new forms of music, such as the concerto. Larger spaces
required instruments that had a brighter and more powerful tonal color, as exemplified
by the violin replacing the viol. Virtuoso musicians adjusted their style so that those
seated in the last row could hear the delicacy of their interpretation. In recognition of
spatial acoustics, for example, the celebrated flutist of early eighteenth century Johann
Joachim Quantz (1966) taught musicians to be aware of, and to incorporate, the effects
of acoustic space on their performances. The need for adjusting to a space was now an
accepted fact.[3]

Musicians' appreciation of spatial reverberation, especially when performing
nineteenth-century music in nineteenth-century concert halls, is abundantly clear in
the many quotations collected by Beranek (1996a). The renowned violinist Isaac Stern
said that "as the [violinist] goes from one note to another the previous note perseveres
and he has the feeling that each note is surrounded by strength. When that happens,
the violinist does not feel that his playing is bare or 'naked'—there is a friendly aura
surrounding each note. . . . The effect is very flattering. It is like walking with jet-assisted
take-off" (Beranek, 1996a). Reverberation is even more critical to organ music, which,
because an organ pipe's valve is an on-off device with no intermediate intensity,
sounds dreadful without it. Unlike pianists, organists cannot produce gradual changes
in loudness by varying the velocity or pressure on the keys, and they have no equivalent of the sustain pedal. They must therefore rely on reverberation to produce smooth
decay and mixing. The famous organist E. Power Biggs, wrote that "an organist will

take all the reverberation time that he is given, and then ask for a bit more, for ample reverberation is part of organ music itself.... Certain French music depends so completely on long periods of reverberation that, no matter how well played, in acoustically dead surroundings it falls apart into disconnected fragments" (Beranek, 1996a).

During the prolific period of classical music compositions, spatial acoustics, from the perspective of the composer, became a recognized extension of music, and from the perspective of the performers, spatial acoustics became an extension of their instruments. Musical compositions explicitly specified the instruments' voices, and implicitly specified spatial acoustics. Both music and space were dynamically responding and adapting to each other in a mutually beneficial embrace, a marriage without any means for an amicable divorce.

The seeds of future discontent were sown with the rapid proliferation of concert halls. If they were to accommodate the passions of musical purists and idealists, future aural architects would need to design a musical space for each genre of music from each historic period: anything else would be an "unacceptable" compromise. But, constructed at great cost and effort, concert halls become as inflexible as natural caverns, flattering some genres and disparaging others. Although conductors and musicians do their best to adapt to spatial acoustics, adaptation has its limits.

As we will see shortly, the implications of music tightly integrated with spatial acoustics would haunt the twentieth-century arts. Social changes unrelated either to music or to acoustics would alter aural architecture in a way that inadvertently damaged the legacy of classical music. And then, still later, with the development of inexpensive artificial reverberators, each musical genre could indeed have its own ideal spatial acoustics. Thus technology first undermined and then supported the marriage between music and space.

Industrialization Creates New Aural Attitudes

As noisy machines and devices permeated society during the nineteenth century, the bucolic soundscape was replaced by an industrial one. Machines of this period produced noises at intensities well beyond that of hand tools, making our modern urban environment appear quiet by comparison. This industrial soundscape modified the way people experienced sound and space. Unintended and unplanned reactions to this new environment changed the concepts of music, musical space, acoustic arenas, and aural architecture. Public acoustic arenas, which had served the traditional role of facilitating social cohesion, shrank as noise overpowered this common resource. Private acoustic arenas became an important alternative to public ones. Sound became something that could be owned and controlled.

At the same time that society was losing the use of public acoustic arenas and focusing on creating private ones, post-Edison technology also created a new form of acoustic arena by connecting locations that were remote in time and distance. Electro-acoustics was the means for capturing, controlling, storing, displacing, expanding, and distributing sound. No longer would sound be only local and ephemeral. Like the printing press, which memorialized fleeting oratory on inexpensive paper for mass distribution, the mechanical microphone converted fragile sound waves into permanent wiggles on wax cylinders, which could talk to future generations. Never before could the passion and immediacy of music and oratory be kept in a desk drawer; never before could sound be separated from its source.

Eventually the unity of a single space for performance and listening, such as a meeting hall or a concert hall, split into separate performance and listening spaces. Microphones and loudspeakers then enabled physically separate spaces to be fused into a new type of acoustic arena. When extended to twenty-first-century technology, acoustic arenas became global, binding people with common interests and shared cultures regardless of their geographic location.

Neil Postman (1993) commented that even Marx understood that "technologies created ways in which people perceive reality, and that such ways are the key to understanding diverse forms of social and mental life." Not only did technology change the aural experience of social life, encouraging active manipulation of sound and enlarging the concept of aural architecture, but it also influenced spatial cognition, sensory perception, and social dynamics. The industrial revolution was also a sensory awareness revolution.

Once electroacoustics began to permeate society, the soundscape was no longer an indirect consequence of people engaged in their day-to-day life. Like a newly discovered natural resource, soundscapes became a new frontier for commercial development and exploitation. By the early twentieth century, a generation of acoustic and audio specialists created multiuse auditoriums, noise control barriers, public address systems, recording studios, broadcast networks, and portable sound equipment. These twentieth-century aural architects replaced the artists, builders, and religious visionaries, who in earlier centuries created theaters, concert halls, and cathedrals. The new technologies of aural architecture changed society's relationship to sound.

Industrialization created new disciplines for analyzing, creating, and controlling sounds. Amateurs experimenting with noise control and sound absorption became professional acoustic scientists and audio engineers. Other professionals nurtured a growing industry of audio inventions for broadcasting, telephone, and phonograph. Although sound was understood as a physical process with a rational explanation, manipulating sound remained arcane, magical, and mystical. Audio and acoustic designs were accessible almost exclusively to specialists. A new subculture of auditory knowledge monopolies could therefore exert increasing influence over aural architecture in real spaces, and especially in virtual ones. To the extent that auditory expertise was now socially important, audio engineers and acoustics scientists became the shamans of aural experience. "Those who cultivate competence in the use of a new technology become an elite group that are granted unreserved authority and prestige by those who have no such competence," observed Postman (1993).

Industrialization Changes the Soundscape

In his study of the soundscapes in Victorian England, John M. Picker (2003) described shifts in people's aural experience as society became increasingly industrialized. In the same century that the explosive eruption of Krakatoa was heard around the world, the birth of the microphone made the sounds of a walking fly audible. In densely packed urban environments, narrow streets amplified sound. Mechanical noises contributed to, and in some cases dominated, the sounds of nature. Increasingly, people were living in soundscapes of their own making, albeit as an unintentional artifact of industrial machines. Cheaply manufactured hand organs symbolized industrialization. Unlike earlier periods, soundscapes were now so varied as to defy the traditional experience of sounds.

Foreground sonic events, especially those having symbolic meaning, are most apparent when they appear against a background of relative silence. At the onset of industrialization, the blast of a whistle from a locomotive chugging through the countryside competed with the barking of dogs on a traditional foxhunt. In the city, the cries of vendors competed with a sea of organ-grinders making a living as entertainers. Somewhat later, industrial sounds became so dominant they become sonic background rather than isolated foreground sonic events. The change was rapid enough—within a generation—to attract political attention.

As if they had just discovered a new sensory modality, scholars, intellectuals, and the literary giants of this period began to comment on sound from this new perspective. As social commentators, they recognized two issues. First, the soundscape now included sonic by-products of machine technology, and second, there was an increasing understanding of both the physics of sound waves and the psychology of auditory perception. Intellectuals recognized both factors. Charles Lamb's "Chapter on Ears" and Wordsworth's "On the Power of Sound" were typical examples from this period, although Wordsworth's poem also retains an age-old mystical attitude toward sound, an attitude that the new soundscape only intensified.

On the Power of Sound (first stanza)
Thy functions are ethereal,
As if within thee dwelt a glancing mind,
Organ of vision! And a Spirit aerial
Informs the cell of Hearing, dark and blind;
Intricate labyrinth, more dread for thought
To enter than oracular cave;
Strict passage, through which sighs are brought,

And whispers for the heart, their slave;
And shrieks, that revel in abuse
Of shivering flesh; and warbled air,
Whose piercing sweetness can unloose
The chains of frenzy, or entice a smile
Into the ambush of despair;
Hosannas pealing down the long-drawn aisle,
And requiems answered by the pulse that beats
Devoutly, in life's last retreats! (Wordsworth, 1835)

Historians who studied this period observed that auditory tranquillity was no longer the inalienable right of the ruling class. Silence had become a precious commodity. The combined assault on all the senses by a profusion of crowds, animals, machines, and a vast array of moving objects became a source of civil discomfort. The loud, coarse music of seedy organ-grinders provoked the wrath of the upper classes. As if their private property was being confiscated, the ruling class responded socially, politically, and emotionally. This pollution of the public acoustic arena motivated the invention of a private one based on the technology, however primitive, of soundproofing. As a symbolic testimony to sonic warfare, a private library in urban London was designed with double-hung walls, skylights rather than windows, and muffling air chambers. Acoustic isolation barriers were the only means for escaping the penetrating noise of densely packed humanity equipped with noisy machines.

Advocates of the old soundscape also waged political battles to quiet the rowdy terrains of urban life. The fight over rights to the soundscape, in keeping with the period, was a class struggle that paralleled the earlier conflict between peasants and aristocrats over land. Exemplifying this conflict, figure 3.6 shows a distraught violinist being overwhelmed by street noise entering through his open window, presumably in the summer when ventilation was needed. Famous intellectuals such as Charles Babbage, the inventor of mechanical computation, and the prolific author and commentator Charles Dickens actively contributed to the battle against the noises of the urban life. Did the multitude of street musicians have rights to the soundscape, which could, without special and expensive efforts, invade the sanctity of the privileged classes? For the elite, the answer was clearly no. Over the next century in many countries, a series of legal battles attempted to regulate the right to generate sound, but most such attempts proved futile.

From the struggle to combat noise, the concept of the private acoustic arena gained prominence among the professional classes. When acoustic arenas became private, owners could then select which sounds would enter and by what means. Aural architecture focused on privatized arenas. The industrial engine, in polluting the public soundscape, also created the radio, telephone, and phonograph as private sound sources. Such devices gave direct control over sound propagation and genera-

Figure 3.6
Hogarth's *Enraged Musicians*. Courtesy of Graphic Arts Collection, Department of Rare Books and Special Collections, Princeton University.

tion to any individual with access to these new inventions. Sound would now enter parlors only by invitation.

These new devices created and reproduced sound without the visual presence of the original source. Captured on wax cylinders, disembodied voices of the dead hinted at humans' ascendancy over mortality. Where early humans heard the voices of spirits, humans of the Victorian age heard the voices of those now dead. The famous RCA picture of the dog Nipper listening to a gramophone, labeled "His Master's Voice," was as much a political and philosophic statement as it was an advertising slogan. The picture embodies a much earlier concept of sound as a means of giving commands that must be obeyed. The dog sits obediently listening to an ethereal voice without a body. Under the overlay of scientific and engineering progress, the mystical experience of sound without a visible source reappeared.

As technology advanced during the early twentieth century, public soundscapes continued to shrink, and private acoustic arenas continued to proliferate as private refuges. With efficient manufacturing and distribution of mechanical goods, individuals readily acquired the means for contributing yet more noise to the public soundscape. Emily Thompson (2002) described the Roaring Twenties as even noisier than early industrial England. The living sounds of natural life, wagons making deliveries and children playing after school, were now overwhelmed by the sounds of speeding engines, honking automobiles, and clacking typewriters; some industrial devices also produced disturbing low-frequency infrasounds. Doctors and legislators expressed concern about the dangers of high noise levels, and a small group of reformers organized the Society for the Suppression of Unnecessary Noise. Whether people experienced modern noises as energizing or as enervating, they could not ignore the cacophony of a society now embracing mechanization as the religion of progress.

The formation of acoustic engineering as a profession was, in part, a response to the perceived need for taming the pervasive noises in public streets and private offices. In this environment, the Acoustical Society of America was born. The need for sound control spawned an industry of acoustic building materials, and a generation of service industries applied these new materials to buildings and machines. Manufacturers developed a range of acoustic building materials with such names as Keystone Hair, Rumford Tile, Audiocoustone Plaster, Acoustifibroloc, Insulite Acoustile and Armstrong Corkustic. In addition to shielding interior spaces from external noise, soundproofing dampened noises that were internally generated. The mantra "absorb and remove all sound" became the prevailing response to high noise levels. Society was redefining sound quality as pure and direct sound, without any influences from the environment. Echoes, reverberation, and resonances were all viewed as a kind of amplified noise.

Buildings and spaces constructed specifically for the new electronic arts of broadcasting and recording reflected this new demand for greater quiet in aural architecture. They were monuments to a new acoustic era, which could be summarized as spatial designs that emphasized foreground sounds in a background of silence. Soundscapes were to be stripped of living sounds and spatial acoustics, thereby weakening the aural connection of spaces to the social fabric. Studios for recording and broadcasting used this new criterion for spatial acoustics, which was a radical break from traditional concert halls.

As the new demand for ever quieter spaces permeated society, professional architects kept lowering their recommendations for reverberation time, from 3 seconds in 1923 to 1.5 seconds in 1930. The new aural architecture divorced itself from the traditional acoustics of Beethoven symphonies in nineteenth-century European concert halls. The open-air Hollywood Bowl amphitheater, as shown in figure 3.7, is a modern version of the ancient Greek amphitheater, with neither reflecting surfaces nor reverberation. Constructed in 1922, it remains the largest natural amphitheater in the United States,

Figure 3.7
Hollywood Bowl. Courtesy of Hollywood Bowl Museum.

seating nearly 18,000 patrons. Its outdoor acoustics became a model for those of enclosed spaces, thus returning to the Greek concept of space. Auditorium walls were nothing but "a necessary nuisance" to support the ceiling and isolate the space from a noisy soundscape. Moreover, as F. R. Watson (1926) suggested, in the ideal concert hall, musicians would have only modest reverberation and the audience none. The perfect auditorium was thought to be a space that replicated the listening experience of a radio in a parlor, or an orator in an open-air theater.

Radio City Music Hall, as the name implies, was a performance space married to the new sounds of radio and the new aural architecture. It was built using the fruits of an energized industry of sound suppression. The sweeping arches over the stage, as shown in figure 3.8, were themselves constructed of a thousand tons of Kalite sound-absorbing plaster, and the rear walls of the hall included a thick blanket of similar sound-deadening materials. Although none of the initial acoustic measurements are

Figure 3.8
Radio City Music Hall. Courtesy of Museum of New York.

currently available, scientists estimated the reverberation time to have been less than 1.0 second for an audience of 6,000. The stated goal was to make the listening experience as pure as possible by designing a system that could transparently communicate the sound from the stage and loudspeaker directly to the listener. When soundproofed studios were combined with soundproofed auditoriums, the path from sound source to listeners would be as "untarnished" as listening to a performance in an anechoic chamber.

At the opening ceremonies of Radio City Music Hall, Roxy Rothafel, the director of the new auditorium, commented: "I think we have made sufficient progress in the science of acoustics to eliminate all possibility of error in reverberation and absorption" (Thompson, 2002). He was confident that acoustic experts had created a space as ideal as possible for sound *transmission*, and that the electroacoustic amplification was so perfect that listeners would not be able to detect its existence. The artistic and aesthetic function of reverberation in enclosed spaces was now replaced with "cleaner" electroacoustics. The evils of spatial acoustics had been banished. By removing reflections and reverberation, and by providing sufficient loudness using electroacoustic amplification, the designer of the hall had created a space where every seat would have the acoustics of the home living room. It was perfect democracy, a perfect Greek open-air theater without spatial acoustics, available to all citizens, not just to the elite.

Radio City Music Hall was the poster child for the idealization of the modern performance space. Other examples include Eastman Theater in Rochester, Chicago Civic Opera, Severance Hall in Cleveland, Kleinhaus Music Hall in Buffalo, and numerous performance spaces in schools and public buildings. Beranek (1962) likened hearing a performance in these auditoriums to hearing a recording in a carpeted living room, and Forsyth (1985) named these spaces "hi-fi concert halls." Not only had artistic attitudes changed, but these performance spaces also had an enlarged seating area, which made traditional shapes impractical. With their weak reverberation and strong intimacy, acoustically dry spaces were judged favorably by the standards of their time.

In response to the new acoustics, musical compositions of the period could support faster tempi, more delicate transients, stronger rhythms, and discordant harmonies. Smoothing the transition among successive notes, a role automatically assumed by reverberation, was provided by musicians using extensive portamento. The auditory arts of this period were as much an adaptation to dry acoustics, as were Gregorian chants to reverberant cathedrals, and Bach's Masses to small Protestant churches. In all cases, the acoustical properties of the space originated from social forces unrelated to the aural arts—industrial noise in one case, religion in the others. Composers simply adapted to the prevailing spaces.

Radio Engineers and Entrepreneurs Redefine Acoustics

In the late nineteenth century, when Thomas Edison invented the phonograph and Alexander Bell invented the telephone, they became the first of a new generation of technologists functioning as aural architects. Their inventions converted sonic vibrations of the air (sound waves) into another medium, wiggling wax grooves and electric signals. If audio information embedded in a physical medium could be converted back into sound waves, sound could be stored, transmitted, manipulated, duplicated, and most importantly reproduced at distant times and places. Telephone and phonograph technology would forever change our experience of space and time.

Although the phonograph was originally intended as an office dictation device, it soon became apparent that the real market for recorded sound was as entertainment. With the proliferation of home phonographs and phonograph records, entertainment sound became portable, permanent, and widely distributed. Similarly, the microphone, which was invented for telephone communications, would enable radio to distribute entertainment to remote spaces in real time. In all these inventions, three common elements were involved: capturing and converting sounds to a new medium, transmitting or transporting the medium from a source to its destination, and reconverting the information in the medium back into sound.

The first prototype of the phonograph was not much more than an acoustic horn, a large conical funnel that concentrated sound at a wiggling stylus, which created a permanent replica of the sound in a wax surface. To reproduce the sound, the needle tracking those wiggles vibrated a diaphragm to create sound waves that corresponded to the original sound. Because the process was crude, inefficient, and noisy, and because there was no means for electronic amplification, to have your voice recorded, you had to place your head close to or in the cone. Like a magnifying glass, the horn both focused and magnified (amplified) sounds that would otherwise have radiated in all directions. The first phonograph was also the first application of closely placed microphones; to achieve sufficient loudness, singers had to place their heads into the horn. The approach worked well enough for a soloist but was, of course, impossible for a chorus, let alone a large orchestra, although Edison constructed an exponential horn 125 feet long in a futile attempt to record one.

Designed to intensify sound and suppress noise, the *close microphone* captured only direct sound, not reverberation and spatial acoustics, these being of lower intensity than direct sound. Close microphones suppressed all forms of background sound, whether desirable or not. Even with the introduction of the carbon microphone, a transducer that converted sound into electrical signals, there was a need for close microphones because carbon microphones and their amplifiers were still noisy. Indeed, close microphones have remained the norm, even though they are no longer required for increasing loudness and suppressing noise.

Through continuous exposure to, and intense marketing of, music recorded with "dry" (deadened or suppressed) acoustics, the listening public came to accept this new concept of music quality. The same dynamic took place during Prohibition when brewed beer was no longer aged: after a decade of exposure, the public accepted this weaker beer as the desirable norm. With repeated exposure, sensory expectations adjust to what is familiar, regardless of its intrinsic attributes. Dry acoustics and weak beer both become matters of habit and custom.

As part of the marketing campaign to advance this new concept of spatial acoustics, test audiences were subjected to "Tone Testing"—asked to make blind comparisons between a live singer and a prerecorded version of that singer (Welch and Burt, 1994). Both singer and phonograph were hidden behind a curtain, and virtually none of the listeners could distinguish between them. At its peak around 1920, there were more than 2,000 such demonstrations, including one at Carnegie Hall (Thompson, 2002). Sound engineers, in striving for precise recordings that would pass these Tone Tests, avoided recording any room resonances, echoes, or reverberation, achieving a truly dry recording. In both cases, the audience heard only the acoustics of the reproduction space. These demonstrations were enthusiastically received by the public, which became a validation of their "precision," meaning the absence of noise, distortion, and spatial acoustics.

As with many marketing campaigns, what began as a technical limitation was transformed into a desirable selling feature. Direct sound was considered as more "precise" than sound modified by spatial acoustics. And to enhance this new sense of quality, even the earliest recording studios were treated to remove the effects of acoustics. A generation of listeners was raised on these recordings, and learned to value direct sound. In the home parlor, a recording of a piano was expected to sound the same as the piano in the parlor. Contemporaneously with the phonograph, the player piano, which played prerecorded paper tapes, appeared as yet another means by which great musicians could be brought into the home (Holliday, 2000). For millions of listeners, who were now being exposed to professional music, the parlor was, not an alternative to a concert hall, but rather the preferred listening space. There was no good reason to bring the acoustics of a foreign space into the warmth of the parlor.

For the nascent, yet high-growth audio industries, winner-take-all capitalism was the major force that drove acceptance of the new inventions. The owners of these industries controlled expensive assets: studio facilities, transmitters, manufacturing plants, and artists under contract. These investments were intended to produce an economic return, which entailed a high sales volume. As is the case today, the corporate mandate was to market available technology and techniques as being the highest possible quality. In this respect, David Sarnoff, the entrepreneur who transformed the primitive technology of radio into the broadcast giant RCA, influenced aural architecture without any intention of doing so. These entrepreneurs determined the listening experience, not just the sound content, but also the acoustics. As sound and auditory art became a commercial commodity, spatial acoustics were removed from performed and recorded music alike.

Unlike earlier applications of technology to the musical arts where the aesthetic component was central, the inventions of phonograph and broadcasting were driven by the passions of entrepreneurs: inventors, engineers, capitalists, and managers. Some of the more charismatic figures of the period included such famous individuals as Thomas Edison, Alexander Graham Bell, Emile Berliner, David Sarnoff, Guglielmo Marconi, Lee De Forest, Edwin Armstrong, and James Jensen. Only a few of them had a personal interest in the musical arts. In every case, the army of technical innovators was financially supported by the consumption of their products and services. Listeners to radio and phonographs were rapidly becoming a market that was orders of magnitude larger than any historic audience. By 1927, a million phonographs were playing more than 100 million records. Musical audiences had grown from the relatively few in the private music chambers of the seventeenth century, to the many who crowded into the public music halls of the nineteenth century, to the vast numbers of listeners within reach of twentieth-century radio.

What appeared to be a new concept of musical space was, from a different perspective, only a different application of one of the oldest paradigms in music and listening spaces. The ancient Greeks had constructed immense open-air amphitheaters to allow their educated citizens to participate in public democracy. Medieval bishops had constructed cathedrals for large congregations, using music, voice, and acoustic spaces as tools for religious purposes. Twentieth-century broadcasting networks now used music and the dry acoustics of their studios as tools for commercial purposes, to reach the widest audience of consumers possible. Only powerful institutions had the necessary resources to invest in large theaters, large cathedrals, or large radio networks. Audience size is everything, suggesting that little has changed over twenty-five centuries.

Designed and constructed in the early 1930s, Rockefeller Center was the palace fortress of the radio aural arts, and Radio City Music Hall was its public temple of unsurpassed size and grandeur. In the early 1930s, the NBC studios were adorned with glass museum cases that displayed modern electroacoustic inventions as if they were holy relics. The sonic priests working in their sanctuary would describe the new NBC studios as a "temple to glorify the radio voice" and a "gigantic cathedral of sound." And far from this new temple, the home living room with its honored radio at the center had become the equivalent of the local parish church.

Especially during the Depression, the importance of radio was not lost on musical and political communities. By 1930, after enthusiastic response to operatic segments performed in radio studios, NBC negotiated the rights to transmit the New York Metropolitan Opera performances. Instead of competing with live performances, radio broadcasts glorified and democratized them. In response to rave reviews of radio opera programs, people who had never heard opera before saved money for trips to New York in order to attend a real opera. For 43 years, Milton Cross, the host of the Saturday afternoon Metropolitan Opera broadcasts, would be a household name (DeLong, 1980).

Although the decision to broadcast classical music into the family parlor would prove very popular, the initial motivation had little to do with dispersing the auditory arts to a wider audience. As Donald C. Meyer (2000) noted, the regulation and financing of radio were still under discussion in the late 1930s. One possibility was the English model of a semigovernmental organization supported by a yearly license fee levied on every radio set. Another possibility, preferred by the radio industry, was to support radio through private enterprise, using advertising and royalties. NBC demonstrated that it served the interests of the wider population, and especially the educated elite, by providing cultural opportunities—orchestral music, classical opera, political discussions, and international news. There was no need, it contended, for the government to intervene to ensure that radio would be used for public purposes. Music in the new medium served political and economic purposes, which only then enabled the technology to transform the aural arts into mass entertainment, which contained its own versions of the aural architecture of musical space. Aural architecture was a hidden passenger on this technology train.

The important legacy of radio and phonograph was the commodization of a previously limited art form. Radio became a growth industry. NBC hired 92 musicians and contracted Arturo Toscanini to conduct a new radio symphony orchestra. Never before or since had a broadcast network created a symphony orchestra of this quality, stature, and renown. At the age of 70, Toscanini was already an internationally respected conductor of great skill and knowledge. His radio orchestra would also make phonograph records for almost two decades, which would be distributed and preserved for future generations. Thomas A. DeLong (1980) described Toscanini's power to define the musical norm when he commented, "an average broadcast in the 1930s reached more people than the total of all those who attended, in person, the four thousand or more concerts given by the New York Philharmonic in its first 90 years." Radio music was *the* music in this period, and the musical style and spatial acoustics of these broadcasts became a universal reference.

Radio music had its own unique artistry and spatial acoustics, in part, because radio studios were not designed to mimic live performances in a concert hall. Studios were designed to be flexible tools of the sound engineering industry producing radio broadcasts. Except for NBC's studios, there is only a limited amount of historical information about early radio studios. Rockefeller Center included 27 studios, representing the best that could be built when the facility opened in 1933. Studio 8H was the largest in the world, with a volume of almost 9,000 cubic meters (320,000 cubic feet) and a height of three stories, a size that could accommodate a full orchestra (Hanson, 1932). Although the volume approached that of a small concert hall, the spatial design was dominated by the requirement for a large number of performers and a variety of program formats. This studio would be used for traditional orchestra music, but it was designed to have a short reverberation time. Sound engineers enjoyed modest acoustic flexibility with movable panels that could be placed over some of the sound-absorbing surfaces. But practices of the period required that the space be acoustically dead even when the panels were placed in their most reverberant position (Thompson, 2002).

Because these concerts reached millions of listeners, and because many, preserved on phonograph records, became classics, the acoustics of Studio 8H were to become famous, or more accurately, infamous. Although Arturo Toscanini favored little vibrato in the strings, crisp timpani with hard sticks, and prominent woodwinds and brass, it was the absence of reverberation in Studio 8H that produced music without depth. Nevertheless, for millions of listeners, Studio 8H was their model of a musical space. Dead acoustics were the cultural norm.

Even though radio broadcasts devalued acoustics in musical spaces, reverberation retained a niche presence as a special effect in radio dramas. Dedicated effects rooms were designed as sonic tools. Rain and thunder machines were used to create the aural impression of a storm, and reverberation chambers were used to create the impression of such unique spaces as caves and haunted houses. On the other hand, using special effects for music was the exception rather than the norm. Performance studios were normally the acoustic equivalent of a hospital, scrubbed clean of everything that could create an aural personality.

In one of the earliest discussions of studio acoustics, Hope Bagenal and Alex Wood (1931) challenged this extreme and, from our modern perspective, untenable view that dry recordings were a synonym for precise reproduction. As renowned acoustic architects in England, they correctly argued that spatial acoustics should be included when recording music written for a concert hall. Their view followed the traditional values of musical artistry, rather than the entrepreneurial values of a nascent American audio industry striving for profitability. However reasonable this view may have been, recording producers and audio engineers faced several harsh realities: recording natural acoustics was difficult if not impossible; artificial reverberation was primitive and inadequate; commercial recording enterprises could not afford to build high-quality reverberant spaces; and the listening public was already educated to consider deadened acoustics as synonymous with quality.

Artistic and Social Conflicts in Concert Hall Designs

If the listening public had never experienced eighteenth- and nineteenth-century classical music in concert halls, the revolutionary changes in radio acoustics might have become permanent. And if musicians had performed only newly composed music written for such spaces, acoustically dead spaces would have become the unchallenged norm. But the music of the grand masters, which was written with the expectation of a performance space with appropriate reverberation, was too valuable and too appreciated to be abandoned. Sophisticated listeners who had once experienced Bach's organ compositions in a church and Beethoven's symphonies in a concert hall came to recognize that they sounded flat, almost inadequate, in the aural architecture of recording studios. At first, while interest in broadcasting ran high, the inconsistency between the acoustics of a concert hall and those of a recording studio remained in the background. But later, even as studio music was penetrating all corners of society, interest in live performances arose and began to grow.

Rather than replacing classical concert performances, broadcast and recorded music increased listeners' interest in live performances. Once exposed to classical opera and symphonic music for the first time through their treasured radio, farmers in rural America sought to attend live performances. As small towns became large cities, there was a pressing need for performance spaces to host lectures, theater, dance, folk music, chamber ensembles, popular music, and classical symphonies. Generic auditoriums were constructed to accommodate expanding audiences with an interest in a wide variety of artistic and social functions. Thus the multipurpose auditorium, an awkward spatial compromise having modest reverberation, came to replace the high school gym and the single-use concert venues designed for a particular musical genre.

By the middle of the twentieth century, the practice of acoustics had split into two separate branches, live and recorded, with two different concepts of aural architecture. As knowledge of acoustic science advanced over the ensuing decades, and as digital signal processing eventually overtook the limitations of analog methods, aural architecture divided still further. On one branch, by modeling famous concert halls and simulating their acoustics, researchers advanced their understanding of the physics and perception of such spaces. On the other branch, by designing artificial reverberators and spatial synthesizers to create virtual spaces, audio engineers transcended the physical properties of sound waves in enclosed spaces. Existing in parallel, the two branches were at once competitors and collaborators. Sharing insights and technology, listeners, artists, and scientists embraced spaces on both branches. Today, the two branches still influence each other. Audio engineers attempt to re-create the experience of a live concert hall in an automobile and a living room, and popular musicians attempt to replicate the sound of a recording during a live performance. The dynamic interplay between the two branches has had different consequences for each, but the basic issues have been similar.

To understand the interplay between the two branches of acoustics, let us now return to the classical concert hall, whose recent history manifests conflicts among art, science, economics, and social forces. Thompson (2002) chronicled the tortuous relationship between acoustic science and spatial design in the nineteenth century. It was a frustrating century because scientific achievements created the illusion that acoustic science was contributing to aural architecture. "As science and architecture parted ways, the subject of architectural acoustics fell into the gap that opened between them" (Thompson, 2002). Furthermore, the artistic and subjective aspects of architecture, which transcend science, were difficult, if not impossible, to evaluate using physical measures.

At that time, the artistic aspect of acoustics—its influence on music—was often taken as an immutable given, and not a proper subject for artistic creativity. There is no better demonstration of this attitude than the story of how Sabine, who would become famous as the world's first acoustic scientist, went about designing the new Boston Symphony Hall in 1900. A music reviewer at a local newspaper later observed that Sabine's calculations held a single objective constantly in view: creating a performance space for the Boston Symphony Orchestra that matched its musical repertoire, which at the time, was predominantly European.

Before examining how Sabine embedded an acoustic personality into his spatial design, we need to examine the process by which a design approach was selected. The historical record detailed the role of tradition, conservatism, and the general unwillingness to engage in artistic and architectural experimentation. When Henry Lee Higginson, the financier, philanthropist, and owner of the project, commissioned design proposals, he received three choices: an elliptical shape, a rectangular design modeled after the Neues Gewandhaus in Leipzig, and a semicircular design modeled after the Greek amphitheater (Stebbins, 2000). The architects favored the semicircular design, which they explored in some detail, but the building committee, wishing to minimize acoustic risk, selected the Gewandhaus design, that celebrated hall being a space of admirable acoustics for the typical repertoire of the eighteenth and nineteenth centuries—and also the repertoire of the Boston Symphony Orchestra at that time.

But even though the sponsors had chosen to replicate the acoustics and overall design of a known symphony hall, in keeping with the democratization of music, the seating capacity of the new hall was to be enlarged by at least a thousand. Changing the model hall's size to accommodate a larger audience, while still preserving its aural personality, was a high-risk goal that required Sabine to invent both theoretical and empirical techniques. His application of science and engineering to the design process allowed the essential aural architecture of one space to be transferred to another. Boston Symphony Hall would eventually be judged a brilliant success. The sponsors had chosen a good reference model, and Sabine successfully replicated its aural personality.

Yet for the first few years after its inaugural opening, reviewers, listeners, and musicians kept up a steady stream of criticism. They kept comparing the aural personality of the new Boston Symphony Hall with that of the old Boston Music Hall. The new acoustic space did not sound like the old space because that was never the goal. The new aural personality was simply unfamiliar to those who had spent years listening to performances in the old hall. Hostility toward the new space, followed shortly by ambivalence, should have been expected because it takes time and experience to reset sensory and artistic expectations. Thus it should have come as no surprise that the public interpreted the changed acoustics of the new hall as a deficiency. The sharpest critics emphasized that Sabine was only an amateur musician who should not be expected, or allowed, to have imposed an aural personality into the design. Responding to the virulence of the criticism, Sabine applied his training as a physical scientist in a futile effort to determine the acoustic parameters that corresponded to cultivated musical taste (Sabine, 1906). That, of course, assumed that artistic taste, which is neither stable nor reliable, but evolves with experience, was subject to scientific investigation. As an acoustic engineer with the mandate to replicate the acoustics of the Neues Gewandhaus, in the new Boston Music Hall, Sabine was not responsible for the decision to choose an aural personality that differed from the old hall's—nor was he an aural architect.

As time went on, most everyone came to appreciate the aural beauty of the new concert hall. And it is now considered one of the world's three best halls, along with

Amsterdam's Concertgebouw and Vienna's Musikverein (Rybczynski, 1996). How did this change in attitude come about? Just as musicians require practice to become familiar with a new musical instrument, they also require practice to become familiar with a new space. For the same reason, visiting orchestras often do not perform as well as a resident orchestra. Every space and every instrument has a unique personality. Orchestra arrangements must also be adapted to a space. In fact, Beranek (1988) suggested that the Boston Symphony Hall gained acoustic maturity only when the orchestra size was increased to over a hundred performers: a larger orchestra was required to fill the larger new space.

If not the acoustic engineer, who, then, chooses the aural personality of a concert hall? Although it is primarily a venue for listeners, musicians, and composers, these acoustic stakeholders are often disenfranchised from the decision-making process. Those who do make the decisions, economic and political sponsors, may not have interest in, much less knowledge of, music or spatial acoustics. Decision makers often find that the visual experience of a space is more compelling than the acoustics, especially since scale models allow visualization but not auralization. Although not his intent, Beranek's acoustic reviews (1962, 1996a) of the world's major concert halls also illustrated the creative variety in visual designs of concert halls. Visual impact is a value encouraged by sponsors who covet a flagship symbol of their home city, and by architects who covet a signature design to highlight their portfolios. These concert halls are visually unique, and compromises between aural and visual requirements are inevitable.

The story of every concert hall is as unique as the biographies of famous individuals. Having considered the Boston Symphony Hall, let us now turn to the Berlin Philharmonic Hall, shown in figure 3.9 and conceived by architects Hans Scharoun and Edgar Wisniewski in the 1950s as a symbolic statement of Germany's cultural rejuvenation and political rebirth in its war-ravaged former capital. For these visual architects, the new space was a visionary statement of German Expressionism, both within and without: heavily laden with geometric symbolism having religious overtones and with the repeated use of the tripled pentagon to represent the unification of space, music, and people (Wisniewski, 1993). Based on a circular model from classical architecture their design made a social and political statement: the raked tiers of the auditorium enveloped the center stage, blurring the spatial distinction between audience and musicians. Although visually intense in every respect, Berlin Philharmonic Hall was not, however, particularly interesting as aural architecture.

Neither of its architects was known for his acoustic expertise. The world-renowned scientist Lothar Cremer had been called in as an acoustic consultant, but his role was advisory and supporting, to ensure against the undesirable and unintentional consequences of architectural innovation. Acousticians must not be allowed to design the musical spaces of concert halls, Wisniewski insisted; otherwise, every concert hall in the

Figure 3.9
Berlin Philharmonic Hall. © 1993 Gebrüder Mann Verlag, Berlin.

world would be based on the risk-free shape of a rectangular shoebox. As its two architects saw it, the visual experience of the Philharmonic would serve to prepare listeners for their auditory experience. From the architects' acoustically naive viewpoint, the critical distinction between live and recorded music would be lost if the visual component did not dominate the aural experience. The unintentional consequence of their visual choices, however, was a delay of 50 milliseconds between direct sound and the onset of reverberation, more than twice the desirable values of 20 milliseconds or less. Their visual creativity had an aural cost: severely reduced aural intimacy.

Avery Fisher Hall at Lincoln Center was one of the more infamous cases where a combination of political, economic, and scientific requirements produced an acoustic disaster that was disliked by reviewers, listeners, and musicians alike. To enlarge the seating capacity, architects designed bulging concave sidewalls, which created focusing echoes. Faced with limited funding, they opted for smooth surfaces, devoid of the adornments whose irregularities would have produced much needed sonic diffusion. When neither adjusting the small reflecting panels on the ceiling nor enlisting a parade of consultants remedied the acoustic defects, the sponsors hired Cyril Harris to completely rebuild the interior, using the proven shoebox shape.

The story of the Avery Fisher Hall is instructive, not because the original acoustic architect, Leo Beranek, was world renowned, nor because the remedy was based on the proven shape of a shoebox, nor even because its acoustic failures stimulated fresh scientific research. The story is instructive because of the preconditions imposed by Harris before accepting the project. "You will agree that acoustics will have priority over aesthetics. If the architects and I should disagree, I'd have to win the argument" (Bliven, 1976). Harris demanded to be an aural architect, not merely a consultant whose recommendations might be ignored. In this respect, he was acknowledging, without explicitly saying so, that Beranek had been a victim, however willing, of a political process he neither understood nor controlled.

A spatial design becomes a fantasy when the sponsors require that its acoustics be appropriate both for lectures, which emphasize speech intelligibility, *and* for Romantic symphonies, which require extensive reverberation. To make the fantasy even more absurd, economically motivated sponsors often insist that a hall have a large seating capacity while retaining a feeling of intimacy. For example, the Royal Albert Hall in London was originally required to have over 6,000 seats, with the additional requirement that every seat have high-quality acoustics and intimate sight lines to the stage. Though nominally part of an artistic process, acoustic design is overpowered by such conflicting requirements.

In summary, throughout history, and certainly in the twentieth century, aural architecture has resulted from the dynamic interplay of visual, aural, social, political, economic, and religious forces. In almost every case, however, nonaural forces were sufficiently powerful to constrain the aural personality of a space. Aural architecture has been subservient to unrelated, yet evolving, cultural values, and remains so to this day.

Artificial Reverberation as a Flexible Compromise

During the mid-twentieth century, audio engineers used close microphones in acoustically dead radio studios, but placed microphones high above the audience area at live performances. Now, you might believe that microphones were elevated above the audience in order to capture the ambience and reverberation at live performances. But given the prevailing negative attitudes toward spatial acoustics, other explanations are more likely. Positioning numerous microphones close to musicians in a live performance would have required that mixing faculties be nearby to create a single channel from the multiplicity of microphone channels, which was simply not practical at that time. Recording studios, on the other hand, were explicitly designed to support multichannel mixing.

Using elevated microphones at live performances provided remote listeners with the same aural experience as those who were attending the performance, including the sounds of shuffling feet, muffled coughs, premature applause, and even rustling candy wrappers. Although clearly a defect in a recording, where the unwanted environmental noise repeats at the same point each time the recording is heard, with live broadcasts, background sounds were desirable: hearing them, home listeners could readily imagine themselves as being part of the audience. From our twenty-first-century perspective, we find it odd that people once considered spatial acoustics as noise, but coughs as signs of life. Attitudes toward "noise" are fleeting reflections of the current culture.

By the middle of the twentieth century, aside from occasional broadcasts from concert halls, there were few spatial choices for recording other than acoustically dead studios. Yet more and more, the desirability of hearing music in reverberant spaces came to be recognized. Like a pendulum, tastes in acoustics swung back to an earlier period. In a classical concert hall, acoustics add reverberation to the end of each note, and soften its initial attack. Reverberation blends a sequence of isolated notes into a continuous musical flow. As mentioned earlier, because an organ note begins and ends abruptly when a key is pressed and released, organ music becomes very flat without some form of reverberation. People came to perceive reverberation, not as noise, but as part of the musical arts. When Decca recorded many of the new rock-and-roll musicians in the 1950s, they used the ballroom at the Pythian Temple in New York because of its acoustics. Without an audience, and with its large volume, its reverberation was perfect for this kind of popular music. In fact, this particular space became the signature sound for many famous performers of the period. Inspired by popular music with its adventuresome interest in novel spaces and sounds, acoustics had returned to music.

Audio engineers and serious musicians soon abandoned the aberrant idea that dead acoustics were a sign of quality in live and recorded music alike. In recognizing the need for ambient acoustics, recording and broadcast engineers actively searched for practical alternatives. How could spatial reverberation be incorporated into broadcast and recorded music? Their initial answer was the reverberation chamber, an enclosed space with a loudspeaker for injecting dry recorded sound, and a microphone for sensing the resulting spatial reverberation. As part of their studio complex, NBC included three such chambers with rectangular shape and an average volume of about 50 cubic meters (1,750 cubic feet), comparable to that of a small bedroom (Hanson, 1932). But with insufficient volume, regular geometry, and strong resonances, these chambers were inferior alternatives to concert halls. Within a few years, however, larger reverberation chambers and recording studios with better acoustics were being built (Rettinger, 1957, 1961; Volkmann, 1942). Nevertheless, designing a high-quality space suitable for classical music from a variety of centuries was still difficult and expensive. A better solution was needed.

Simultaneously with the advances in reverberation chambers, Laurens Hammond (1939) introduced a home organ that included reverberation. How else could the sound of an organ at home resemble that of an organ in a church? Using electromechanical delay-line technology that had already been developed at Bell Laboratories, R. L. Wegel (1932) conceived of the first helical spring reverberator as an intrinsic part of the instrument. The spring reverberator contained one transducer that converted an electrical version of sound into a mechanical vibration, and a second transducer that converted it back again. The time for a mechanical vibration to traverse the spring was an audio delay. And if the spring was configured to reflect mechanical vibrations internally at its ends, then, much like the sonic reflections off the walls of a room, the result was a primitive form of reverberation.

Over time, what had been a few isolated examples of reverberation as special effects became an industry producing a wide range of adjunct devices for popular musicians, notably guitarists. These devices included the Binson Echorec, Uniton Swissecho, Selmer Echo, and Watkins Copycat, many of which were nothing more than portable tape recorders configured with multiple heads and feedback loops. By combining such echo devices with spring reverberators, popular musicians acquired a full repertoire of effects. From our perspective, these devices were all crude and ambiguous mergers of musical instruments and spatial acoustics, both of which were energy decay processes. There was no conceptual difference between the vibration of a plucked string and the vibration of a coiled spring. When combined, the instrument simply had two vibrating elements.

The designers of spatial synthesizers and reverberation devices struggled with one central problem: finding a medium that could hold audio information for many seconds without being degraded by additive noise, nonlinear distortion, and frequency limitations. The medium must transmit sonic information even when represented in another physical format, such as a bending wave in a plate or vibration in a spring. Candidates for such media included air, metal, magnetic tape, optical film, and electronics. Once a medium for holding sound was selected, the inventive focus then, and only then, turned toward emulating physical spaces.

The worldwide race to invent high-quality spatial emulators of concert hall acoustics took place in Germany, Austria, Great Britain, and the United States. In 1960, the winning device was announced—a large electromechanical steel plate, which required a forklift to move (Kuhl, 1958). After extensive research on the wave mechanics of metal plates, it acquired the necessary quality to be used for broadcast and recorded music. Shortly thereafter, under pressure to reduce size and weight, a Viennese company perfected spring reverberation using random etching of a coiled wire (Fidi, 1970), and a few years later, a German company replaced the steel plate with gold foil (Kuhl, 1970).

These devices became the mainstay of audio studios because reverberation was no longer dependent on a physical space; high-quality reverberation could easily be added in the mixing studio. Close microphone techniques, which suppressed background noise and provided artistic flexibility, no longer produced sterile music. Moreover, reverberation devices included control over the most basic spatial parameter: reverberation time. Audio engineers, who used or designed these devices, acquired a new role: aural architects of synthesized and virtualized musical spaces.

The search for even better reverberation solutions accelerated with the digital and computer revolution. Digitized audio, signals in the form of bits, could be stored indefinitely in a memory device without any degradation. The perfect delay medium was now at hand, and with supporting elements such as analog-to-digital converters and arithmetic computation, a complete audio delay system was possible. When Manfred R. Schroeder and Benjamin F. Logan, Jr. (1961) demonstrated their computer simulation of reverberation, it became clear that digital audio could finally provide a collection of high-quality basic components that could implement any audio algorithm. The initial reverberation algorithm, which was very primitive by today's standards, was a recreational project (Schroeder, pers. comm.). Yet the advent of audio signal processing by computer, which was truly revolutionary, stimulated the visionary dreams of many other acoustic possibilities, reverberation being only one such example (Schroeder, 1970).

It would not be long before an all-digital reverberator took its revered place in the studio alongside traditional audio equipment (Bäder and Blesser, 1977). A new competitive race began among those with skills in digital technology, including EMT, Lexicon, Ursa Major, Quad-Eight, AKG, and Quantec. Unlike earlier developments in audio technology, however, the computer industry controlled the rate of progress of digitized audio. This technology required an investment far greater than that which could be supported by the audio industry itself. Audio engineers simply had to wait patiently until a digital infrastructure produced the necessary power, quality, and cost reduction. Early frustration at poor results was often overwhelming because technology did not yet offer the ability to replicate the sound of natural acoustics, which was fundamentally a very difficult problem.

By the end of the twentieth century, after 25 years of intense research and development in acoustic simulation, and after corresponding advances in computer technology, the collective efforts of hundreds of innovators finally produced systems that approached, and in some cases surpassed, the acoustic quality of real concert halls (Blesser, 2001). It took a full generation for the technology to transform an engineering problem into an artistic tool. The next step in the evolution of reverberation occurred when algorithms took advantage of the processors found in personal computers, which were now powerful enough to implement even the best algorithms in the living room.

At this point, artificial acoustics have merged with electronic music to become a unitary concept, creating a new art form that is not restricted to sound waves in acoustic spaces, within instruments or without. Artists are now able to treat acoustics as fluid and amorphous. Each instrument can have its own attendant reverberation. The sonic timbre of a musical note merges with the effect of its space, both in location and reverberation. Musical voices need not be spatially consistent. Within the same composition, a violin in a small space near the listener can exist together with a trumpet in a

cathedral at a distance. A musical voice can appear at multiple locations within a dynamically changing space. The twentieth-century revolution in musical space existed
in parallel with the more apparent revolution in musical styles. By the twenty-first century, they have converged.

The histories of virtual spaces and concert hall spaces are like two branches on the
same tree, but with different social and technical forces influencing each branch to produce different artistic results. Most notably, virtual spaces are inexpensive and flexible,
whereas concert hall spaces are expensive and inflexible. These differences, more than
the specific properties of the actual aural architecture, have dominated the artistic
implications of space. It is trivial to adjust parameters of a virtual space and almost impossible to modify the acoustics of a concert hall space. The nearly immutable aural architecture of a concert hall, a physical legacy lasting centuries, provides artistic stability
and consistency. In contrast, virtual spaces are ephemeral, with novel spaces appearing
almost every moment. Like almost everything in our culture, changes are accelerating.
The economic and political constraints found in concert halls disappear when using a
personal computer at home—the final democratization of music in space.

By working with flexible and inexpensive tools, aural architects of virtual spaces became similar to instrument designers, composers, arrangers, performers, and listeners,
all of whom can readily experiment with variations of the sonic arts. Every listening
environment and every sample of recorded music is a variant of a musical space.
When adjusting spatial simulators, audio engineers perform the roles historically performed by acoustic architects and engineers. Or rather, they become real aural architects with the means to make space—virtual, artificial, or imaginary—an artistic tool.
No previous generation of aural architects had such freedoms. Aural architects of concert halls are rarely given the freedom to invent a new aural personality. In comparison
to the variety of virtual spaces, the subtle differences among concert halls, though recognized by connoisseurs and musicians, are minor. Moreover, there may be only one
or two concerts halls in a major city, but there are thousands, if not millions, of virtual
spaces in the vast repertoire of recorded music.

Paradoxically, the ease of inventing novel sounds and new spatiality, which is an
intrinsic property of modern signal-processing algorithms, has a subtle aesthetic cost.
Over the centuries, artists have embraced and incorporated new tools and techniques,
such as horsehair bows, metal woodworking tools, and pneumatic organ valves, but
years were then spent refining and mastering those tools and techniques. Generally, a
period of incremental refinement followed each technology-induced artistic revolution. Now, however, as the speed of technological advances continues to accelerate,
the refinement period becomes ever shorter. On the other hand, it still takes years to
master the artistic nuances of a modern studio reverberator, with its hundreds of presets, each having hundreds of parameters. The objection that synthesized music and
space break with past traditions of acoustic instruments and real concert halls is not

well taken: artistic traditions are *always* evolving, although it is fair to note that the new rules of synthesized spatiality have not yet been refined. Aural artists have acquired an elevated taste for spatial novelty, which itself is an artistic revolution that parallels the technology revolution. Novelty now competes with refinement.

Musical space in the twenty-first century will be the story of what artists and engineers do with their newly acquired freedom. But even as the present brings new aural art forms, the past is still with us. Ethereal musical voices in virtual spaces recall the floating voices of cave spirits from earlier times. The history of auditory spatial experience, aside from changes in technology and social values, is more circular than linear. As happened with the shamans of pretechnical societies, cultural values still determine which spatial choices will be revered and which discarded.

4 Aural Arts and Musical Spaces

Music and architecture have the common property of putting us inside a sensorial whole different from that we ordinarily live in.
—Violet Paget, 1932

Spaces intended for music are ubiquitous because music is found in all human societies, and all music must exist in a space. Early human hunters were choosing a musical space when they moved their ceremonies from a forest to a cave. And today, acoustic engineers design concert halls and other auditoriums. Spaces used for music are one of the most frequent applications of aural architecture. Over the millennia, musical spaces evolved in parallel with changes in musical styles and aesthetics,[1] just as social spaces evolved with the changes in aesthetics, social class, and attitudes toward power and privacy. Regardless of the period or culture, when used primarily for music, a space becomes a musical space. In such spaces, the attributes of social, navigational, and aesthetic spatiality become less important, or even irrelevant.

Musical spaces, especially modern concert halls, are a very narrow application of aural architecture in a very specific context. When we look at the aural architecture of musical spaces as an extension of music, we observe a tighter binding between music and spatial acoustics than between musical spaces and the other applications of aural architecture. Even though the principles of physical acoustics remain universal, their application to musical space is dictated by the values of the aural architects who are supposed to represent the composers, conductors, and musicians, as well as the listeners in the audience.

How listeners experience reverberation depends on whether the environment is primarily a social, navigational, aesthetic, or musical space. In a specific situation, reverberation may be desirable, disagreeable, or irrelevant. There are no universal criteria. In a musical context, musical rules dictate the desirable properties of reverberation, implicitly specifying reflection patterns, spectral balance, decay rate, initial delay, and onset shape. Such details are all potentially important to the experience of music. But in a social space, only the ratio of early to late sonic reflections influences the size of the acoustic arena.

Because music is the primary sonic illumination in a musical space, and because the historic repertoire of Western classical music embodies expectations about the role of spatial acoustics, aural architects have little flexibility in choosing the properties of a musical space. As acoustic consultants, they advise visual architects *how* to achieve specific acoustic properties, but not what properties are appropriate for music. Although this distinction is true for all applications of aural architecture, the contrast is more apparent with musical spaces.

However vast and comprehensive the literature on the physical and perceptual acoustics of concert halls, it rarely challenges the accepted goal: to design new spaces whose acoustics match those of the spaces that hosted famous European composers of the seventeenth through nineteenth centuries. The aural architecture of historical musical spaces, even if long since demolished and even if designed without regard to acoustics, dictates the aural architecture of new musical spaces. For example, Bach composed the Saint Matthew Passion for the Thomaskirche in Leipzig, and inflexible connoisseurs would like to listen to this composition performed with the same instruments in the same kind of acoustic space. Indeed, because the acoustics of musical spaces have become frozen in time, most modern concert halls do not have aural architects.

The two-way relationship between music and space is actually a pair of one-way relationships. Just as the designs of modern spaces are constrained by the need to host the historical musical repertoire, which is taken as an inflexible requirement, so composers write their music for immutable spaces that already exist, perhaps having been built a century earlier. There are a few rare exceptions. For example, Richard Wagner, with considerable effort, was able to raise the funds needed to build the theater at Bayreuth according to his artistic vision. Most often, however, composers assume a fixed aural architecture. At best, they can choose or recommend particular performance spaces for their compositions. Thus, because music and its spaces are rarely created at the same time, the two-way relationship between music and its spaces seldom exists in practice.

From the perspective of a composer, who is manipulating many musical dimensions unrelated to space, spatial acoustics is just one of the ways to achieve a given musical goal. Even if inflexible, space is still part of the musical language. Composers and musical scholars actually demonstrate an unconscious understanding of the aural architecture of musical spaces by how they discuss and manipulate space. At a minimum, for example, tempo and reverberation must be matched; a rapid tempo in a cathedral produces musical soup. We can thus extract the composers' implicit understanding of aural architecture for musical spaces by examining their viewpoint, even if their specialized musical language disguises that understanding.

Rather than creating a distinct language to describe the aural architecture of musical space, composers, designers, scholars, and researchers borrow from their respective host disciplines. Aural architecture is subsumed into the auditory arts using the language of music, absorbed into physics using the language of acoustics, treated as psychology using the language of perception, or hidden within the misapplied language of visual architecture. To describe musical spaces over the millennia, from the remote past to the unknowable future, we need a broader, overarching language of spatial concepts. Such a language must communicate issues arising from music, acoustics, perception, and even the signal processing of virtualized spaces.

Expanding the scope of aural architecture to include twentieth- and twenty-first-century computer algorithms that simulate spaces transforms musical space into an abstract concept, disassociated from physical reality. What we end up with as the language of aural architecture for musical spaces, whether real or virtual, is musical spatiality.

From one perspective, musical space is a physical environment—a concrete reality, a place where listeners and performers congregate, such as a concert hall. From another perspective, space is an ill-defined abstraction that relates only to our perception of spatiality, which also exists in the imaginary spaces created by computers. To appreciate space as having attributes that are both tangible and abstract, by analogy, consider money, which is both the physical coins and bills—cash—in a purse or pocket and a bookkeeping entry in a computer database. Cash and computer entries are the concrete and virtual versions of the same monetary abstraction. Similarly, electronics expanded the concept of musical instruments from vibrating strings and resonant enclosures to signal-processing algorithms. Advanced technology enables the design and construction of virtual objects (synthesized instruments in virtual spaces), which replace or coexist with concrete objects (real instruments in concert halls). Initially, as an alternative embodiment of a physical entity, a virtual object shares the same properties as the concrete object that it replaces, but eventually, new freedoms allow a virtual object to acquire its own properties. Spatial concepts therefore evolve as the properties of the virtual versions diverge from concrete versions. Moreover, for the first time in human history, the introduction of virtual spaces allows music and space to be designed simultaneously in a truly two-way relationship. The real revolution in musical space may, in fact, be that space becomes a real-time artistic activity. Becoming an aural architect of music space thus becomes possible, as well as important, unlike previous periods where space was a building that lasted centuries.

Musical Space as an Artistic Abstraction

In the distant past before recorded history, music was a communal activity in a shared environment; everyone participated without a clear distinction between active performers and passive listeners. Somewhat later, in some cultures, those with special skills became musicians (performers), whereas others became the audience (listeners).

When participants split into two groups, their respective spaces were separated: performers sat in one area, the audience in another; musical space split into a performance space and a listening space.

As listening areas expanded to accommodate larger audiences, progressing from small chambers to large concert halls, the enlarged seating area no longer had uniform acoustics. The space under a deep balcony, for example, is a distinct subspace that is only partially coupled to the main seating area because the balcony casts an acoustic shadow over the space under it. Similarly, the first row of the orchestra has different acoustics from the last row. Just as the location of an observer determines the observer's visual perspective, the location of a listener determines the listener's aural perspective. Although we speak of a concert hall as a single space, more accurately, it is multiple coupled subspaces with similar but subtly different acoustics. The stage and the audience area are coupled subspaces. Depending on the design constraints, coupling may be weak or strong, which then determines the degree of acoustic similarity among subspaces.

With the advent of recording and broadcasting in the early twentieth century, which placed listeners in a large number of diverse and private locations, the number of aural perspectives increased still further. The direct coupling between performance and listening spaces was broken. Musical space became two *de*coupled spaces, with a transport mechanism used to bring the music from one space to the other. When music is reproduced electronically, a listener actually experiences a hybrid comprising at least three sets of spatial attributes: the acoustics of the performance space where the music was recorded (recording studio or concert hall), the acoustics added during the mixing process when the music was prepared (spatial synthesizer), and the acoustics of the listening space where music actually heard (living room). Each of these spaces, unrelated and unique, influences the final experience.

Neither a performance space nor a listening space need be physical; either or both can be created with electronics. By using closely placed microphones, multitrack recorders, and studio electronics, music can be created without a live performance and without an actual performance space. Similarly, by using headphones and binaural processing, listeners can experience music without a listening space. In the extreme, musical synthesizers feed a spatial synthesizer whose output is converted to sound only at the listener's ears. In this case, spatial acoustics are no longer the result of sound reflecting from surfaces, but from computers using algorithmic rules to manipulate digital signals. Spatial acoustics do not have physical constraints if there are no physical spaces to provide those constraints. When sound is disconnected from a physical environment, new concepts are required to discuss the evoked perception of an imaginary space.

Who are the aural architects of a musical space that is not physical? With concert halls, we can identify the aural architects, and we can evaluate their creations. With electronic reproduction, however, many individuals privately contribute to the spatial experience. Acoustic engineers determine the physical properties of the recording environment; design engineers develop the recording and reproduction equipment; recording engineers place the microphones; mixing engineers prepare the final musical product for distribution; interior decorators select furnishings for the listeners' acoustic space; and listeners position themselves and the loudspeakers within that space. Often acting independently, these individuals are members of an informal and unrecognized committee of aural architects who do not communicate with one another. With their divided responsibility for the outcome, they often create the spatial equivalent of a camel: a horse designed by a committee.

Periodically during the twentieth century, members of this committee designed systems that attempted to replicate the spatial experience of a concert hall in a living room. Although early attempts failed because of inadequate technology and knowledge, multichannel reproduction systems are now able to re-create the aural perception not only of a specific concert hall but also of a specific seat in that hall, whether real or virtual. Such technology presents aural architects with a clear choice: to duplicate a specific aural experience modeled after a real space or to create an artistically meaningful one in a virtual space without a physical counterpart. This choice has important artistic implications.

The possibility of this second choice forces us to reexamine the assumptions underlying auditory spatial awareness. By our original definition, auditory spatial awareness is the *internal* experience of an *external* environment. A physical space exists in the world, and the experience of that space exists in the listener's consciousness. We seldom think about the distinction between internal and external spaces because they are so tightly linked. When using spatial synthesizers, however, we have only an internal experience. Audio engineers who design these devices call them "virtual space implementations," and the external reality is only an arcane algorithm. Looked at another way, the experience of a concert hall is also internal and the physical acoustics of an enclosed space only a means for creating the internal spatiality.

To appreciate more fully the nature of a virtual space, consider the visual analogue of computer-generated cinema. When we view an image of a humanoid from another planetary world, we are seeing a virtual person. Virtual objects function as if they were real but they do not actually exist, and often, could never have existed. Semantically, *virtual* is a fancy word for "illusory," but with an intrinsic ambiguity. For example, an image of a hypothetical person and an image of a person with three eyes and a transparent body are two distinct concepts. One is indistinguishable from a real person, whereas the other is a science-fiction fantasy. Both are images of a virtual human.

The same two cases apply to virtual space synthesizers. In one case, a synthesizer simulates or replicates the listening experience at a particular seat of a realizable concert hall. In the other, a synthesizer creates a spatial fantasy as an aesthetic extension of the musical arts. Initially, the distinction between the two cases symbolizes different concepts, but when taken to its logical conclusion, the distinction disappears: in both cases, someone must select the attributes of spatiality that are musically relevant and desirable. Whether modeling reality or creating a fantasy, the creator of a virtual space is an aural architect. The fact that musical spatiality had been created by physical spaces is just a historical artifact of older technology.

A simulated spatial reality can be understood as a surrogate implementation: real spaces use sound waves, whereas virtual spaces use signal-processing algorithms. If the virtual space closely mimics a real space, a listener will, in effect, perceive the real space. Using an engineering criterion, we say that the simulation is "perceptually equivalent" to that of the real space, creating a compelling illusion. Replication is an engineering function based on technical knowledge. In contrast, creating a spatial fantasy is the artistic activity of designing musical spatiality.

Replication preserves the immutable legacy left by those architects who designed the original spaces. When the designers of spatial simulators blindly mimic the acoustic defects in specific concert halls, they are abdicating responsibility for the properties of virtual spaces. Acoustic defects in real concert halls may arise from economic constraints, from favoring visual aesthetics or human comfort at the expense of acoustic integrity, or simply from inadequate knowledge of acoustics. Those same architectural defects and unfortunate compromises, typical of most concert halls, are then included in the simulation even when they need not be. By abandoning the goal of pure replication, however, and accepting responsibility for optimizing, refining, and improving their simulations, designers relegate the original concert halls to the limited role of initial models. Eventually, with enough enhancements, the simulations evolve into new artistic interpretations of space.

While physical space can be described using the physical language of acoustics, virtual spaces do not have a natural language. Because of the tight linkage between auditory spatial awareness and the acoustic properties of real spaces, the language of spatial perception almost always refers to external attributes, such as size, distance, orientation, texture, shadows, and so on. In order to discuss the experience of virtual spaces, we need an alternative concept that assumes neither the completeness nor the consistency of a real space. Chapter 2 introduced the concept of *spatiality* to represent spatial attributes that describe internal experiences when there are no external references. Yet, because spatiality is entirely an internal concept, for which there is no natural language, we must still borrow words and concepts from acoustics, perception, and the musical arts. Unfortunately, borrowing overloads the meaning of words, creating unavoidable ambiguities and confusions. For example, *reverberation* refers both to an acoustic process in an enclosed space (or in a signal-processing algorithm) and to the aural experience of that process.

During the last century, we witnessed two revolutionary shifts in the way we conceive of musical space. First, the recording technology of the early twentieth century split musical space temporally, spatially, socially, and artistically—partitioning *what*, *when*, and by *whom* music could be heard. In doing so, it added many new participants to the process of creating spatial experiences. The computer technology of the late twentieth century then virtualized the space *where* the music would be heard, *how* it would be created, and *who* would be responsible for its aural architecture. As this pattern of virtualization continues, musical *space* will merge ever more completely with musical *instruments*, with both becoming ever more abstract.

Enclosures Produce Temporal and Spatial Spreading

Which attributes of aural architecture are relevant to music? To be useful to both architects and musicians, our answer must encompass a wide range of both physical and virtual spaces, yet also be consistent with the language of spatial acoustics and musical aesthetics.

Although the abstract concepts of spatiality exist apart from real spaces, there are four good reasons to begin our discussion of spatiality with a concert hall. First, having been around for many centuries, concert halls have been extensively studied by artists and scientists alike. Second, most readers are likely to have experienced a concert hall. Third, by focusing on a concert hall, we have less need to create a separate vocabulary to distinguish physical and perceptual attributes. And fourth, a concert hall lets us explain signal-processing algorithms, which are even more arcane than spatial acoustics, in terms of simple analogies.

The basic physics of sound in a large enclosed space are relatively simple. A sound source, such as a short note from a flute, radiates an ever-expanding spherical wave of sonic energy in all directions, like the circular wave of cresting water when you drop a pebble into a pond. When a sound wave encounters an object, a secondary sound wave is then reflected back into the space. Reflected sounds, which are now traveling back into the space, are reflected repeatedly as they arrive at other surfaces. The process continues indefinitely, with each sonic reflection producing a multiplicity of new reflections until there are an infinite number of tiny overlapping reflections arriving from all directions. In fact, the word *reverberate*, originating from Latin, means "to throw back"; sound is thrown back into the space by being reflected from the interior surfaces: walls, ceiling, floor, and objects within the space. Because these sound reflections are so many and so low in intensity, they become a perceptual unit, called "reverberation," which is a major aspect of spatiality. Listeners first hear the direct sound when the spherical wave arrives at their ears, and they then hear a multiplicity of diffuse sound reflections as the reverberation continues.

We aurally experience an enclosed environment such as a concert hall by the two primary attributes of musical spatiality: *temporal spreading*, the way that reverberation extends the duration of a sound; and *spatial spreading*, the way that it broadens the direction of arriving sound waves. Though unrelated concepts, temporal and spatial spreading are tightly coupled when produced by a given physical space. Consider a flute note in our concert hall. Temporal spreading elongates the note; when the flutist stops blowing, the note does not end but continues to reverberate, often for many seconds, before gradually fading away (gradual decay). Spatial spreading transforms the flute note, radiating it from a single location on stage, into a multiplicity of sound waves radiating from multiple reflecting surfaces distributed throughout the space, and creating an enveloping sound field with a broader and more diffused location.

To appreciate temporal spreading, consider the opposite case of a flute in an open space with no walls or floor to reflect sound back toward the listeners. The original spherical wave expands from the flute, and once it passes the listeners' ears, it then vanishes. The moment the flutist stops blowing, sound ceases, as if an imaginary acoustic vacuum cleaner had removed every trace of the previous note. An open space is thus like a bucket with no bottom, impossible to fill no matter how much sound you pour in, whereas an enclosed space is like a leaky bucket, filling easily and emptying slowly. Enclosed spaces are storage containers for sonic energy. A highly reverberant space, like a bucket with a tiny hole, is readily filled with sound, and once filled, it stays filled for a long time. And a cathedral, whose reverberation time can be as long as 10 seconds, is like a mammoth bucket with a minuscule hole.

Spatial spreading is more difficult to conceptualize than temporal spreading, even with a scientific background. Consider the visual analogy of a candle burning in a lightproof room. If the room has flat, perfectly nonreflecting interior surfaces, you see the direct light radiating from the candle and nothing else. If, however, the interior surfaces are covered with an infinite number of mirrored dots, reflected light arrives from all directions. You see the candle *and* the infinite number of microscopic reflected images of the candle fused together as background illumination, uniformly spread throughout the space. The direct candlelight in the nonreflecting room is like direct sound in an anechoic chamber; the diffused candlelight from the mirrored dots is like the spatial spreading of direct sound in a reverberant space. Acoustic and visual reflecting surfaces produce spatial spreading—enveloping sound and diffuse illumination.

Even though direct sound produces reverberation, they are each perceived as distinctly different sounds. Part of the distinction arises from our ability to localize the origin of direct sound but not of reverberation. As a single coherent spherical wave, direct sound is perceived as having originated from a well-defined location, such as from a stage. As a diffuse sound field composed of waves arriving from all directions, reverberation appears to float in the air without an origin. Direct sound and reverberation have different spatial properties—perceptually localized at a point in space versus diffused throughout the space. Spatial spreading transforms the former into the latter.

In a large enclosed space, because both temporal and spatial spreading of sound originate from a single process in a single space, they are inextricably coupled together, with the geometry of the enclosure and the properties of its surfaces determining the nature of both types of spreading. In contrast, lacking an enclosed volume of air to hold the sound and interior surfaces to distribute it, an open space produces neither temporal nor spatial spreading. Although a concert hall and an open space represent the two extremes, there are many intermediate cases with varying degrees of temporal and spatial spreading.

Because our inventory of music originates from a variety of cultures and periods, we need a variety of musical spaces with the appropriate temporal and spatial spreading. Ideally, both types of spreading should be flexible and independent components of the auditory arts. As they arise in a given concert hall, they are difficult, if not impossible, to manipulate separately, but in a virtual space when we use a spatial synthesizer, they become independently adjustable artistic parameters. We can analyze any physical or virtual space in terms of its ability to produce both types of spreading, which are the initial abstractions in our language of musical spatiality.

Large Auditoriums Create Metainstruments

Over the centuries, innovative artisans have invented countless novel musical instruments, starting with almost any object that made or changed sound. In the sense that it, too, changes sound, we can consider the musical space of a concert hall to be, not simply a place for musicians and listeners to gather for performances, but also an extension of the musical instruments played within it. Although the origin of musical spaces can be traced to the need to shelter the audience from the weather and external noise, the role of space in music became an artifact of that need, one that indeed now dominates the art form. Thus our analysis of aural architecture acquires a new meaning.

As a simplification, a musical instrument contains two acoustically coupled elements: an active energy source, and one or more passive elements that create a unique timbre. For example, a violin has vibrating strings, an active source of sonic energy when the violinist pushes a bow across the strings. Muscular energy is converted into acoustic energy at the point of contact. Vibrating strings then couple their acoustic energy to the violin body, which acts as a passive resonant enclosure and surface radiator. Drums, organ pipes, cymbals, and even singers can also be modeled as active energy sources coupled to passive resonators. Both the passive and active elements of an instrument may serve either or both of two separate functions: shaping the spectral timbre, and creating temporal spreading.

Modeled acoustically, an instrument in a concert hall is a primary resonant enclosure coupled to a secondary resonant enclosure, the concert hall, which makes a strong
contribution to the temporal spreading of the instrument's sound. When the violin's
body transmits its sound to the concert hall's enclosure, the violin becomes a metaviolin; without a secondary resonant enclosure, the violin remains a protoviolin. The difference between these two concepts of violin arises from the existence of the secondary
resonant enclosure. Performance spaces create a new class of musical instruments:
metaviolin, metaclarinet, metaoboe, and so forth. Although musicians sit inside the
secondary resonant enclosure, rather than holding it in their hands, this difference is
irrelevant from the perspective of creating music. A resonant enclosure is a resonant
enclosure.

The long history of instrument design illustrates the wide range of resonant structures that have been used to create artistically meaningful instruments. Depending on
its size, structure, materials, and coupling, each instrument's small resonant enclosure
possesses an aural personality; each produces distinct pitches, timbres, sustained resonances, tonal filtering, and combinations thereof. Large resonant enclosures, such as a
concert hall or cathedral, have thousands of resonances that blend, without a characteristic frequency, but still produce temporal spreading. (These resonances cannot be
perceived individually because of their high density.)

If a musical instrument has only a small primary resonant enclosure, which does
not produce temporal spreading, the addition of a larger secondary resonant enclosure
serves a critical role. Consider a wind instrument, such as a clarinet with one adjustable
air column; its resonance frequency (pitch) is changed by selectively closing holes. A
clarinet has no temporal spreading. A clarinetist can produce only a sequence of discrete notes, not chords.[2] However, when a secondary resonant enclosure adds sufficient reverberation to the clarinet, previous notes are extended while the clarinetist
plays a new note. Consider the notes C, E, and G, as shown in figure 4.1. When played
in a dry space, these sixteenth notes are heard as a sequence (left). But when played in
a highly reverberant space, their reverberation exists at the same time, thus creating
the aural impression of a C major chord (right), composed of the pitches from notes
C, E, and G. A metaclarinet can play a chord, a protoclarinet cannot.

Figure 4.1
The effect of spatial acoustics on the perception of three rapid sixteenth notes C, E, and G. Left:
In a dry space, they are heard as a discrete sequence. Right: In a highly reverberant space, they
may be heard as a half note C major chord.

Even for those instruments with the ability to produce simultaneous notes and chords, the attack and the ending of a note may be too abrupt. For example, to control airflow, a pipe organ uses binary valves, which are either full on or full off. Musical notes appear and disappear with sharp transitions that are musically unpleasant. Blending and softening the transitions between notes is possible, however, when a secondary resonant enclosure produces reverberation. Because a large resonant enclosure only gradually fills and empties, listeners hear the slower leading and trailing edges of notes. In addition, this same flywheel mechanism allows the organist to acquire some control over intensity by regulating the duration of the notes. A short note does not last long enough to completely fill the space with sound. Partial filling produces a quieter note. Spatial acoustics are so integral to the sound of an organ that it is generally tuned and customized for the properties of a specific performance space. If there had never been such a performance space, organ builders would have had to construct and transport a very large resonant enclosure, indeed, to achieve a similar effect.

Without being explicit, composers write music to be performed with specific instruments—a protoclarinet or protoviolin for outdoor venues, and a metaclarinet or metaviolin for enclosed venues. The *meta-* or *proto-* prefixes are implied, although not formally articulated by the choice of performance space. For example, Schubert's *Trout* Quintet might well have had the title "Quintet for Metapiano and Metastrings Having a Secondary Resonant Enclosure with a Reverberation Time of 1.2 Seconds." Similarly, music directors are selecting instruments when they exercise their discretion by choosing performance spaces. Moving the orchestra from an outdoor to indoor venue is equivalent to replacing protoclarinets with metaclarinets. Moving the orchestra from a concert hall to a small chamber is equivalent to replacing one metainstrument with its smaller cousin, having a smaller secondary resonant enclosure. Marching bands use only protoinstruments.

Although the designers of instruments focus on musical aesthetics, the need for portability plays an overriding, albeit hidden, role in their designs. With a very few exceptions, the resonant enclosures of musical instruments have been limited to a size that could be readily transported. Musicians traveled to large resonant enclosures (performance spaces), but carried small ones (musical instruments) with them.

Although resonant enclosures as large as a concert hall are not of course transportable, a designer can use an electronic synthesizer to *simulate* a large resonant enclosure while retaining the size and weight of a small musical instrument. By compressing the effective volume of an instrument's resonant enclosure by at least seven orders of magnitude, the portability criterion becomes obsolete. A synthesizer can simulate the large resonant enclosure of a Gothic cathedral as easily as the small one of a clarinet. Or it can simulate one, two, three, or any number of resonant enclosures with varying degrees of coupling among them, thus creating a new class of pseudoinstruments with no physical counterparts. Despite their differences, synthetic and physical resonant enclosures serve one and the same function: temporal spreading.

Although all forms of temporal spreading with comparable decay times are relatively similar, the size of the object producing energy decay has a strong influence on perception. Listeners do not confuse the sound of a protoguitar (string vibration) with that of a metaguitar (spatial reverberation), even when these have identical decay times and frequency spectra. The distinction resides in the parameters that determine the fine structure of the decay process. For a protoguitar note, each of its many partials (multiples of the fundamental pitch) decreases with a relatively smooth and consistent envelope, and their phase alignment remains coherent and slowly varying. In contrast, for the note of a metaguitar, each partial decays with random amplitude and phase alignment. The randomization of spectral components is musically very important—and readily perceived by listeners.

Many physical and perceptual parameters arise from the size and shape of a resonant enclosure. In addition to differences in randomization, size influences the perception of resonances. In a large resonant enclosure, such as a concert hall, resonances are so dense that they fuse into a continuum without changing the timbre, whereas in a small resonant enclosure, such as an instrument, resonances are individually perceptible, contributing to the instrument's characteristic sound. Before considering intermediate-sized resonant enclosures, we can first decouple the parameters of these resonant enclosures from the enclosures themselves. Such parameters, which include randomization, resonance density, attack rate, linearity of decay, and so on, need not be locked together as they are in physical resonant enclosures. With virtual instruments and spaces, especially when merged into a single computer implementation, the parameters of resonant enclosures become artistic choices: we can have full randomization with discrete resonances, or vice versa. The more musically innovative and aesthetically pleasing choices are likely to give rise to a whole family of identifiable virtual pseudoinstruments, each having its own name.

With popular music, mixing engineers often adjust parameters on their spatial synthesizers to create virtual resonant enclosures that could not be produced with sound waves in real ones. The label "spatial synthesizer" is misleading in two important respects: the selected parameters need not correspond to a space, nor need they convey the feeling of spatiality. In one such application, engineers enhance a protosinger's weak voice in order to create an aesthetically pleasing metasinger's voice. Without such enhancement, the protosinger's voice would not have the required appeal of today's fashion in popular music. The additional resonant enclosure adds aural substance and body to an otherwise aurally inadequate oral resonant enclosure, without creating the impression of a performance space. This creates the illusion that the singer has a larger head, mouth, and vocal tract, thereby adding depth, mass, and impact to the voice. In this application, the synthesizer creates an auxiliary resonant enclosure that fuses with nature's oral one.

There are several conclusions to be considered. First, even if produced by a performance space, temporal spreading is conceptually part of the design of musical instruments. Second, when listeners attend a concert hall, they are placing themselves inside the large resonant enclosure of metainstruments. Third, concert halls are a constrained means for implementing aesthetically pleasing sounds, whereas synthesizers transcend those constraints. Finally, virtual instruments and virtual spaces are both based on the same concepts: synthesized virtual resonant enclosures that create temporal spreading.

Artistic Relevance of Temporal Reverberation

Virtuoso musicians master their favorite protoinstruments through years of practice, and they master the corresponding metainstruments only when performing in specific concert halls. During a rehearsal, a conductor walks into the audience area to hear the blending of metainstruments, which are unique to that space. Musicians are adapting to their metainstruments by sensing the details of temporal spreading.

Reverberation has a complex temporal behavior, rapidly increasing in loudness in the onset region, holding steady in the sustain region, and slowly decreasing to inaudibility in the decay region. Spatial geometry determines the actual shape of these three regions. Some spaces have a fast onset, some have a long sustain, and others have a rapid initial decay. Each note has four temporal regions: the region of its direct sound, and three regions of reverberation: attack, sustain, and decay. Depending on the music, the locations of the listeners, and the aural architecture of the space, each region makes a different contribution to the listening experience.

These four temporal regions interact in a complex fashion. On stage, the direct sound is so loud that it may mask the early part of reverberation that immediately follows. In contrast, at the rear of the concert hall's auditorium, the early and sustaining parts of the reverberation dominate the direct sound, which is much quieter than at the front; whereas in the center of the auditorium, all four versions of an isolated musical note are distinguishable.

To hear the complete reverberation of a note, there must be silence after the note terminates, hence the name "stopped reverberance." Music must stop in order to avoid masking the ongoing reverberation. Enter a large church, clap your hands, and you hear stopped reverberance. Most untrained listeners think of the long decay as being the essence of reverberation, even though its low intensity makes it vulnerable to masking by other loud sounds. Without intermediate intervals of silence, music is typically heard as a connected sequence of discrete notes, with the current note masking the reverberation decay of earlier notes. Only the earlier parts of a reverberation process, before significant decay, are audible with continuous music, hence the name

Figure 4.2
The first six notes from the C major scale, C, D, E, F, G, and A.

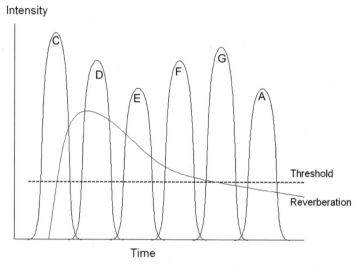

Figure 4.3
The effect of modest reverberation on the previous six note sequence. The reverberation of note C is strongest when the direct sound of note D appears, but inaudible when the direct sound of notes F, G, and A appear.

"running reverberance," which describes our aural experience of most music we hear in public musical spaces.

When listening to music, we actually experience a complex mixture of the four versions of multiple notes. Consider a hypothetical sequence of short notes. At a given moment, there is the direct sound of the current note, the onset region of the previous note, the sustaining region of the note before that one, and the decay of many earlier notes. The composite is aesthetically pleasing if there is an appropriate balance between these stages.

To illustrate the blending process of reverberation with the direct sounds, consider the sequence of six notes from the C major scale, as shown in figure 4.2. In a concert hall, each note produces reverberation with its three regions, onset, sustain, and decay. Figure 4.3 graphs the loudness (intensity) against time of the reverberated note C,

Intensity

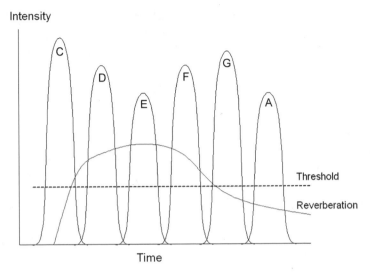

Figure 4.4

With reverberation having a longer sustain region than in the previous example, the reverberation of note C is loudest for the direct sound of note E, and still modestly loud for the direct sound of notes D and F.

which exists simultaneously with the direct sound of notes D through A. In actuality, the other five notes also produce reverberation, but they are omitted for the sake of clarity. When note D appears, the reverberation from note C is still relatively strong. As it decays still further, it produces less blending with each successive note. Eventually it falls below the masking threshold (dotted line) to become inaudible.

The degree to which a note blends with the reverberation of a previous note depends on the shape of the reverberation process, the intensity of the two notes, and the time between them. In this example, running reverberation is audible during the time interval that happens to include notes D and E. The composer influences blending by selecting the tempo. With a faster tempo, notes F though A would also appear while the reverberation from note C was still audible, whereas with a slower tempo, note E would occur when the reverberation was inaudible. Musicians also control blending by the loudness of their playing. If, for example, note F appeared at a much higher intensity, it would mask the reverberation of the notes preceding it. Because the spatial design of a musical space determines the shape of the reverberation process, an aural architect affects blending. Figure 4.4 graphs the same six notes in a space that has reverberation with a slower onset and a longer sustain. Unlike our example in figure 4.3, the reverberation from note C is now relatively uniform during notes D, E, and F, and they all blend with a relative constant amount of reverberation.

Although average listeners do not notice such nuances, professional musicians and music connoisseurs hear them. If the shape and intensity of reverberation during the first few hundred milliseconds are grossly inappropriate for a particular kind of music, listeners may become aware of a subliminal defect without being able to articulate what is wrong. Elegantly crafted reverberation is like the tonality of a violin by Stradivari. Crudely crafted reverberation is like the sound of a student violin mass-produced in a factory. Many can hear the difference under controlled circumstances, but only a few appreciate the difference when enjoying music.

In a concert hall, listeners hear a mixture of the *direct* sound, the initial wave front as it would have appeared in open space, and reverberated sounds held by the enclosure. The ratio of the two depends strongly on the distance between sound source and listener. Whereas reverberating sound is relatively uniform throughout the space, the intensity of the direct sound, like all spherical waves, decreases rapidly with distance. A tenfold increase in distance between the first and last row of the concert hall auditorium corresponds to a tenfold decrease in the intensity of the direct sound. Relative to the direct sound, the reverberation is soft in the front row and loud in the last row.

Unlike the subtleties of reverberation, the ratio of direct to reverberant sound is noticeable even to those without training. Figure 4.5 graphs both the direct sound and reverberation of the six notes in figure 4.2 as heard in the last row of a very reverberant auditorium. Reverberating notes are labeled c through a, and the corresponding weak

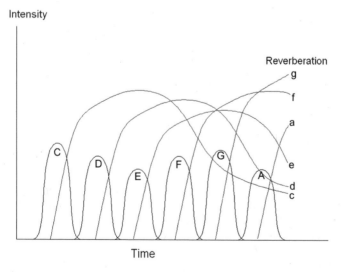

Figure 4.5
At the back of a highly reverberant space, such as a cathedral, the reverberation from these six notes overpowers the direct sounds of the individual notes.

direct sounds, C through A. Notice that when note A appears, reverberation from the previous five notes still dominates its direct sound. In such circumstances, listeners would experience soupy music or aural mud, rather than the clarity of distinct notes. Even if the reverberation decreased rapidly after the long sustain region, running reverberance would already have done its damage.

When choosing your seat at a concert hall, you are selecting the intensity of the direct sounds. In a shoebox-shaped hall, such as Boston Symphony Hall, seats in the first third of the second balcony provide an almost perfect balance, at least to my taste, between direct and reverberant sound. And the balcony extends into open space, with neither obstructing shadows nor sound absorption from densely packed human bodies. But because such seats face across the hall rather than toward the stage, they are visually undesirable and among the least expensive in the hall. Thus, even at musical concerts, visual rather than aural perspective determines desirability.

Some concert halls have such pleasing reverberation that recording engineers welcome the opportunity to record in those spaces, and they judiciously select the optimum locations for their microphones. Even under ideal circumstances, however, recording spatial ambience is still difficult. Microphones, even if placed above the audience, also detect background noises: coughs, whispers, the roar of airplanes, and the rustling of candy wrappers. Placing microphones near musical instruments avoids these noises, but the direct sound then overpowers the reverberation. Recording in a concert hall without an audience changes its acoustics because the audience makes an important contribution to absorbing sound. For example, the reverberation time of the Boston Symphony Hall is 1.8 second when full, but 2.6 second—aural soup—when empty (Beranek, 1996a).

Spatial synthesizers were born out of a historical need to supplement or replace the acoustics of a performance space when recording. Once having replaced physical acoustics, the mixing engineer enjoys the freedom to adjust a wide range of parameters. These include steepness of the onset, amplitude and duration of the sustain, rate of final decay, intensity of direct sound, density of reflections, and dozens of other technically abstruse parameters. An engineer can, without exaggeration, choose among hundreds of premade virtual spaces or create thousand of variants. Instead of a fixed physical space that lasts for a century, a synthesizer instantly virtualizes a disposable space. But as typically applied, a synthesizer remains a surrogate for a concert hall—a common resonant enclosure shared by all metainstruments.

When each individual instrument is given unique spatial parameters, a synthesizer becomes more than a surrogate for the resonant space of a concert hall. Creating the secondary resonant space for a metaguitar is no longer constrained by the properties of such a space for a metaclarinet or metapiano. The secondary resonant space of a metaguitar can be modified, improved, or changed into a variant without affecting those of other metainstruments. In fact, the family of pseudo-guitars includes the synthesized protoguitar, metaguitar, as well as many other variants.

Spatial acoustics and musical instruments merge to become metainstruments, even though they are physically separate objects. From a conventional perspective, the design of musical sounds involves assigning some properties to portable instruments and others to performance spaces. These assignments are actually artistic, social, economic, and technical artifacts, most of which relate to the need for instrument portability and sheltered seating. Modern technology allows other assignments as long as the temporal spreading is consistent with the musical composition. As Wallace Clement Sabine (1922) commented while designing the Boston Symphony Hall, cultures that gathered in enclosed spaces for music developed musical forms dominated by melody and harmony, and cultures that gathered in open spaces for music developed musical forms dominated by percussive sounds with strong rhythms. Our heritage of classical music arose from a limited set of parameters. A wider range of parameters now opens up possibilities for novel musical forms.

Listening Environment Determines Spatial Spreading

Let us now turn to the other principal attribute of musical spatiality: spatial spreading. Whereas temporal spreading merges with the acoustic properties of instruments, spatial spreading makes a more direct contribution to listeners' sense of their external world. To appreciate how the spatial distribution of sound creates a sense of the world, first consider the directly opposite case where all sounds emanate from a single source—the complete absence of distributed sound. You can easily experience what this would be like by listening to a recorded symphony orchestra performance from a single high-quality loudspeaker located in the middle of an open field. Even if the music were originally performed by a world-renowned orchestra in a preeminent concert hall, the experience is aesthetically uninspiring. Music radiating from such a loudspeaker sounds dull, flat, feeble, and remote. You hear the musicians and recorded spatial acoustics as if they were embedded inside the loudspeaker box, or as if the box were an open window into a distant concert hall. Regardless of the amount of reverberation embedded in the recorded music, you still experience yourself as being in an open field. In other words, reverberation radiating from a point source automatically separates the listener from the performance space.

Spatially distributed sound, or lack thereof, conveys the difference between here and there. Our example of a loudspeaker in an open field is stark because the respective acoustic spaces are so different: a concert hall versus an open field. The orchestra is *there* in its concert hall; as listener, you are *here* in your open field. Whereas temporal spreading is a property of the performance space, spatial spreading is a property of the listener's space. When spatial and temporal spreading are consistent within a single space, the performance and listening spaces fuse into one. An outdoor environment has neither type of spreading, and a concert hall has both. Our loudspeaker in an open field reproducing music from a concert hall splits a single acoustic space into two.

Sensing the location of a sound source and sensing the experience of an enclosed space are polar opposites. On the one hand, localizing a sound source involves decoding a single spherically radiating sound wave before it is contaminated by reflections. On the other hand, perceiving an enclosed space arises from millions of sound waves emanating from an infinite number of surface locations. With reverberation, a sound wave is equally likely to arrive from the left, right, front, back, above, or below. When a sound field is fully diffused, uniform, and random, you cannot detect the location of its source because your auditory experience is identical in all directions. The difference between localized sound and enveloping reverberation parallels the difference between a protoviolin and a metaviolin. With localized sound, the instrument is external to the listener, whereas with enveloping reverberation, the listener is inside the sound generation process, within the mass of air holding the sonic energy.

Because orientation is not possible, a fully diffuse sound field creates an enveloping feeling unlike any other listening experience. A visual analogue would be the fully diffuse lighting inside a spherical room with surfaces of frosted glass uniformly illuminated from behind. Such a space would be devoid of all visual details, which would prevent spatial orientation. A visual analogue of the directly opposite case, direct sound from an instrument, would be light from a single bulb in a conventional room containing other objects. The source of illumination has a defined location, and objects create shadows and reflections.

Paradoxically, our idealized description of reverberation assumes that the listener cannot hear the enclosing surfaces as such, yet the enclosure must be present in order to create reverberation. To use a water analogy, compare the experience of floating in a bathtub with swimming in a lake. Although you can experience being immersed in water in both cases, the experience is altogether different because the container holding the water dominates in one case and disappears in the other. There is no awareness of a lake having a bottom and sides even though it is clear that it must have them in order to hold water. Conversely, the boundaries of the bathtub dominate that experience, offering physical support, spatial orientation, and constraints on the water dynamics. Listening to ideal reverberation is like swimming in a lake of sound with no awareness of its container. Experiencing a container and experiencing the reverberation that it creates are perceptually unrelated, even though they are physically equivalent. In the context of music, aural architects avoid the perception of the container (enclosing walls) whenever possible, while emphasizing the water contained within it (spatially diffused reverberation).

Unlike listeners in aural navigational or embellishment spaces, ideally, every listener in a concert hall space should experience the identical enveloping reverberation. Listeners should not hear the local acoustics of such elements as an overhanging balcony, an open entry door, the materials of local walls, the geometry of the seating area, the proximity of a large granite column, or the volume of a secondary alcove. Although removing the balcony in a concert hall is economically impractical, aural architects of musical spaces would do so if they could. With our idealized concept of reverberation, every seat should be suspended in air to avoid the negative influence of a floor, walls, and sound-absorbing neighbors.

A personal anecdote illustrates the aesthetic consequences of a surface with strong aural attributes in a musical space. I once heard a concert while seated in the last row of Kresge Auditorium at MIT. That seat was located against a back wall that had been extensively treated with a massive amount of sound-absorbing material. Because there was no sound reflecting from that surface, perceptually, the wall was the virtual equivalent of open space. Throughout the concert, I had the feeling of being pulled into the void behind me, as if I would fall out of the auditorium when leaning backward. Even though the wall was visually and tactilely present, my aural awareness undermined my sense of physical security. From an aural navigational perspective, the acoustics of the rear wall should have provided a clear sense of my proximity to the back of the auditorium. In a musical space, a totally sound-absorbent wall is a defect.

In contrast to the case of concert halls, achieving spatial consistency and uniformity with recorded and broadcast music is impossible. Although audio mixing engineers have complete control over temporal spreading, which is determined by the location of microphones in the performance space or by the adjustment of spatial synthesizers, they have no control over spatial spreading in the listener's environment. The uniqueness of each listener's private environment dwarfs the small differences among most seats in a concert hall.

As a listener, you have a vast range of choices for the way you can listen to music. First, when using stereophonic headphones, you hear the space entirely within your head. The violin might be located behind your left eye and the clarinet behind your right eye. Rather than you being in the performance space, the performance space is located, aurally at least, inside your head. Second, with two stereophonic loudspeakers in your (typical) living room, you perceive the music as being external to you. Two channels cannot reproduce enveloping reverberation because sound comes only from one direction, the front. The space is equivalent to a large open window into a concert hall. Finally, if you were to install a standardized surround-sound system in a carefully designed room, you might well experience the sound field the audio mixing engineer intended. A multichannel audio delivery format would provide enough spatial information to reproduce the engineer's intent, but only if you elected to use those channels in the prescribed fashion. The audio mixing engineer would then be the aural architect.

If you enjoy being your own aural architect, rejecting the artistic intentions of the audio mixing engineer, you can create your own experience of space. You need only basic acoustic knowledge, which is readily available in the literature, and the economic means to acquire a spatial synthesizer, to design the environment with the correct acoustics, and to place multiple loudspeakers in the correct spherical configuration. A stereophonic recording will create the illusion of your being in the choicest seat of the best concert hall. In an ideal listening space, especially with your eyes closed, you will be transported to another space. The illusion is compelling. Regardless of the technology or the criteria for quality, spatial spreading must be explicitly created or re-created in the listener's environment.

As well as being a member of the committee of aural architects, a listener is also a member of other social systems. Most listening spaces, be they concert halls or living rooms, are subject to social constraints that often conflict with the acoustic requirements for listening to music. Enjoying music is also part of a larger social context, such as relaxing in the living room with the family, or attending a concert with friends. Whereas an aural architect might focus on the aural quality of the experience, a listener often focuses on music as a vehicle for social cohesion. Living rooms are not just musical spaces; they are mostly social spaces subject to social rules.

Most recreational spaces are acoustic accidents, with sizes and properties that range from large, upholstered living rooms of the affluent to the small and barren dormitory rooms of struggling college students. Only a few individuals have the freedom and economic means to design a space optimized solely for its aural architecture. Similarly, a concert hall is expected to be visually elegant, and to enhance social and visual communications among the participants, especially between the musicians and the audience. Sight lines are important. The organic vitality of a shared emotional experience creates social cohesion. That, in turn, enables listeners to express their appreciation with a standing ovation at the end of a virtuoso performance. A concert hall provides a total experience that depends on all the senses.

Rather than being an activity that exists in isolation, aural architecture is part of, subordinate to, and competes with, a broad range of other human needs. It has always been so, and will always be. Aural architects could create an acoustically "perfect" concert hall or an "ideal" music reproduction system if all social constraints were removed and the space were limited to a few individuals. There is more than enough technical and acoustic knowledge available to achieve that goal. The concert hall might be ugly, the reproduction system might require a carefully designed and dedicated listening space, and the participants might be socially uncomfortable, but spatial acoustics could approach some idealized definition of perfection.

In fact, in our diverse culture with its emphasis on individuality, there is no universal definition of an ideal listening space. Although recorded music began as an attempt to faithfully replicate live performances, reproducing recorded music is now almost entirely decoupled from its live counterpart. Because live and recorded music are supported and sponsored by different subcultures, each with their own artistic expectations, their respective attitudes toward an ideal space also diverge. The subculture of popular music listeners shows more appreciation for spatiality attributes that are creative and novel without requiring the spatial consistency of a physical space, whereas the subculture of live jazz and classical music listeners shows more appreciation for the spatial tradition of quality concert halls.

When spatiality is considered as abstract art form, there is no requirement for the listener to aurally visualize a real space. Indeed, when the spatial attributes of a simulation are incomplete, inconsistent, or contradictory, a listener has no possibility of aurally creating a spatial image. Aural architecture for reproducing recorded music therefore contains a dilemma, or even a paradox. Do listeners want to aurally visualize a real environment from the acoustics in the reproduced music, or do they only want to enjoy reproduced music containing artistic nuances of spatiality? The answer depends on the aesthetic attitudes of the listener. Even when spatial attributes are an accurate version of a real space, most listeners have neither the ability nor the interest in hearing a musical space as a space. Rather, they simply enjoy music that is enhanced by spatiality. Based on ubiquitous acceptance of recorded music with overtly contradictory acoustics, we must conclude that spatial accuracy is not a significant criterion for much of our musical culture. Nevertheless, some audio scientists, mostly those invested in classical music, still *assume* that most listeners appreciate spatial consistency, and that listeners are subliminally disturbed by spatial contradictions. In contrast, producers of popular music assume that listeners are surprised and delighted by weird spatial effects.

To illustrate an example of spatial inconsistency, consider a recording engineer who provides each musical voice with its own reverberation, each corresponding to a unique acoustic space. One musical voice might have the long reverberation of a cathedral; another might have the short reverberation of a small chamber; and a third might lack reverberation, as in an open desert. Under these assumptions, theoretically, the musicians could not be sitting together in a musical ensemble. A listener cannot aurally visualize a space that is simultaneously large and small, enclosed and open. Therefore, spatial consistency is itself an artistic and aesthetic judgment, rather than a prerequisite for musical quality. If music is pleasing, listeners are unlikely to be bothered by such spatial inconsistencies.

The epigraph to chapter 1 quoted Winston Churchill remarking that "we shape our buildings, and afterward our buildings shape us." This idea is as valid for concert halls as it is for electronic *reproduction* of music. The cultural shift that enabled electronics to replace buildings shifted the ownership and control of musical space from one group to another. These new owners-controllers now shape our concepts of space. Hypothetically, had spatial synthesizers existed before enclosed spaces, the evolution of music would have been shaped by audio, mixing, recording, and broadcast engineers creating virtual spaces, rather than by builders and aural architects creating preeminent concert halls, opera halls, cathedrals, and other musical spaces.

Abstract Dimensions and Artistic Concepts

Having established a basic language for describing musical space based on temporal and spatial spreading, let us move the discussion to a higher level of abstraction. Abstract concepts arise when consistent patterns emerge in a few cases, which are then generalized to explain other cases. Although they are often the hallmark of scholars and academic researchers who dazzle their colleagues with their intellectual prowess, abstractions also have pragmatic value. For our purposes, abstractions create a universal framework for discussing, analyzing, and extending the concept of musical spatiality.

The earlier discussions made a distinction between an enclosed acoustic space and the acoustic objects that were enclosed by it. When subjected to a more careful examination, however, the distinction blurs and then disappears. A section of a wall surface can function both as acoustic object and as part of a spatial boundary; the body of a violin is as much an enclosed resonant enclosure, albeit small, as it is an acoustic object that radiates energy.

A more relevant attribute of acoustic objects involves controllability—musical playability. The enclosure of a clarinet is playable by placing various fingers over selected holes. And if the ceiling of a concert hall also had holes, a giant could play the space with his fingers. In part, the property of playability originates from size: large objects are difficult to manipulate. This is illustrated by a pipe organ, which requires mechanized assistance to convert finger motion on a human-sized keyboard into valve motion of pipes distributed over a wide area. But with electronic virtualization of acoustic attributes, musicians or their computer surrogates can manipulate every parameter. By controlling all parameters, a virtual giant could play any virtual resonant enclosure of any size, even one as large as a concert hall.

In the second half of the twentieth century, contemporary musical artists initiated the merger of space and music. When Edgard Varèse was invited to collaborate with Le Corbusier for the Philips pavilion of the 1958 Brussels Expo, he used 350 loudspeakers to gain complete control over where each sound entered the pavilion (Ernst, 1977). The historical language of music and space was inadequate to describe Varèse's composition.

Let us now consider three proposed perceptual dimensions as a language for describing a wide range of musical and spatial experiences. No doubt, there are other abstract dimensions that could (and will) be added, but these three provide an initial foundation. To a modest degree, these dimensions are independent.

Dimension 1: Primary versus Slave Sonic Events

Let us assume that music is the result of interactions among acoustic objects—physical materials that have acoustic properties—and that every acoustic object is either an active sound source or a passive sound modifier, either injecting energy or changing energy already injected. This discussion extends the earlier concept of a protoinstrument (without any enclosing resonant space) and a metainstrument (within a large resonant space).

First, we apply this concept to musical instruments. The vibrating strings of pianos, violins, and guitars are active energy sources, as are the vibrating reeds of clarinets, saxophones, and oboes. After being energized by a human action—breath, impact, and movement—these vibrating objects produce sonic events (musical notes) that radiate and couple their energy to passive acoustic objects. The housing of a piano, the body of a violin, and the tube of a clarinet are passive modifiers of that injected sound energy. They resonate, filter, reflect, amplify, and radiate the injected energy, thereby creating a unique pitch and timbre. Instrument bodies are passive acoustic objects.

Next, we apply this same concept to the musical space of a concert hall. The sonic energy sources are the protoinstruments; they inject sound energy into the concert hall. The passive acoustic modifiers of that injected energy are the physical objects that comprise the acoustic space: walls, statues, people, doors, and balconies. These acoustic objects reflect, disperse, shadow, diffuse, and absorb sound that was injected by the instruments. To indulge the poetic license of analogy, the acoustic architect designs an acoustic space as if it were the body of an instrument, and composers and conductors manipulate musicians as if they were vibrating strings and resonant air columns. Space and the objects within it are passive acoustic objects.

The distinction between active sound sources and passive sound modifiers is, however, more theoretical than real because energy sources and acoustic objects are tightly bound together. With a violin, strings are mated to the violin body, and with a drum, the membrane is mated to its enclosed resonant enclosure. Similarly, according to the intentions of the composer, musicians are mated to specific spatial acoustics: a string quartet matches those of a small chamber; a symphony orchestra matches those of a large reverberant concert hall, Gregorian chant matches those of a cathedral, and a military marching band matches open-air acoustics.

After careful analysis, the distinction between passive and active blurs to the point of becoming useless because, in isolation, a vibrating string or membrane is almost inaudible. A pure energy source, an irrelevant abstraction, always requires passive acoustic objects to efficiently radiate sound. In fact, we can model music as a sonic energy package that progressively passes through a series of passive acoustic objects, each of which then radiates and couples energy to other acoustic objects, and eventually to listeners. Initiated by muscular energy, vibrating strings produce sound, resonating air enclosures produce sound, which is then radiated by sound-reflecting surfaces and alcoves.

What, then, is an *active* acoustic object? It is nothing more than the particular acoustic object that is *perceived* as having produced the *initial* sonic event, called the "primary (independent) sonic event." With many passive acoustic objects radiating sound initiated by a single energy source, the acoustic object that produces the earliest and loudest sound is usually identified as being the source of the primary event. For the violin, the primary event is the direct sound from its body, although this definition will be enhanced during the later discussion on fusing. The concept of a primary event, which is based on perception, replaces the concept of the protoinstrument, which is a physical description.

Other passive acoustic objects also radiate sounds. When their sounds fuse with the primary event, they are not perceived as being distinct sonic events. For example, when the direct sound from violin and its reflection from the back wall of the performance stage perceptually fuse together, the composite is a single primary event. But when the direct sound and its reflections are perceived separately, for any of many possible reasons, they become the primary (independent) and the secondary (dependent slave) sonic events. For the violin, the slave sonic event is its reverberation. The concept of a primary and secondary sonic event replaces the concept of the sound from a metainstrument. The earlier distinction between active and passive acoustic objects is now a distinction between primary and slave events—a distinction that is purely perceptual. In a large resonant space such as a concert hall, a violinist produces a primary sonic event (direct sound) and a slave sonic event (reverberation). But in a smaller resonant space, such as a bathroom where spatial resonances fuse with the direct sound, we aurally perceive only a primary event whose overtones are selectively amplified by the space.

There are other examples where a passive acoustic object may or may not produce a recognizable sonic event. For example, an early sonic reflection fuses with the direct sound, whereas a later sonic reflection becomes a perceptibly distinct echo. As a rule, spatial attributes either modify the primary sonic event or create one or more distinct secondary sonic events. When a secondary sonic event exists, and when it is bound but not fused to the primary sonic event, the secondary event is experienced as a *slave* event rather than as a new primary event. Slave sonic events are perceived as being caused by their respective primary sonic events.

Primary and slave sonic events are more useful abstractions than the concepts of proto- and metainstruments for exploring how musical events bind to each other and to the musician. Let us begin with reverberation as a slave sonic event. Concert hall acoustics and spatial synthesizers are processes and devices that create these slave sonic events, which retain their spectral and temporal coupling to the primary sonic event, and hence to the musician. Similarly, some popular musicians use what is called a "fuzz box" to produce unusual slave sonic events. By extracting pitch or any other parameter from an electric guitar, these devices produce novel sonic events with bizarre temporal and spectral envelopes. Without meaning to trivialize this discussion, a spatial synthesizer is a spatial fuzz box. Both produce slave sonic events.

To extend the concept of slave sonic events still further, consider the following hypothetical situation. An electronic keyboard controls two instrument synthesizers, one of which first produces a clarinet sound from stage left, and the other then produces a delayed cello sound from stage right. Although the two musical events differ in timbre, decay pattern, attack time, and spatial location, they are still tightly bound to each other and to the musician. Because of the temporal order, the clarinet note is experienced as the primary sonic event and the cello note as the slave sonic event. In contrast, with an individual cellist and clarinetist, the two musical events are unbound; there are two primary sonic events. It is doubtful that any listener would confuse a real cellist and a real clarinetist with a single musician controlling a synthesizer producing the sounds of both instruments. Listeners can readily perceive the degree of binding between sonic events.

From this perspective, a metapianist is actually playing two tightly bound instruments: the protopiano producing primary sonic events, and the concert hall resonant space, producing slave sonic events. Control of the two instruments could be separated, however, with a pianist playing a protopiano and another skilled musician, as accompanist, playing reverberation from a spatial synthesizer. Reverberation then becomes its own primary sonic event because the binding between the two sounds is now loose. Binding between two sonic events can vary. Bound sonic events (primary and secondary) and unbound sonic events (two primary) are two extremes of a continuum rather than categorical choices. With the merging of electronic instruments and spatial synthesizers, the degree of binding can vary over a wide range.

Dimension 2: Localized versus Diffused Sonic Events

An acoustic object, be it a musical instrument or a spatial property, can be perceived as having size, distance, and direction. Sizes range from infinitesimal to infinite. At one end of the range, a point source produces a single coherent sound wave that radiates in a spherical shape. As the wave traverses their ears, listeners can sense the size and direction of the source. At the other end of the size range, a sonic process that produces a multiplicity of similar sound waves traveling in all possible directions results in a uniformly diffused sound field of infinite extent and without orientation. Large acoustic objects radiate sound over their entire extent, just as the atmosphere produces a luminous blue sky, as a diffused field without location. Every sound field exists between these extremes.

To appreciate the distinction between localized and diffused sound sources, consider two arrangements for singing. From the front row of a concert hall, a listener perceives a soloist as existing at a location on stage. Alternatively, if a listener is encircled by a ring of singers who are synchronized in the same musical register, the chorus has no perceptual location. From within the circle, singing is an enveloping sound within which the listener exists—inside the sound generation process. At a distance from the chorus, it becomes a source that can be localized—outside the sound generation process.

Separating performers from the audience, a long-standing tradition, places listeners outside of the musical creation process, thereby allowing them to aurally localize all musicians. That need not be the case—listeners could exist inside a musical process. A few listeners could sit within the orchestra, or the musicians could be distributed in the concert hall auditorium. Hypothetically, a sound system could reproduce the sound field that exists inside the resonant enclosure of a violin. A listener would then experience the violin from inside the sound generation process. The perception of the violin is transformed from a point source on stage to an enveloping sound.

Just as a listener can be remote from a soloist or inside a chorus, spatial acoustics provides analogous cases of a discrete sound reflector and diffused sound reflectors. An echo from a planar surface is the aural equivalent of the visual reflection of a candle in a mirror. Both the original candle and the reflected candle produce a coherent wave radiating from a single point in space. Enveloping reverberation, composed of millions of sonic reflections distributed throughout the space, is equivalent to a large chorus. It is experienced as an infinite number of sound sources, aurally analogous to an infinite number of mirrored dots placed on all surfaces of the space. Just as an observer becomes visually immersed in the light from these dots, and the illumination is a collective property of the space, not of a visually localizable image, so does a listener become aurally immersed in enveloping reverberation. Even though an echo and reverberation are both slave sonic events that arise from a primary event, they differ along the dimension of localized versus diffused.

Having examined the extremes in size, ranging from an infinitesimal to infinite, let us now turn to intermediate cases. For example, when spread over an area 20 meters by 10 meters (65 feet by 35 feet), a large chorus on stage produces a wall of sound, which we hear as a large, but finite sound source. Similarly, when listening to reverberation through an open window, we also hear a wall of sound. In both cases, sound is neither a point source nor infinitely diffused. The relative size of an acoustic object is proportional to the listener's distance. At great distances, all sources appear as if they originated from a single point in space. From outer space, the atmosphere of the earth is a point light; on earth, the atmosphere is enveloping luminance.

In addition to these examples, spatial acoustics can also change the apparent size of a musical source located at a single point in space. Acoustic scientists call the experience of aural size the "apparent source width." The early sonic reflections, perhaps within the first 75 milliseconds, fuse with the direct sound rather than becoming distinct slave sonic events. But in fusing, they increase the perceived size of the primary sonic event. Because of these sonic reflections, soloists in concert halls sound larger than their counterparts in open fields without sonic reflections. With its ability to create early reflections that are spatially distributed, a spatial synthesizer becomes another means for controlling the perceived size of sound sources. And if used as part of the performance, size becomes an artistic parameter.

This discussion explains how listeners perceive the size and location of two kinds of acoustic objects: active sound sources such as musicians, and passive sound reflectors such as wall surfaces. There is a third kind of acoustic object, one that removes sound by absorbing or shadowing. For example, a nearby column or overhanging balcony creates an acoustic shadow, and an open window or a fiberglass mat absorbs sound. These, too, are acoustic objects because they produce spatial inconsistencies in the nearby sound field; any nonuniformity in the spatial distribution of sound implies the presence of an acoustic object or geometric anomaly, which is equivalent to an object. Even a resonant enclosure embedded in a wall becomes an acoustic object if it selectively amplifies specific frequencies at one region of the acoustic space.

To illustrate the properties of an absorbing acoustic object using a visual analogy, replace a mirrored surface with a black panel, substituting light absorption for reflection. The panel is visible to an observer as an object, not by the light that it reflects, but by the light that it removes. The panel is a localized visible object with a size and location. Similarly, a listener perceives an absorbing or shadowing acoustic object at a location if the ambient sound in that region is missing or perturbed, as if an acoustic vacuum cleaner had removed acoustic dust from only one region of a uniformly dusty wall.

If, however, the black panel is replaced with thousands of black dots scattered on all the surfaces, it becomes diffused with infinite size and no location. Acoustically, a single listener in the audience is equivalent to a black dot in that the listener absorbs sound at one location. And a large audience is then similar to a sound-absorbing wall, which removes expected sonic reflections from the ground surface. Sounds, especially those with high frequencies, are weakly reflected from the auditorium when it is densely filled with people. To transform the audience, a single surface having size and location, into a diffused property of the entire space, listeners would need to spread out on every surface, including the ceiling.

Traditionally, the musical arts treat musical instruments (active acoustic objects) as the only kind of acoustic object that should have size and location. All other acoustic objects, be they reflective, absorbing, dispersing, or resonating, should be fully diffused with infinite size and no location. Failure to achieve this goal is viewed as a spatial defect—an artistic judgment without intellectual foundation. This need not be so. There is no reason why artists could not manipulate any attributes of any acoustic object as part of their musical art. Although such an approach is impractical with physical acoustic spaces, virtualized music makes such artistic manipulations possible, and perhaps even desirable.

Dimension 3: Fused versus Decoupled Sonic Events

As a simplification, music is the aesthetic art of combining simultaneous and sequential sonic events. This leads to the central question: which sonic events bind together into a perceptually fused musical element and which remain as separate elements? When addressing this question, composers and musicologists usually consider only the primary sonic events—notes—originating from the primary acoustic objects, called "instruments." When composers include space as part of the composition, they focus on where instruments are placed, and they assume that the notes originate from these locations. However, by broadening the concept of sonic events to include slave sonic events, especially those which are spatially distributed, music and space combine into a single concept. Popular, contemporary, and virtual music often take advantage of slave sonic events, both localized and spatially diffused.

The merging of music and space forces us to review the concept of sonic fusion. When do sonic events bind to each other so tightly that they fuse into a single perceptual unit? When do they bind but without fusing? And when do they remain perceptually independent? As a general principle, pitch, timbre, timing, and directionality determine the degree to which sonic events will bind. A chord, for example, results from the fusion of multiple notes at specific pitch ratios, synchronized onsets, and locations. Similarly, a chorus is the fusion of many voices singing ensemble. A sonic event that results from fusion becomes a new sound, different from the sonic events that created it. Consider a food analogy. Gazpacho, a soup made of pureed vegetables, is the fused equivalent of a salad. Each vegetable contributes to the final taste of the soup, but they are too fused to be experienced individually, whereas a salad is a collection of vegetables that are bound but not fused.

Conversely, when attributes are too dissimilar, listeners hear each sonic event as an independent sound stream, as, for example, the sound of a clarinet and that of an airplane. A singer who begins too early or at the wrong pitch does not fuse with the other singers in the chorus. A musician who sits at the rear of the auditorium, even one playing with perfect pitch, does not fuse with the other musicians on stage because of a difference in location and timing. Determining when or if multiple sonic events fuse together is a complex question that depends on a wide range of physical and perceptual properties. A full explanation of the artistic rules that govern fusion in music is beyond our discussion. We can, however, explore specific cases of musical fusion with attributes that arise from spatial acoustics. In the context of large enclosed musical spaces, fusion of spatial reflection is relatively well understood by acoustical scientists, but less so by composers, who treat spatial fusion as a subset of musical fusion.

The distinction between an echo (distinct sonic reflection) and reverberation (multiple sonic reflections) is a consequence of perceptual fusion. An echo is a slave sonic event with a coherent direction and delayed arrival; two echoes are likely to remain distinct events when they are separated in time and direction. But when the rate of sonic reflections exceeds a critical threshold, perhaps 10,000 per second, they fuse completely into a new slave sonic event: reverberation. Fusion destroys the perceptual size and directionality of individual reflections.

The concept of reverberation as the fused composite of all sonic reflections is too simplified and inaccurate. Early sonic reflections also fuse with the direct sound rather than only with later sonic reflections. Consider a virtual space that produces only early sonic reflections within the first 100 milliseconds of the direct sound. Rather than creating the perception of reverberation, these sonic reflections enlarge the perceived size of the acoustic object. Conversely, if the direct sound is very weak, early sonic reflections fuse with the late sonic reflections as reverberation. Fusing with both the direct sound and reverberation, early sonic reflections serve as a smooth and continuous bridge between a primary and slave sonic event. Perceptually, however, the early sonic reflections are more important for their role in the perceived size of the primary sonic event.

Having established that early sonic reflections fuse with the direct sound and that late sonic reflections fuse to one another to become reverberation, we are in a position to revise the previous definition of primary and slave sonic events. The primary sonic event now includes the early sonic reflections, those within 100 milliseconds of the direct sound, and the slave sonic event includes the remaining late reflections. This definition is solely perceptual. Had we not modified the definitions, a primary sonic event would remain the direct sound from a source even though the direct sound is almost inaudible at the rear of a concert hall auditorium. Instead, listeners hear a primary sonic event just because early sonic reflections contribute a significant amount of energy to the otherwise weak direct sound.

This new conceptualization treats spatial acoustics as two perceptual processes: modifying the direct sound to produce a primary sonic event with increased size and intensity, and creating a separate slave sonic event that is experienced as enveloping reverberation. Space plays a critical role in both processes. But the two processes are distinctly different even when they originate from a common physical process. With this new definition, we are now in a position to explore the role of primary and slave sonic events in music.

Musically, fusion becomes even more interesting when we consider the relationship among multiple slave sonic events, which originate from different primary sonic events. Consider a time-ordered sequence of notes, each of which produces reverberation that exists concurrently with the reverberation from earlier notes. Sequential becomes simultaneous—slave sonic events have a different temporal ordering from primary sonic events. The musical rules for pitch fusion, such as with chords, now apply to slave sonic events because they have the same pitch as their primary events. Slave events fuse in the same way that their primary events would have fused had they existed simultaneously. Many musical rules for melody are the consequence of slave sonic events; reverberation and melody are interdependent.

Intuitively, composers, musicians, and aural architects understand these issues, but mostly in the context of music in physical, not virtual spaces. With spatial synthesizers, we have a wider range of aesthetic choices in designing slave sonic events from sonic reflections. For example, we could have many sonic reflections immediately following the direct sound, but not afterward, thereby creating a primary sonic event but not a slave event. We could have pure reverberation without a primary event, which makes reverberation its own primary event. We could have reverberation without a bridging region to weaken the binding between the primary and secondary events. We could construct multiple slave sonic events; or replace a single echo with a dense cluster of sonic reflections. In fact, the degree and type of perceptual binding between a primary and a slave sonic event is an artistic parameter. Concert halls have inflexible binding, which varies by seat location. Without the constraints of physical acoustics, there is no limit to creative variants. However, given the lack of seasoned and prolonged experimentation, most invented samples of slave sonic events are aesthetically appealing to at most a few listeners.

Finally, these discussions imply, somewhat incorrectly, that all listeners experience the same degree of fusion from a given sound combination. Fusion is a process of the auditory cortex that is influenced by learning and experience. There are examples where one listener hears the component events and another hears only the fused result. Many of my colleagues hear the coarseness of an inadequate sonic reflection rate from a primitive spatial synthesizer, whereas untrained listeners may hear this as complete fusion. Similarly, a conductor can often hear individual violinists in a string section, whereas average listeners hear them only as fully fused sound.

Application of Spatial Rules to Music

The previous discussions defined three spatial dimensions for musical sonic events: primary versus slave, localized versus diffused, and fused versus decoupled. By incorporating these dimensions into the language of musicology, spatial rules merge with musical rules. The proposed dimensions do not replace or compete with the existing concepts of artists, scholars, acousticians, and audio engineers. Rather, these dimensions enrich our understanding by demonstrating how music and aural architecture interact. These three dimensions are as relevant for a classical symphony in a nineteenth-century concert hall as they are for contemporary music using computer technology to implement electronic instruments and virtual spaces. In both cases, listeners experience the spatiality of musical events.

The traditional analysis of music and architecture errs when it views these two perspectives as distinct. The proposed dimensions use a musical language for the attributes of musical spatiality. Consider two phenomena that, though they originate from the same mechanism in a physical space, are experienced separately: (1) space creates slave sonic events to supplement primary sonic events; and (2) space contributes spatial attributes to both primary and slave sonic events. With synthetic implementations, the composer and acoustic designers have independent control over these phenomena. Regardless of the implementation, every musical sonic event has attributes of size and location; every musical sonic event has a means for controlling those attributes; and every musical sonic event has one or more artists who exercise control of those attributes.

The language of music is partitioned on the basis of who exercises control. Because conventional composers, conductors, and musicians exercise control only over pitch, timbre, dynamics, attack, duration, and tempo—the attributes of *musical notes* produced by instruments—they pay less attention to the spatial attributes, which are assumed and required, but not actively controlled. The direct sound is part of musicology, but listeners hear primary and slave sonic events, not just the direct sound. For those composers and musicologists who recognize spatial rules as an important part of music, and there are some who do, their language is often ad hoc and pragmatic.

From a trivial perspective, music results from the application of artistic rules: creating sonic events, and manipulating the relationship among them. There are rules for the size and location of all musical sonic events, for the creation and properties of slave sonic events, and for the degree of binding between primary and slave sonic events. Spatiality is part of those rules. Individuals from specific disciplines generally focus on a subset of rules without an appreciation or awareness of the others. And within a discipline, many rules are hidden within implementation constraints or left unexamined as immutable traditions. Concert hall architects and musical composers rarely share their respective rules with each other, even though both rule systems influence the musical experience of the audience.

We now arrive at the central questions about musical space: how are the rules for spatiality applied, who applies them, and when are they applied? Because the answers vary with culture, musical style, artistic prerogatives, and the choice of real or virtual space, a detailed exposition is far beyond the scope of this discussion. Nevertheless, if we ignore some exceptions, we can arrive at a few generalizations that illustrate the basic principles.

Composers, conductors, musicians, and mixing engineers apply rules that influence the relationship among primary sonic events when they select or change pitch, tempo, and other musical nuances that are too subtle to have names. Only aural architects and mixing engineers apply rules for influencing slave sonic events when they design real or virtual spaces. Those who create computer music apply rules for all sonic events when they select a collection of algorithms.

Compositional rules implicitly contain an awareness of reverberation as a slave sonic event created by the performance space. For example, a rapid sequence of four primary sonic events (melody fragment) corresponds to a simultaneous mixture of four slave sonic events (chordlike mixture). Unlike primary sonic events, which can be very short and sequentially distinct, slave sonic events usually have a minimum duration and blend together. Slave events are more likely to form a mixture rather than remain distinct. Depending on the pitch, tempo, intensity, and temporal shape of slave sonic events, the resulting mixture may be harmonious or discordant. The composer and conductor adapt to aesthetic blending of slave and primary sonic events, but their control is, at best, indirect and unreliable.

The consequence of a particular spatial rule depends, not just on who applies it, but also on when it is applied. Long before a new composition exists, architects had already applied their rules when they designed performance spaces. Composers apply their rules before the composition becomes music. Musicians apply their rules before the microphone signals are combined into a performance. Mixing engineers apply their rules before their musical production has been released to listeners. Each participant in the artistic chain adapts to the actions of the previous participants and anticipates the actions of those following. Designers of performance spaces are first in the sequence with their spatial attributes memorialized in stone, and listeners are last with their choice of a reproduction environment. All participants are part of a distributed team of aural architects, each contributing aspects of spatial experience.

As implied earlier, spatial abstractions become useful only if they explain many cases. The proposed abstractions are more than adequate to describe a live performance in a concert hall, but they would be burdensome, superfluous, and academic if used only for that one case. The abstractions are powerful because they have broad applicability to all forms of music, especially electronic music of the twenty-first century. The abstractions provide a common language to each member of the distributed team of aural architects.

Using our abstractions, we now return to an earlier question: should spatial attributes be consistent and accurate in order to aurally visualize a real space, or should spatiality be a collection of artistic components that ignore the rules of real spaces? It is clear that the classical music tradition, which has evolved in real concert halls, implicitly embraces the rules created by real spaces. Furthermore, such spaces include additional architectural rules, such as avoiding aural embellishments and strong local acoustic objects. A listener is able to aurally visualize such concert halls because the perceptual rules for modifying primary sonic events and creating secondary sonic events are completely consistent with the physical attributes of the space. Under our abstract concepts, the "rules" for space with classical music are severely restricted, and a spatial synthesizer should duplicate them because the composer assumed their existence when the music was conceived.

Conversely, contemporary composers, musicians, and audio engineers, especially when deploying virtual instruments and virtual spaces, enjoy a much wider range of rules than their historical counterparts. In the extreme, every primary and every secondary sonic event can be created independently with varying degrees of size, fusion, location, duration, timbre, and loudness. There is no distinction between a musical instrument and its space; space no longer serves the role of providing seating for the audience. Space becomes the aural equivalent of Escher's bizarre pictures of impossible objects and spatial geometries: artistic illusions. Perceived locations can suddenly produce strong sonic reflections, absorb specific frequencies, or create acoustic shadows. Spatial inconsistencies, rather than being perceptual contradictions, are then part of the art. Attributes of spatiality can be changed instantaneously. Aural size can suddenly increase and then immediately decrease, as if the walls jumped to a new location. A spatial synthesizer, whose properties can be adjusted to produce an almost infinite number of spatial attributes, becomes a "playable" aural art form. Listeners simply appreciate experiencing the "art of spatiality."

Obviously, with this expanded set of artistic rules, listeners have no possibility of aurally visualizing a virtual space as a physical space. There is simply no space to aurally visualize. In the hands of a brilliant artist, the creative application of spatial rules may produce aesthetically pleasing results. Yet when inventing new rules and applying them in novel ways, an artist is just as likely to create musical experiments that have little enduring value.

The application of aural architecture to cinema is a good example of aesthetically pleasing spatial rules that never presume a space as a real environment. The sound tracks that create the aural image of a space complement the visual image appearing on screen without requiring consistency. The two spaces, one for each sense modality, are neither externally consistent with each other nor internally consistent within themselves. Visual space is constantly switching perspective and distance. Looking at the world through the eye of the camera, a viewer floats through space, penetrates walls, and instantly appears in a remote environment. Meanwhile, for the purposes of creating a mood or telling a story, an aural space may be a mixture of an intimate conversation in a bedroom, soothing music in a cathedral, and natural sounds from an open space. By itself, either the aural or visual space can abruptly change without apology or warning. In fact, unrelated aural and visual spaces often coexist simultaneously, as for example, observing an automobile race while listening to an intimate conversation in a bedroom.

The practice of aural architecture by those who design concert halls, recorded music, and cinema sound diverges because each application has its own clients with their unique rules, goals, and constraints. But with music as sonic illumination, applications involve similar experiences—music in space and space in music—all subject to a common perceptual ability of listeners: auditory spatial awareness. Musical space is just one application of aural architecture, one for which this chapter has proposed a unique language.

As a principle of art, Samuel Taylor Coleridge (1907) asserted that perceptual conflicts and inconsistencies become irrelevant because of a "willing suspension of disbelief... which constitutes poetic faith." Artistic space never represented itself as being real space; it is only the *experience* of space that is real; and achieving artistic impact often requires spatial contradictions. Although art acquires its power by the freedom to selectively transcend reality, each art form transcends physical space in its own peculiar way, with its own concepts of spatiality.

5 Inventing Virtual Spaces for Music

We all know that Art is not truth. Art is a lie that makes us realize truth, at least the truth that is given us to understand. The artist must know the manner whereby to convince others of the truthfulness of his lies.
—Pablo Picasso, 1923

The great Renaissance painters of the fifteenth and sixteenth centuries applied their newly acquired knowledge about spatial perspective and geometric projection to their pictures, representing three-dimensional spaces with what was, at the time, an amazing degree of accuracy. Now in the twenty-first century, aural architects replicating the experience of a concert hall musical space at a remote location are creating a kind of aural photograph, although there are alternative representations of artistic space that ignore realism. During the three centuries following the Renaissance, realism as an approach to visual exactitude gave way to Impressionism, Expressionism, Modernism, Cubism, Futurism, Dadaism, Surrealism, to name but a few of the proliferating painting styles. Faithful representation of physical reality became subservient to a new language of stylized and subjective expressions of experience. Objects and spaces could depict strong emotion and intense symbolism precisely when they were no longer constrained by having to replicate physical attributes. For example, in Salvador Dalí's *The Persistence of Memory*, melting clocks—virtual objects—communicate the concept of subjective time, even though they have no ability to track the passing of time. Time is, after all, relative and personal.

By the early part of the twentieth century, accepted truths about reality were being revised on a daily basis. The belief in physical matter as something substantial was undermined by revelations from atomic physics. Solid objects were actually voids; matter was energy, time and space fused; and eventually, particles became abstractions rather than miniature marbles. As if to prove that experience is not reality, exhibits in science museums and special effects in films routinely demonstrated sensory illusions. Ideas such as chaos, entropy, uncertainty, relativity, and randomness, permeated the arts, sciences, politics, and, in turn, the larger social fabric. Revolutionary changes in the nature of physical reality permeated our artistic view of ourselves and our world. Virtual auditory space was invented in this revised context of subjective space.

Like their visual counterparts, aural artists worked within this radically changed intellectual environment, but they would have to wait until the mid twentieth-century before acquiring the electroacoustic tools necessary to virtualize space and objects. Karlheinz Stockhausen (1959) was one of the earliest composers to recognize that these tools could be used as an aural canvas, something that could create a virtual space: "My idea would be to have a spherical chamber, fitted all round with loudspeakers. In the middle of this spherical chamber, a platform transparent to both light and sound would be hung for the listeners." Contemporary composers and audio engineers soon embraced this new aural canvas as a means to an artistic end. Aural space was no longer limited to physical sounds in a place where we lived, worked, gathered, or performed. Aural spaces became nonobjective, nonrepresentational, and nonphysical, and like their visual contemporaries, aural architects of virtual musical spaces also created surrealistic environments. When technology liberated aural artists by providing them with the aural analogues to paint, brushes, and canvas, centuries of intellectual evolution blossomed in only a few decades.

Prophetic visions of the future are sometimes found in the distant past, especially when brilliant minds anticipate what will be possible without being confined by their immediate reality. When Francis Bacon (1626) described the "sound houses" of his utopian college in his essay *The New Atlantis*, he was prophesying the electroacoustic world of contemporary music of the twentieth century:

We have also diverse strange and artificial echoes, reflecting the voice many times, and, as it were, tossing it; and some that give back the voice louder than it came, some shriller and deeper; yea, some rendering the voice, differing in letters or articulation from that they receive. We have means to convey sounds in trunks and pipes, in strange lines and distances.

Without tools for creating an aural space, spatiality remained subservient to other compositional elements, such as rhythm, melody, timbre, and tempo. But with the evolution of advanced electroacoustic tools, Bacon's seventeenth-century ideas, once merely footnotes to history, would be rendered into sound for ordinary listeners to hear; musical space became increasing fluid, flexible, abstract, and imaginary. This trend was most apparent in the second half of the twentieth century. From the perspective of electronic music, spatial design is an application of aural architecture without assuming a physical space. Musical space is unconstrained by the requirements for normal living, and musical artists are inclined to conceive of surreal spatial concepts.

Like M. C. Escher's painting of an imaginary space with interwoven staircases that simultaneously lead upward and downward, aural artists also have the freedom to construct contradictory spaces. As an analogy to a virtual aural space, figure 5.1 has ele-

Figure 5.1
M. C. Escher's *Relativity* © 2005 M. C. Escher Company–Holland. All rights reserved. www
.mcescher.com.

ments of visual spatiality, but the space itself could not exist. Similarly for an aural
space, we can create sounds that appear to come closer without moving, or a spatial
volume that is simultaneously large and small. Modern audio engineers and electronic
composers, without necessarily realizing their new role, became the aural architects of
virtual, imaginary, and contradictory spaces. Aural spatiality can exist without a physical space.

By abandoning conventional norms defining music and space, modern artists created
contemporary music.[1] Although this class of music is considered by some to be an irrev-
erent and unpleasant form of noise, the new *rules of space* are still worth investigating

because they exist apart from the compositional creations that incorporate them. These rules are interesting both because they predicted the popular music of the late twentieth century and because they suggest future direction for the twenty-first century. Even if some twentieth-century contemporary music has not left an enduring legacy, the new rules of aural space are likely to survive in other aspects of our art and culture.

The rule that requires musicians to perform in a tight cluster on the stage and listeners in predefined seats in the audience area is readily broken, as is the rule that requires both musicians and listeners to maintain a static geometric relationship throughout the performance. Moreover, when knobs on equipment can alter virtual spatial attributes, the rule that requires spatial acoustics to remain constant and consistent during a performance is also easy to transcend. In the world of virtual spatiality, acoustic space and sound location are no longer based on the laws of physics; acoustic objects can change their size and location instantly. Acoustic space and sound location have become as dynamic as the sequence of notes in the composition. As with all artistic rule systems, however, breaking old rules is easier than replacing them with meaningful new ones. A few decades is a very short duration for refining a new art form.

A virtual space is not only a compositional element in music, but also an experience that can be extracted from music and then applied elsewhere, for example, to auditory displays in the cockpit of an airplane, the fictional spaces of computer games, or the dual audiovisual spaces of cinema. In these applications, there may not be consistency among the different sensory modalities. In some sense, with the ubiquitous technology of the twenty-first century, the experience of spatiality frequently dominates the experience of a physical environment. Space is no longer just a geographic framework (near-far, front-back, up-down) for positioning sounds relative to listeners. Space is no longer just a response to the acoustics of the environment. The older definition of cognitive maps of space as the internal representation of an external world, introduced in chapter 2, becomes fluid, plastic, and even more subjective. Aural architects of virtual spaces are manipulating their listeners' cognitive maps.

Artistic Dimensions of Space and Location

Composers have always understood, both intuitively and consciously, that the location of the musicians contributes to listeners' experience of a musical space. The hidden problem with positioning musicians throughout a space is that sound waves move comparatively slowly. Large acoustic spaces produce large delays, which displaces the temporal alignment of music arriving from different locations. Two notes beginning at the same time may arrive at a listener at different times. The spatial manifestation of time is an artistic issue for both listeners and performers, and as in advanced physics, time and space are related and connected concepts.

When musicians are tightly clustered, the time for a direct sound to travel among them is small, and synchronization depends on their artistic skills alone. Conversely, when an orchestra is large and spread across the stage, the sound delay places a limit on aural synchronization. Because musicians separated by 20 meters (65 feet) will hear each other with a 60-millisecond delay, the visual cue of the conductor's moving baton takes over the function of producing temporal consistency. When musicians in a large orchestra are perfectly synchronized in time, neither the conductor nor the listeners hear that temporal alignment because they are closer to some musicians than others. For example, a listener near the stage but far off to the left will hear a musician at the far right side of the stage with a delay after hearing a musician on the left, even though the two musicians are playing the same note at the same time. This problem is exacerbated if musicians are widely distributed throughout a large space.

Composers can compensate for audio delay in several ways. Tight synchronization is not required if the composer includes a temporal gap, perhaps silence, between sounds originating from widely distributed locations. The location of the musicians, which depends on the particular geometry of a space, can then become a compositional component, although when the composition depends on a specific spatial organization, the music is not easily transported to other spaces without having to be adapted. For this reason and because it is less flexible than other options, composers have seldom manipulated the spatial distribution of musicians.

With the advent of electroacoustics, perceived location and intrinsic audio delays were separated. For example, deploying individual microphones and headphones for each musician removes the intrinsic delays when they listen to their colleagues. Unlike air as a medium, electrified sound moves through wires instantaneously. The sound engineer is therefore free to electroacoustically reposition musicians anywhere in the virtual space, without destroying the synchronization among them. Two musicians separated by a distance of 50 meters (165 feet) can still be heard synchronously. Aurally perceived location has nothing to do with actual location; virtual spaces and virtual locations break the relationship between time and space.

Anyone who creates a complete sound field that produces the experience of spatiality is functioning as an aural architect. Traditionally, sound sources from loudspeakers were viewed as injecting sonic events into a listening space, but with the advent of surround-sound reproduction, the sound field includes, and in some cases, replaces the experience of the listening space. This chapter traces the history and evolution of space in music, ending with the aural architecture of virtual spaces.

Incorporating Location within Traditional Music

Many of the spatial ideas found in contemporary music originated from an earlier period when musicians were occasionally distributed within the performance space. There is a long tradition of antiphonal music, a dialogue of call and response among

distinct groups of musicians at different locations, which does not require tight synchronization or simultaneous playing. This style is found in the chanting psalms of
Jews in biblical times, and in early Christian music dating from the fourth century. In
the late sixteenth century, Giovanni Gabrieli extended the tradition of *cori spezzati*
(divided choirs) as an adaptation to the unique architecture of Saint Mark's Cathedral
in Venice (Grout, 1960). The musical space was vast, and it contained two widely separated organs and choirs at opposite sides of the cathedral. Adapting to that uniqueness, composers at Saint Mark's featured a dramatic use of antiphony between the
halves of the double choir. The penchant to divide performers was also part of the
Venetian polychoral tradition, started by Adrian Willaert and culminating with nine
choral groups distributed throughout the cathedral (Mason, 1976). The refinement of
cori spezzati represented a musical revolution, and also appeared in secular music of this
and earlier periods, such as madrigals with echoes (Arnold, 1959).

By the twentieth century, the use of spatially distributed musicians became less unusual and more innovative. Richard Zvonar (1999) cites numerous examples. Charles
Ives, in *The Unanswered Question* (1908), placed the strings offstage to contrast with
the onstage trumpet soloist and woodwind ensembles. He was influenced by his father,
a Civil War bandmaster and music teacher, who had experimented with two marching bands approaching the town center from different directions. Henry Brant then
extended the idea in *Antiphony I* (1953) and *Voyager Four* (1963) with five ensemble
groups placed along the front, back, and sides of the space. Three conductors were
required.

For modern composers, dispersing musical sources throughout a space is no longer
revolutionary; location is an active component of a composition. Antiphony and spatial distribution evolved into a space-time continuum, which Maja Trochimczyk (2001)
calls "spatiotemporal texture." At any time, a musical voice could appear from any direction, and by intentionally sequencing attributes of space, time, pitch, and timbre, a
voice can create the illusion of movement (changing position) and transformation
(changing size). When used in this way, space is a musical dimension. Charles Hoag,
in *Trombonehenge* (1980), used thirty trombones surrounding the audience as an imitation of Stonehenge, and R. Murray Schafer, in *Credo* (1981), surrounded the audience
with twelve mixed choirs. Extending the blending of musicians and listeners still
further, Iannis Xenakis scattered 88 musicians among the audience so that the listeners
are actually inside the music; in another of his compositions, musicians moved
through the space rather than remaining seated.

Based on traditional theory, music has a temporal and pitch structure, and within those dimensions, a composer manipulates musical voices so that they either
fuse into a unitary whole or remain segregated as distinct elements—musical layers.
Contemporary music, however, has added a spatial dimension. Composers now require new rules for manipulating fusion and segregation. The proliferation of composi-

tions that manipulate space signifies a new form of sound imagery (Trochimczyk, 2001).

An analysis of contemporary music is made even more complex by the addition of the two related ideas: incorporating the spatial dimension of voice location, and elevating sonic segregation over fusion and blending. During the last century, even without using space as an artistic element, Western music abandoned fusion as a prerequisite. Layered musical elements retain more of their perceptual identity when not fused. Space has become just another tool for creating musical layers.

Maria Anna Harley (1998) analyzed spatial music in terms of perceptual principles that contribute to segregating musical elements. By drawing on Albert S. Bregman's *Auditory Scene Analysis* (1990), she applied the principles of perceptual psychology to music. Spatial differences between sound sources that result in temporal differences at the ears augment the aurally perceived segregation of musical elements. Like differences in time, pitch, timbre, and attack, differences in spatial location are yet another means to enhance this segregation. In other words, similar but not identical sounds belong to separate musical layers when they are also spatially separated. Disparate locations de-emphasize fusion. Many modern composers, such as Bartók, Boulez, and Stockhausen, intuitively use this principle in their music.

That twentieth-century music drifted away from fusion is consistent with spatial separation of sound sources. As a means of preventing fusion, Brant (1967) used several artistic principles that derive from spatial separation. In one composition, he illustrated his concepts by distributing stringed instruments along the walls on the ground floor of a concert hall, as well as in the first, second, and third balconies, thereby creating a broad and intense wave of sound. Spatial separation preserved the clarity of contrasting layers, especially when different musical elements are in the same register. Because identical or harmonically related notes in two musical layers would typically fuse if not spatially separated, spatial separation afforded the composer greater musical flexibility by permitting increased complexity without concern for unintended confusion.

Placing the performers below, above, behind, or to the side of listeners is not intrinsically interesting. Indeed, serializing the direction of music from a sequence of orientations or choosing an arbitrary geometric shape for performer location is, for Harley (1998), simply a failure to understand the new art. Spatial music is interesting precisely because, and only because, it allows combinations of musical elements that would otherwise be artistically weak without using spatial distribution. As if to prove this assertion, Trevor Wishart (1996) analyzed spatial movement in soundscape art, apart from a musical context, and came to a similar conclusion about space as a segmentation tool.

In her summary of musical space, Harley (1998) concluded that "geometric floor plans and performance placement diagrams are integral, though inaudible, elements of the musical structure—as integral and inaudible as some abstract orderings in the domains of pitch and rhythm." Spatial organization of sound sources and listener locations are components of music. Yet even when the musical score carefully specifies an organization in time and space, the composer is still constrained by the inherent inadequacy of human performers to achieve precision timing when physically separated.

Consider two musicians located at different places but playing the same note on the same instrument. Using the concepts of Pierre Boulez (1971), there are four important cases that differ only in relative timing: simultaneous beginning and ending (fused), delayed onset of one musician's note relative to the other's but still overlapping (conjunctive interval), a small temporal gap between the end of one musician's note and the beginning of the other's (disjunctive interval), and a large delay between the two musicians' notes (distinct sonic events). The fused case corresponds to a distributed choir singing in unison, and the last case corresponds to the historical use of antiphony. The middle two cases are interesting because they have the potential to create the perception of virtual movement, which Boulez calls "mobile distribution" or "dynamic relief." In contrast, a *fixed distribution* or *static relief* represents a static state without kinematics. Timing has always been a critical dimension in composition, but timing combined with space becomes two-dimensional: spatiotemporal.

This extra spatial dimension, in addition to preserving segregation of musical textures, offers other possibilities. A disjunctive interval can produce a sudden change in the aurally perceived location of a musician, and a conjunctive interval can produce smooth transition between the two locations, *space glissando*. However, both effects are fragile, depending on the skill of the musicians to control timing, pitch, timbre, attack onset, and termination. And both effects depend on the location of the listener relative to the musicians. Musical movement is therefore an illusion, or a metaphoric allusion, rather than an imitation of a physical process. In addition to this change in perceived location, true motion of a sound source produces a Doppler frequency shift. Whereas physical motion in physical space has a reality, virtual motion in virtual spaces is an artistic prerogative.

Electronic Presentation of Spatial Music

If loudspeakers, rather than stationary performers, generate the sound field, the audio engineer has the freedom to control the timing, timbre, pitch, and other parameters of each sound source, notably musical spatiality. This permits a wide range of possibilities. At one extreme, each loudspeaker can reproduce one musical voice, which would be located at its respective loudspeaker. The composer can thus place a voice only at loudspeaker locations. At the other extreme, sounds emanating from multiple loudspeakers can be synchronized, collectively producing any particular sound wave. For example, if all loudspeakers in a rectangular array produce the same sound, but with controlled delays, the resulting sound field can simulate a single voice at any orientation and distance. Although each is itself a sound source, when loudspeakers are combined, the perceived location of musical voices is unrelated to the location of the actual loudspeakers. With enough loudspeakers, the range of spatial possibilities is truly infinite, including the experience of spatiality, hence the designer functions as an aural architect.

As mentioned in chapter 4, one of the earliest examples of loudspeaker arrays was introduced at the 1958 World Exposition in Brussels. Philips commissioned Le Corbusier to design a sound pavilion with built-in loudspeakers and sponsored Edgard Varèse to compose "Poème Electronique," an 8-minute multimedia presentation (Treib, 1996). The composition was married to that space, and the space was designed for that composition. This experience became a prototype of a spatial art where the architecture, color, images, voices, and music were all superimposed to create an experience far greater than the components. Few artists, then or since, have been given such freedom.

Willem Tak (1958), the lead sound engineer from Philips working under the direction of Varèse, designed an electroacoustic system with a single artistic goal: "The listeners were to have the illusion that various sound-sources were in motion around them, rising and falling, coming together and moving apart again, and moreover the space in which this took place was to seem at one instant narrow and 'dry' and at another to seem like a cathedral." The virtual acoustics of the space were explicitly embedded in the prerecorded composition. Varèse not only selected and processed each sound element, but controlled its source location. Because the space held 400 visitors, the actual musical experiences depended on each listener's location. It was an art form that allowed for a different experience for each listener.

By today's standards, their technology, however effective, was primitive. The music was recorded on a single track, but with two additional tracks for special effects and reverberation. To achieve spatial control, a second system comprising 180 channels activated switches, motors, and relays to distribute the audio and video to the various projectors and loudspeakers (de Bruin, 1958). Altogether there were 350 loudspeakers positioned throughout the pavilion; because there was neither a scoring system nor a means to envision the auditory experience without first hearing it, Varèse and the sound engineers spent months composing the eight-minute performance by trial and error. It was a labor of love. At the age of 75, Varèse (1998) fulfilled a dream that had begun many decades earlier: "For the first time I heard my music literally projected into space." Before the pavilion was demolished, some two million visitors attended these performances, which was a public milestone in the transition from mechanical to electronic art.

A decade later, at the Osaka World's Fair in 1970, the German pavilion consisted of a spherical auditorium some 28 meters (90 feet) in diameter, containing fifty loudspeakers arranged in seven concentric circles. The audience, sitting on an acoustically transparent floor, could experience sound arriving from any direction, and thus be completely enveloped by Stockhausen's music. A dream that had preoccupied the composer since 1958 was realized with this opportunity to present his electroacoustic spatial music without the constraints of performers at fixed positions (Kurtz, 1992).

Although as many as twenty soloists performed; their music fed the microphone inputs of a special mixing and control console designed to drive the array of loudspeakers. The console included a "sound mill," a giant pan potentiometer with a large handle that could steer the microphone signal to any of fourteen loudspeakers. Stockhausen described how the sound "could make complete circles around people, not only horizontal circles, but vertical circles...or spiral movements of all different loops....Multiple sound sources could be made to swirl along arbitrary trajectories, intersecting and interleaving each other. This polyphony of spatial movements, and the speed of the sound, became as important as the pitch of the sound, the duration of the sound, or the timbre of the sound" (Cott, 1973). The conductor-composer controlled the audience's experience by turning knobs rather than by moving a baton.

Besides the German pavilion, shown in figure 5.2, many other pavilions at the Osaka World's Fair were also presenting music using electroacoustic technology. Xenakis composed his 12-channel composition *Hibiki Hana Ma* for 800 loudspeakers in the Japanese pavilion (Zvonar, 1999). In a somewhat different vein, the Pepsi Cola pavilion was configured as an electronic space that could take on the different personalities of the composer. It used 37 loudspeakers and 8 signal-processing channels, which provided amplitude modulation, frequency modulation, and spectral filtering. The music source was a combination of 16 monophonic tape-recorded channels and 16 microphones for live performers. Electronics were changing, creating, and projecting sounds into the space.

In recognition of the need for a controllable, yet portable, performance venue for the new *art* of electroacoustic spatialization, David Worrall (1989) constructed a 7-meter (25-foot) geodesic dome at the Australian Center of the Institute for the Arts and Technology. It contained 16 loudspeakers on an acoustically transparent surface, thereby giving the composer complete freedom to envelop the audience within a three-dimensional sound field. The listening environment, the dome, disappears from the experience because the actual listening space did not influence the sound field. Worrall (1998) concluded: "The space is *in* the sound. The sound is of the space. It is a space of sound, but there is no sound *in* space." Using electronic tools to synthesize and process the sounds and their location, the composer was also able to both incorporate and embody the experience of space.

By accepting a spatial compromise, Jonty Harrison (1998) and other composers at the University of Birmingham constructed a portable sound diffusion system (spatial control of sound) comprising up to 30 channels of loudspeakers that could be temporarily installed in a traditional performance space. This system, shown in figure 5.3, and its music have been well received in many European countries. The performer-

Figure 5.2
Stockhausen's music presentation environment at Osaka World's Fair. Courtesy of the Archives of Stockhausen Foundation for Music, Kuerten, Germany.

musician operates a diffusion console that provides a real time mechanism to control the output from each of the channels, rendering dynamic sound sculptures into the particular space. By linking the composition to the rendering system, the composer retains control over the way in which the music is diffused into the space. In the view of Barry Truax (1998), "composition and diffusion can be understood as two complementary processes: bringing sounds together, and spreading them out again in an organized fashion." However, because the acoustics of the performance space exist simultaneously with the embedded acoustics, as with normal music, the listening experience retains some of the personality of the performance space. The distributed

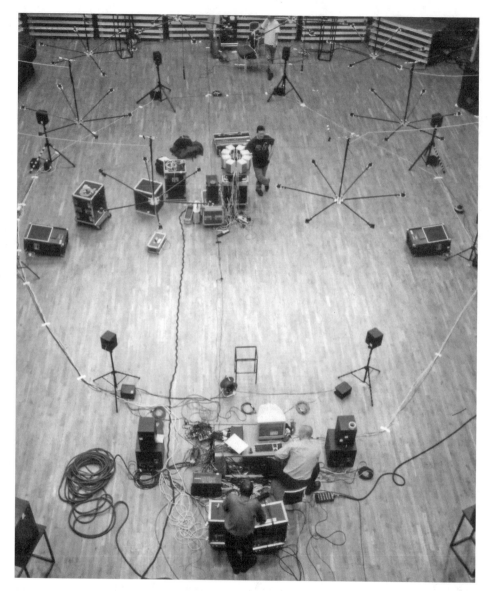

Figure 5.3

The BEAST sound diffusion system creates a spatial experience entirely from 30 loudspeakers. Courtesy of Kevin Busby.

loudspeakers provide complete control of the sound source direction without removing reflections and reverberation.

Other examples of sound diffusion systems have come and gone over the years. These include the Acusmonium of the Groupe de Recherches Musicales (Emmerson and Smalley, 2001), and the Gmebaphone of the Groupe Musique Expirimentale de Bourges (Zvonar, 1999). During the recent decades, thousands of compositions have been presented to the public using such systems. Recently, these specialty systems, custom-designed with great effort, have been replaced by software running on personal computers. By transforming compositional tools into commodities, the process of composing is now accessible to anyone with an interest in the auditory arts. For example, the music-rendering language Csound, developed by Barry Vercoe at MIT in 1985, has become one of the most widely used sound synthesis software programs. Manuals, tutorials, examples, and premade algorithms are readily available (Boulanger, 2000). Although the language has all of the modules needed to synthesize the acoustics of any seat in Boston Symphony Hall, it can just as easily synthesize a science-fiction fantasy of cosmic space. Regardless of the artistic value of contemporary music, composers clearly function as aural architects, creating spatial experiences. Especially with electroacoustic music, aural architecture is clearly no longer limited to physical spaces.

Natural Environments as Spaces for Auditory Art

Under an expanded definition of aural architecture, choosing an existing space for its aural properties is just as much architecture as building an aural space. Selecting and building both imply a social criterion. While some twentieth-century composers were actively embracing computer-generated sounds in virtual spaces as the inevitable evolution of traditional music, others were advocating a return to natural environments. A derivative art form derived from Schafer's concept of the soundscape (1977) gave rise to musical performances in natural spaces such as meadows and woods.

Soundscape music merges with acoustic ecology to become the art of a living world with natural sounds and acoustics. As with very early cultures in previous millennia, the distinction between sound and space once again disappears. Two aural components contribute to a listener's associations to a particular space: its unique sounds and its characteristic acoustics. For example, forests have the sounds of birds and rustling leaves. And forests have particular acoustics resulting from the movement of reflecting surfaces, air turbulence, and thermal refraction. Both its sounds and its acoustics contribute to hearing the space as a forest; either aspect can dominate or complement the other. Those aspects of sounds that convey a spatial association, not as a result of illuminating spatial acoustics but by themselves, are also part of aural architecture. Such a conclusion is valid only when using the broadest definition of aural architecture: any sounds or acoustic attributes that produce the experience of space for listeners.

Early advocates of the genre, now known as soundscape music, were making both a political statement about industrial noise and a cultural statement about appreciating sounds that were otherwise ignored. In its original manifestation, soundscape music was performed in an outdoor environment characterized by natural acoustics, unrelated to the acoustics of enclosed performance spaces. This genre of music emphasized the sounds of life: chirping songbirds, barking dogs, and storming weather. And its environment was meadows, lakes, town halls, parking garages, and any space used by people. From the perspective of these ardent advocates, soundscape music was original and natural music, whereas concert hall music was not; indeed, the structure of the concert hall only served to block out the sounds of nature.

Extracting aural elements of soundscapes has a long history in music—evoking the experience of far-off spaces by mimicking their natural sounds rather than the spatial acoustics. Although it is not possible to imitate the spatial acoustics of a forest, office, battlefield, or church in a concert hall, it is possible to imitate and even reproduce there the characteristic sounds of thunder, typewriters, cannons, or bells. Traditional composers ignored the acoustics of natural spaces because they had neither control over nor interest in them. Imitating characteristic sounds was their only viable alternative for conveying a sense of a specific space. Such imitative sounds were allusions to or illusions of real sounds, and listeners could readily imagine being in the associated space.

Numerous examples illustrate the vividness of such spatial allusions. The hunting horn motifs in the works of Haydn and Mahler evoke a pastoral landscape. Beethoven's Pastoral Symphony is a tribute to country life with the sounds of quail, nightingale, and cuckoo. Jean-Philippe Rameau featured birdsong in *Le Rappel des Oiseaux* (Descent of Birds) and in *La Poule* (The Chicken). Ferde Grofé imitated the clip-clop of donkey hooves in *On the Trail* from the *Grand Canyon Suite*. Majestic ocean waves were portrayed in Debussy's *La Mer*. And in the twentieth-century composition the *Pines of Rome*, Ottorino Resphigi included a recording of a bird from high up at the ceiling. Much earlier, sixteenth-century German organ builders occasionally included a *Vogelsang* (birdsong) stop, implemented with inverted pipes immersed in water to mimic the sound of a bird. Evoking the image of an office space, Leroy Anderson elevated a typewriter to musical instrument for his *Typewriter Song*. Wind machines periodically create the sounds of storms in the vast outdoors. Creating spatial associations by using the characteristic sounds of a specific space is a long-standing tradition in musical compositions.

In contrast to auditory allusions, by moving a performance into a natural space, soundscape music elevates the role of acoustics in conveying a spatial experience. In one such work, Schafer's *Music for Wilderness Lake* (1979), twelve trombones were placed around a small lake. The composition specified that it be performed at dawn and dusk, when the wind is slight and when sonic refraction, the bending of sound by thermal gradients, is most apparent. Schafer used the focusing effect of thermal inversion layers to propagate sound over long distances (Harley, 1994). To set the mood, musicians had to walk through the woods in the dark in order to arrive at the performance destination. As quoted by Harley (1994), Schafer saw that "the solitude and closeness to nature affected their manner of playing and provoked, at times, 'pantheistic' sensations, giving rise to an unforgettable experience." Other examples of natural acoustics include two choral groups at opposite ends of a lake. Musicians were aware of, and responded to, the natural sounds of water, birds, and weather, all of which were beyond their control. The artistic power of an environment is far stronger when it is more than an allusion. For example, Richard Wagner's *The Ring of the Nibelungen*, with its magical world of gnomes and giants along the Rhine, was recently performed in the Amazon rain forest (Rother, 2005).

The essential idea of soundscape music is to enlarge acoustic boundaries, extending the space to the acoustic horizon. And within that expanded area, all sounds of life are included, not just those represented on the composer's score. Yet few of us are likely to travel to a lake in northern Canada or to the Amazon jungle to listen to music in a space without boundaries. As an art form, soundscape music attempts to bring that remote experience to urban listeners by including its three central components: environmental sounds, their location within the environment, and the acoustics of a natural unenclosed space. For advocates of the soundscape, aural architecture is an unenclosed space, limited only by the acoustic horizon.

In considering spatial frames of reference, Simon Emmerson (1998) conceptualized musical space as progressively expanding circles, from the smallest to the largest: sonic event, performance stage, acoustic arena, and soundscape. The sonic event occurs at the instrument; traditional performers are located on a performance stage; and the audience sits in the acoustic arena bounded by the wall of the space; but with a soundscape, the space is far larger and determined only by the acoustic horizon. Even while sitting in an enclosed space, the composer can remove the aural experience of the acoustic boundaries created by walls by including spatially distributed sounds at remote locations, creating the illusion of boundless space. Replacing reverberation with the acoustics of air turbulence and thermal inversion layers adds to the illusion of unbounded space.

Whereas traditional music is the art of creating sound from instruments, soundscape music is the art of the aural environment. It is a shift in emphasis, from space as the container of the art, to space as the art of the container. Soundscape art creates a sonic landscape, whereas traditional music emphasizes foreground performers. With landscapes and soundscapes, the background is also important.

What sounds should be included in a particular soundscape? At one extreme, a composer can wander through a natural environment with tape recorder and microphone to capture "found" sounds that have the potential to represent that environment. At the other extreme, those sounds can be "abstracted" by elaborate signal processing using digital audio workstations (Truax, 2002). Both found and abstracted sounds, however, retain some aspect of their identity even when merged with the sounds of traditional musical instruments. For example, in the *Cricket Voice*, Hildegard Westerkamp recorded the sounds of a cricket in the stillness of the desert. But she then processed the sound by changing its pitch, timbre, duration, and reverberation, all of which created an alien but still recognizable sound of an enlarged insect. By including breaking branches, reverberating caves, honking horns, falling bricks, and clattering jackhammers, soundscape composers are more inclusive, less conservative, than their traditional colleagues. Thus spatial environments have always been a source of inspiration, literally and figuratively.

Composers, conductors, and performers often search for new performance spaces, manipulate familiar spaces in new ways, or include electroacoustics to augment natural acoustics. Deep caves and abandoned grain silos become alternatives for concert halls. On the one hand, deep caves allow contemporary listeners to hear music as their prehistoric ancestors did. On the other hand, such spaces excite the imagination of artists jaded by the infinite flexibility of electronics. For thousands of years, we have looked to nature as a source of sensory inspiration. Searching nature for its spatial artifacts is easier than inventing spaces, and exploring what has already been found is one of the oldest artistic traditions. Nature, with a little assistance from innovative artists, has bequeathed a range of interesting underground spaces as artifacts of geological history. Some of these subterranean environments are artistically and emotionally provocative, if for no other reason than their sonic uniqueness.

Jennifer Berezan (2000), songwriter and folksinger, had the unique opportunity to use the Oracle Chamber in the Hypogeum at Hal Salfieni, Malta, as her recording studio, where she recorded *ReTurning*. The space is a large underground tomb carved into the limestone for use by an ancient civilization for burial ceremonies. She described the experience as "singing inside the earth, in a place that had been used thousands of years for ritual, for oracles, for prophecy. It was obvious that the people who had built it had an incredible understanding of acoustics and of the value and power of sound for healing.... Initially, I described the experience as being in an altered state [as if] the whole place vibrated with our songs; it was exquisite, and beyond words" (Berezan, 2000).

Using the stalactite grotto of Jeita near Beirut, shown in figure 5.4, a glistening subterranean world of natural acoustics, in 1969, Karlheinz Stockhausen performed several of his postmodern compositions (Kurtz, 1992). The musicians, located on a platform constructed over the abyss below, were illuminated with spotlights in an otherwise darkened environment. The largest chamber, the Red Chamber, had a volume approaching 100,000 cubic meters (3,500,000 cubic feet), large by any standards.

Figure 5.4
Grotto of Jeita in Lebanon, where Stockhausen presented his concert. Courtesy of the Archives of
Stockhausen Foundation for Music, Kuerten, Germany.

Unlike a normal concert hall, listeners were as much as 80 meters (260 feet) from the performers, which allowed the natural acoustics to dominate the direct sound. Underground rivers could be heard in the background. The effect of the space was total, and multisensory. Visitors gained access to the cave by walking for 15 minutes through a tunnel and smaller caves until they reached the main grotto. As part of the mood, the access path contained hidden loudspeakers rendering Stockhausen's music. A Catholic priest said of the performance of *Stimmung*, "It was the longest prayer I have ever known and the happiest" (Kurtz, 1992). The music in the tunnel set the mood in the same way that a grand entrance foyer of a concert hall does.

Michael Kurtz (1992) reported that "the peculiar acoustics of the grotto had their own role to play; the sounds reverberated for a very long time and sometimes seemed to roll physically for hundreds of meters along the rock walls, accumulating in a manner never experienced in normal concert halls and stimulating the musicians to shape time and dynamics in a special way." Stockhausen, in an interview with Jonathan Cott (1973), said of the acoustics: "the caves made the music sound both prehistoric and also like something out of science fiction." Although the reverberation time was as long as that of a cathedral, perhaps 8 seconds at middle frequencies, the acoustics were not that of a single large space. Because the space was actually a combination of multiple connected spaces, the onset of reverberation was delayed and softened, reaching only a modest intensity. In addition, with dramatically larger sound absorbtion at high frequencies, reverberation was significantly reduced, thereby removing any sense of harshness.

Before civil unrest forced the closing of these caves, François Bayle also performed in the grotto of Jeita. More recently, as stability returned to the region, the Lebanese artist Sami Makarent hosted a musical evening of poetry and music, and the piano prodigy Guy Manoukian performed there as well. All concerts were sold out. Apart from the acoustics, the visual impact is so striking that these caves are a featured part of organized Lebanese tours. If the caves were more accessible to the listening public, they would most likely have become major musical venues with a repertoire of compositions written for them. Moreover, had such music achieved popularity, scientists would have studied its acoustic properties in order to duplicate the spatial experience in other locations.

Even though underground natural enclosures are infrequently used as alternatives to concert halls and recording studios, musicians have performed and recorded music in numerous caves and caverns around the world. The list includes the Cathedral Cavern in Australia, the Mammoth Cave in Kentucky, Caves of the Iron Mountain in New York, the Beit Govrin Caves in Israel, the Chiselhurst Caves in the United Kingdom, the caves of the Faroe Islands, and the Buddhist caves of Pitalkhora in India. Modern performers continue to release commercial music recorded in these spaces, each of which is acoustically unique. Selecting a performance space is also a form of aural architecture.

New Ways of Listening to Electroacoustic Music

The new rules for musical spaces are themselves part of the larger new ways of listening. Art derived from a new set of rules is, by definition, unfamiliar. As Nicolas Slonimsky (1965) pointed out, "music is an art in progress, and... objections leveled at every musical innovator are all derived from the same psychological inhibition, which may be described as Non-Acceptance of the Unfamiliar." A few of the many quotations collected by Slonimsky illustrate the breadth and intensity of critical reactions. In 1855, Wagner's *Lohengrin* was described as having "no more real pretension to be called music than the jangling and clashing of gongs and other uneuphonious instruments." Stravinsky's *Le Sacre du Printemps* was called the "Massacre du Printemps." Brahms's *Requiem* was heard as "so execrably and ponderously dull that the very flattest of funerals would seem like a ballet." Gershwin's *An American in Paris* was "nauseous claptrap, so dull, patchy, thin, vulgar, long-winded and inane." Saint-Saëns's *Danse Macabre* was dismissed as no more than the "clatter of the bones of a skeleton" (Slonimsky, 1965). As mentioned earlier, even the acoustics of the Boston Symphony Hall, today considered one of the world's great concert halls, were intensely criticized during its first few years. Hostility to change is the norm rather than the exception.

How do listeners experience the new aural architecture of virtual musical spaces, when vision and hearing of a unified space are decoupled? Although the literature does not directly address this subject, there is an extensive commentary on spatial application to contemporary music, and to auditory imagery in film. Let us consider two cases: an aural space devoid of a visual counterpart, and an aural space subservient to its visual counterpart.

Because we experience space as a unified entity composed of input from all the senses, not as segmented aural and visual spaces, the relationship between the two sensory modalities determines the resulting experience of space. Even with the emphasis on aural space in a traditional concert hall, listeners hear the space illuminated by music, and they hear music in the space. But they also see the space and the musicians. The most desirable seats are determined by sight lines as well as by aural experience. When there is something to view, vision plays an important role. In contrast, for many twentieth-century technology-based art forms, aural space exists without a visual counterpart, or the experiences of aural and visual spaces diverge, complement, or contradict each other. This is a radical change.

With contemporary music, there is often no visual counterpart to what is being heard. Harrison (2000) described how the German concept of *elektronische Musik*, the electronic replacement of acoustic instruments, and the French concept of *musique concrète*, prerecorded music created before the performance, became *acousmatic music*. This name refers to the Akusmatikoi, pupils of Pythagoras, who were expected to sit in absolute silence listening to their master hidden behind a screen without any visual contact (Emmerson and Smalley, 2001). Radio, telephone, and off-screen sound in film are also acousmatic, because there is no visual anchor (Chion, 1994). Acousmatic music is ideal for free-form auditory imagery, sound metamorphoses, and virtual spaces. It is designed to be unconstrained by, as well as decoupled from, visual cues that would otherwise dominate or undermine the validity of the intended auditory images, including the image of a particular space.

In contrast to acousmatic music, spatial concepts in film sound and multimedia are simultaneously presented through both visual and auditory inputs. The sound source and its spatial environment may or may not match a corresponding visual image, and the relationship between sound and sight is a free-form artistic dimension. Film sound has been extensively analyzed, not by scientists, but by those artists who have spent their professional lives mastering the complexities of the medium. The brilliant composer, filmmaker, and critic Michel Chion (1994) identified three kinds of listening states: casual, semantic, and reduced. He contended that how we listen to film sound is determined, in part, by the artistic context. These three states, while subjectively defined, also apply to other forms of listening besides film. Aural architecture, especially in its manifestation as a virtual space, also involves the same three listening states.

With *casual listening*, the most common of the three, listeners attend to sounds in order to gain awareness about the source or the causal event, and this type of listening merges with other information: vision, memory, imagination, emotions, and social context. Casual listening is prone to the influences of these other factors, and deceptively, it creates the illusion of acoustic properties that may not actually be present. For example, while watching a dog fight, a listener will perceive any barklike sound as a dog's bark even if it does not have strong barklike attributes. In a battle scene, a popping balloon is a cannon. Similarly, the auditory perception of large space, as determined by reverberation, can be undermined by visual cues of a smaller space. Casual listening is very plastic because it fuses concrete experiences, external events, and a listener's expectations. We hear what we expect to hear even if it is not actually represented in sound, and we do not hear what we do not expect.

Semantic listening attends to sounds as a sequence of codes and symbols, which serve to communicate a message. The listener hears the codes, not the sounds that carry the coded information. There are obvious examples. Spoken language is a sequence of phonemes used to represent words, phrases, grammar, semantics, and meaning. Similarly, music is a sequence of notes that represent the melody, beat, harmony, and so on. In semantic listening, perception transcends the sounds themselves even though they carry the information. Semantic listening is flexible, as for example, in a noisy environment when lip reading supplements the auditory information in speech that would otherwise be ambiguous or unintelligible. Although somewhat of an exaggeration, in a religious context, reverberation can be a symbol of God's home or the cosmic universe, not spatial acoustics.

Reduced listening, according to Chion, focuses on the sounds themselves independent of their origin or meaning. "Reduced listening takes the sound—verbal, played on an instrument, noises, or whatever—as itself the object to be observed instead of the vehicle for something else" (Chion, 1994). To focus this intensively on sound takes training and mental discipline. Unlike the other two listening states, reduced listening has no natural perceptual language. With this type of listening, a sound reduces neither to a sonic event, such as a dog bark, nor to a semantic code, such a musical melody, but remains a pure sound for which there is no linguistic or experiential representation. Skilled acoustic scientists with experience in concert hall auditoriums, as well as audio engineers designing spatial synthesizers, are able to focus on selected acoustic attributes while mentally excluding others. They can hear, for example, contradictory statistics in the reverberation tail. There is no natural language for reduced listening.

Composers of acousmatic music expect reduced listening because there are no visual or event anchors to connect sounds to musical instruments. Similarly, aural architects of virtual spaces also expect reduced listening because there is no space associated with the aural experience. The electronic disconnection of sounds from their familiar origins presented listeners with an unfamiliar experience of music and its space, and with the cognitive burden of acquiring an appreciation for virtual spaces. After a half century of exposure to these new art forms, many listeners now have sufficient experience to accept virtual spaces as the norm. But as a by-product of this acceptance, many popular performers who gained fame through their recorded music cannot create a compelling live performance when a real acoustic space is reintroduced. Although Bacon had alluded, albeit unwittingly, to spatial sounds of the twenty-first century, he would have found them strange because he lacked familiarity with these new experiences. In this new art form, the aural architecture of a virtual space is more important than the architecture of a real space.

Artistic Implications of Technical Issues

In their quest for novel spatial acoustics, musicians and composers search for unusual spaces that have artistic potential. And every space has at least the potential to be an auditory inspiration. Underground caves, which arose from geologic processes, are acoustically unique and artistically interesting. Man-made spaces built for other social purposes, such as religious ceremonies or political gatherings, can also provide aural inspirations for those who discover their interesting acoustic properties.

This raises an interesting question. How would an audio engineer, functioning as an artist with unlimited freedom and resources, design, build, and experiment with an artistically interesting virtual space? By "interesting," I mean a space with acoustic attributes, or combinations of acoustic attributes, unlike any that already exist. The answer is historically consistent: select from among acoustic accidents. Modern audio engineers create such "accidents" by tuning, changing, adjusting, adding, and removing components in their signal-processing algorithms, iteratively creating a range of accidents as they acquire intuition and experience about the relationship between the parameters of their algorithms and the sounds they produce. Contrary to the hubris of some experts, few designers of virtual spaces can predict the aural properties of a novel algorithm unless they have already had experience with something similar. We might call this design process the "art of inventing and evaluating controlled accidents," a process in which the discards are never made public.

With each new generation of technology, this process repeats itself. Hence, during the last century, rapid advances in computer and sound technology resulted in a continuous and organic process. As technology moved faster and faster, the rate of discovery increased from a new acoustic space per century (discovering a cave), to a new virtual space per day (modifying an algorithm). As scientific understanding and engineering sophistication accelerated, modern technology also increased the number of acoustic accidents. Yet creating and evaluating them still require special skills.

Just as virtuoso violinists do not construct their own violins, and just as violin builders do not play virtuoso violins, so audio engineers are usually neither composers nor performers. Acoustic architects building concert halls and engineers designing virtual-space simulators function as craftsmen with special skills and expertise. In the early days of electroacoustics and contemporary music, many functioned in both roles. Eventually, however, the skills of composing and engineering diverged. But even when compartmentalized within their respective professions, artist and engineer are still dependent on each other, albeit often with conflicting goals, rewards, visions, and knowledge.

Previous discussions have suggested that aural artists, in combination with audio engineers, have the possibility of creating arbitrary sound fields to match their imagination. That is an ideal based on theory, not practice. Artists, often without significant economic resources, rarely have access to such powerful systems. Audio engineers, typically working within budget constraints for commercial firms, rarely have a mandate to design an artistically innovative system. And limited knowledge often prevents both groups from achieving their personal goals. The quality of the art of virtual space depends on understanding the properties of those tools available. The history of virtual spaces is therefore the story of an evolving relationship between sophisticated audio engineers, creating spatial tools, and impatient artists, incorporating such tools long before they are fully refined.

Because technology advanced so rapidly during the second half of the twentieth century, the art of virtual space never had a stable period of quiet contemplation. Contrast the rapid changes in virtual-space technology with the relative stability of concert hall acoustics: revolution versus evolution. Even at the beginning of the period of accelerating technology, change was opportunity. Although early audio rendering and presentation systems were far from ideal, scientists and musical artists immediately recognized and embraced their potential as artistic tools.

Four decades ago, Stockhausen could have rendered the aural experience of the Jeita caves using the sound system at the Osaka World's Fair. The aural experience of a real cave would then have become a virtual cave. While Stockhausen and Varèse were introducing millions of visitors to new aspects of spatial music, scientists at BBN, the famous acoustic company founded by Leo Beranek and his colleagues, designed a research tool to simulate concert hall acoustics. Within an acoustically dead space, Thomas R. Horrall (1970) constructed a twelve-channel system that simulated the directional response of Boston Symphony Hall based on measurements from a scale model. I had the opportunity of bringing a class to a private demonstration where, with our eyes closed, we could easily imagine being in the symphony hall. I was told that even experts familiar with the hall could identify the seat that had been used for the simulation. From the extensive written records of the twentieth century, it is evident that experimental spaces and spatial simulations are more the norm than the exception.

Over the next few decades, with the rapid advances in acoustic science and computer systems, many experimental virtual-space systems will appear throughout the world, each with its own goal. Yuji Korenaga and Yoichi Ando (1993), for example, created a generic simulator for auralizing particular seats in concert halls. Like selecting a seat based on its sight lines, attendees could select a seat after previewing its acoustics. A far more ambitious project at Helsinki University by Lauri Savioja and colleagues (1999) involved the creation of a complete auralized environment that included a virtual orchestra to render the music, a virtual space containing a virtual listener, and the means of presenting the experience to a real listener. In every respect, this system created an imaginary world, fully disconnected from the real world. Although the system was designed as a scientific tool to study real spaces and the perception of such spaces, at the time, it was the Stradivarius of virtual-space synthesizers.

Yet when Ingolf Bork (2000) evaluated sixteen different spatial simulators, he found that none of them were sufficiently accurate to duplicate a designated target space. In part, these systems were still limited by insufficient computational power to duplicate a detailed model, and in part, they were limited by the burden of modeling the sound dispersion of the large number of small nooks and crannies found in real spaces. These systems were, however, more than adequate to enable artists to create imaginary spaces. Even though artists and scientists may use the same technology in similar ways, their goals and attitudes are fundamentally different from each other. Artists use such systems to discover interesting acoustic accidents, whereas scientists use such systems to replicate a specific set of properties.

Systems for Creating Personalized Sound Fields

Whether intended for an artistic or a scientific purpose, virtual spaces (simulation systems) and performance spaces (concert halls) have one overriding problem that has no ideal solution: the size of the "sweet spot." Because the sound field varies throughout either a real or a virtual space, each listener hears a different sound in each part of a space. Only a limited area of the space approaches some aesthetic ideal. A real concert hall has a relatively large sweet spot, but seats far under the balcony, in the first row, or far off to the side are all outside it. Most virtual spaces have a smaller sweet spot, which is relatively equidistant from all loudspeakers. Although artists and engineers strive to enlarge the area of the sweet spot, either by choice of music or technology, there are intrinsic limits that depend on how the geometric arrangement of surfaces, loudspeakers, or both influences sound fields.

Given the onerous trade-offs in designing a virtual space with a large sweet spot, engineers have one set of choices, artists another. On the one hand, an engineer can simplify the task if the sweet spot is allowed to be small, or an artist may ignore the fact that location determines the listeners' aural experience. On the other hand, both artist and designer may require a large sweet spot in order to accommodate a large audience or to allow listeners to move about. In terms of the area of the sweet spot and aural properties of the virtual space, the range of choices is, indeed, immense. Engineers, building such systems, readily admit that the problems are beyond an engineering criterion, becoming artistic and social issues (Kleiner, Delenbäck, and Svensson, 1993). Who chooses the properties of the space, and who is the aural architect of a musical space? The answer varies from case to case.

Like the acoustics of a particular performance space, the properties of the sound presentation system are difficult to change once constructed, especially when codified into an industry standard. The artistic implications of choosing a 2-channel stereo system or a 350-channel spherical loudspeaker array parallel those of choosing a small chamber or a large cathedral. Music prepared for one presentation system does not readily transfer to another without expensive format conversion. And once the presentation means have been selected or constructed, artists and listeners must then live with, and adapt to, their aural properties. Composers writing music and sound engineers mixing recorded music assume, but do not control, the playback process.

Aural architecture of virtual spaces becomes the design of a spatial experience for each individual listener, not the aural architecture of the composite space. Space is individualized, with listeners having the same individual control over their listening environments that audio engineers have over their spatial synthesizers. Space becomes an individual experience, rather than a common environment with relatively uniform properties.

Just as an acoustic architect must understand building materials, a virtual-space architect needs to understand sound presentation systems—the tools for creating a listener's sound field. One such tool, headphones, illustrates the complexity of creating spatial experiences. In several respects, headphones are the simplest form of sound presentation—two signals, two transducers, inexpensive, and well suited for private listening. But even this simple tool is complex. Beginning with headphones and ending with massive loudspeaker arrays, the following discussion illustrates many of the implications of using sound systems to create spatial experiences.

This discussion also highlights the interdependence of the artists, the engineer, and the scientist; the aural architect is actually a committee, not an individual. Just as artists must have a rudimentary appreciation of the perception of sound as created by presentation systems in order to adapt their art to the properties and constraints of sound presentation, so scientists and engineers must have a comparable appreciation of the needs of artists who use their spatial tools. However technical and abstract the details of sound presentation, the complex choices that must be made have very real implications.

Headphones create a spatial experience for a single individual. But for all their simplicity, when you listen to a stereophonic recording intended for loudspeakers, headphones destroy your perception of external space and location. The source location and spatial acoustics exist entirely *inside* your head, between your ears, not outside in the world. The violins may be located behind your left eye, the clarinets behind your right, and the soloist behind your nose, a phenomenon referred to as "in-head localization."

The explanation for in-head localization is both clear and mysterious. When you listen in a real environment, the sounds at both your ears are almost identical, but with critically important small differences in the amplitude and arrival time of sound waves as they move past your head. These differences are necessary and sufficient to place the sound source at a specific external location. But when you listen with headphones, especially to music prepared for stereophonic loudspeakers, the differences between the sounds at your right ear and those at your left violate the relationship that would have existed between them for a real sound field in a real space.

Recording engineers can take advantage of the two loudspeakers in a stereophonic presentation to create distributed sound sources, for example, violins on the left and trumpets on the right. With loudspeakers, both your ears still hear the violin and the trumpet without violating the perceptual rules for externalizing sound. However, with headphones, only your left ear hears the violins. The human brain simply does not understand the isolation of the right from the left ear. How could a violin be heard only by one ear? If a violin existed in the external world, the right ear would also hear it.

The auditory cortex is an evolutionary adaptation to the physical rules of external sounds, which always appear in both ears. In-head aural localization is therefore an evolutionary *artifact* of listening to sounds that contradict those rules.

It should nevertheless be possible to create headphone sounds that correspond to the principles for externalizing sound sources. To a large extent, the principles that describe the relationship between the sounds at the right ear and those at the left are already known. When such rules are implemented in signal-processing algorithms, a stereophonic format can be converted into headphone presentation, called "binaural audio." Alternatively, rather than converting a sound field captured by traditional microphone placement into a binaural format, audio engineers sometimes use a technique called "dummy-head recording." A life-sized model of a head with miniature microphones inside the dummy's ear canal is placed at a particular seat in a concert hall, as if the dummy's head were that of a real person. To the degree that the model head has the same acoustic properties as a real head, these microphone signals match the sounds at the eardrums of a real person. When we listen to such binaural recordings through headphones, sound sources are externalized, and we have a sense of spatial acoustics.

The binaural technique is, however, incomplete. For both a real person and a dummy head, sound sources located anywhere in the median plane, front, above, behind, or below, always produce the same sound in both ears. There is no difference in amplitude or time arrival that would allow the brain to determine the orientation in the median plane. How do people discriminate among these orientations? The answer is the external ear, the pinna, whose nooks and crannies are actually a spatial filter that changes the frequency content of the sound depending on direction. For example, the pinna filter might contain a small frequency dip at 4,400 Hz for frontal sounds, but a 5,800 Hz peak for sounds arriving from the rear. By recognizing these minor spectral perturbations, the brain decodes the correct orientation in the median place. Yet because each person has unique pinnae, correct auralization requires that the pinnae in the dummy head match those of a particular person.

Henrik Møller and colleagues (1996) showed that, because the differences in pinnae among listeners are quite large, using a generic pinna, one shape for everyone, resulted in significant auditory errors, predominantly confusions in front-back and above-below aural locations (Wenzel et al., 1993). As a consequence of thousands of hours of normal listening, each brain acquires a model of *its* particular ears. Rather than using a dummy head, which is an approximation, Pauli Minnaar and colleagues (2001) placed microphones into the ear canals of a real person. Spatial accuracy improves when a listener listens to a binary recording made with the listener's own ears. Obviously, it is impossible for a mixing engineering in a recording studio to individualize the pinnae for music intended for a general population.

Even if binaural headphones provided an accurate auralization of an acoustic space, listeners do not expect the world to move when they turn their heads (Thurlow, Mangels, and Runge, 1967). A binaural presentation with headphones locks the sound field to your head, not to an external frame of reference. When you hear the sound of a trumpet, for example, directly in front of your nose, it remains there, not fixed in space, even when you turn your head. In contrast, under normal conditions, when you turn your head, the sound at your ears correspondingly changes. Your brain, having invoked the muscles to reposition your head and sensing your head's new orientation, compensates for the expected change in sound, keeping the image locked to the external world. Your brain preserves a static external world by understanding that sound *should* change with your head movement. When, however, you use an externally mounted head-tracking sensor, a device that provides a signal to indicate the direction of your head, changes in the position of your head can be used to produce a corresponding change in the sound at your ears, thus mimicking a sound field fixed in external space (Karamustafaoglu et al., 1999). The world remains static if head rotation compensation is sufficiently rapid (Wenzel, 1997).

To create a compelling experience of acoustic space using headphones, a great many intricately interrelated factors need to be considered (Begault, Wenzel, and Anderson, 2001). When reverberation is included, for example, it lowers elevation errors and decreases the probability of in-head localization. Head tracking reduces front-back confusion by as much as half, but does not contribute to accuracy of aural localization or to aural externalization, and contributes nothing if the listener's head is not rotated. Individualized pinna filters improve performance but only if the sound has significant energy above 4,000 Hz. As a general rule, a given factor becomes highly significant when other factors are also making a significant contribution. A robust system must include all factors to faithfully mimic a real experience.

When individual pinna filters, head tracking, and virtual acoustics are combined into binaural recording, headphones become a wonderful scientific tool. But they are not yet practical as a means for listening to musical space. Nevertheless, without any special processing, headphones are entirely adequate for listening to music *as music* if there is no concern for the loss of aural localization. Even with expected advances in technology, which might make such systems economically feasible, measuring an individual listener's pinnae is procedurally complex and beyond the interest of the average listener. Therefore, headphone auralization systems have limited artistic application for the general public.

If an auralization system creates an accurate sound field around the listener's head, perhaps within a volume of 1 cubic meter (35 cubic feet) rather than just inside the listener's ear canals, many problems in creating a spatial experience with headphones disappear. In this approach, sound is both locked to the external world, avoiding the need for head tracking, and filtered by the listener's pinnae, avoiding the need for individualization. Two or more loudspeakers are carefully placed relative to the listener at predefined locations, and each loudspeaker signal is carefully processed with a special algorithm that creates the correct binaural sound field near each ear, a method called transaural ©. The algorithm is concerned only with a small area of the sound field, near a single listener.

Ideally, a transaural loudspeaker system would have the advantages of headphones but without its disadvantages. Extending the earlier work of Manfred R. Schroeder and Bishnu S. Atal (1963) at Bell Laboratories, who attempted to duplicate concert hall acoustics, Duane Cooper and Jerry Bauck (1989) showed that such systems were practical without excessively burdensome constraints, such as using an anechoic space for listening. When properly designed, the acoustics of the listening space play only a minor role with a transaural presentation. After deriving the complex mathematical filters necessary to create a transaural sound field, and after building a basic system with two loudspeakers, Bauck and Cooper (1996) commented that listeners experienced full immersion in the sound field with nary an audible hint of loudspeakers. In this uncanny transaural rendering of spatial acoustics and sound sources, aural reality matches acoustic theory, at least in the controlled settings of the laboratory.

The remaining issue is still the size of the sweet spot, the area where the effect is compelling and realistic. Transaural presentations assume that both the location and the orientation of the listener are fixed, and that the left and right sound signals are correct only in a small volume of space around the listener's ears. Cooper and Bauck (1989) informally observed that a ±30-degree head rotation produced a benign, albeit noticeable, change in the audio perspective. There was more tolerance for forward-backward motion. As listeners move outside of the sweet spot created by a transaural presentation, the sonic image gradually collapses, although not catastrophically, as it would with ordinary stereo loudspeakers.

More recently, researchers have studied the techniques to enlarge the size of the sweet spot. Using a pair of loudspeakers placed at a 60-degree angle about 70 centimeters (30 inches) from the listener, William G. Gardner (1998) measured the shape of the sweet spot, which he defined by the degree of cancellation of the unwanted signal, left loudspeaker to right ear and vice versa. At 10 decibels (dB), where the sonic effect was still pronounced, the sweet spot was about 25 centimeters (10 inches) long but only about 5 centimeters (2 inches) wide, thus confirming earlier observations that such a narrow width would be problematic for normal listeners.

Because the shape and area of the sweet spot depend on the loudspeaker configuration, scientific analysis provides insight into the means for improved robustness. In one such study, José Javier López, Felipe Orduña, and Alberto Gonzáles, (2000) demonstrated that closely spaced loudspeakers, corresponding to a narrow angle, enhanced the area of the high-frequency sweet spot, and conversely, widely spaced loudspeakers enhanced the area of the low-frequency sweet spot. The explanation is straightforward. Wide angles at low frequencies took advantage of the natural acoustic shadowing of the head, while narrow angles at high frequencies avoided the corrupting influence of that acoustic shadowing. The obvious solution to the conflicting requirements is to split the high and low frequencies such that the former drive closely spaced tweeters and the latter drive widely spaced woofers (Bauck, 2001). When this is done, the area of the sweet spot becomes significantly larger, but only at increased cost and complexity.

If such system were to gain widespread applications, other improvements could be added. Gardner (1998) illustrated several examples where the sweet spot could be automatically repositioned by 15 centimeters (6 inches) in any direction. Similarly, additional loudspeakers provide another means of controlling the size and location of the sweet spot. Bauck and Cooper (1996) derived the generalized mathematics for an arbitrary number of loudspeakers and listener locations. Each additional degree of freedom, whether through head tracking or auxiliary loudspeakers, allows for an improvement in the spatial accuracy in the selected acoustic area.

Transaural sound presentations are particularly appropriate when the listener is located in a fixed seat such as while driving an automobile, piloting an airplane, or using a personal computer. In these cases, the seat is fixed by the nature of the control surface, be it a steering wheel or a monitor. As a means of creating a virtual yet general auditory display, this technology has appropriate applications beyond the auditory arts. It has been used as a human interface for presenting critical information to pilots (Begault and Pittman, 1996), and for creating an intimate sense of interacting participants in teleconferencing (Rimell, 1999). Transaural systems are inappropriate for groups of listeners who must occupy a large area, as well as for listeners who need to move about. The transaural spatial experience changes dramatically with position because the sweet spot is *so* sweet that the contrast to other locations is disconcertingly stark.

To conclude: presenting sound with stereo headphones, binaural headphones, and transaural loudspeakers illustrates the high degree to which the spatial experience depends on the means for creating the sound field. And in all cases, the area and quality of the sweet spot are the overriding trade-offs. Implicit in these systems, with their severely circumscribed sweet spots, is the cultural assumption that listening should be private and asocial. Even without any technology to mitigate in-head aural localization, however, headphone listening is socially important simply because it allows individual listeners to maintain the privacy of their acoustic arenas while in a public setting. The experience of space is of little consequence when the issue of who should control the acoustic arena dominates. Indeed, enclosed performance spaces may have arisen from the need to create an acoustic boundary between a smaller private and the larger public acoustic arena.

The Automobile as a Special Environment

Behind the discussions about the art and technology of sound presentation systems, however intellectually interesting, we find a more powerful consideration: the life style of individuals in the culture. Listening to music with headphones while exercising at the local health club is very different from listening to music in a home theater room, or from listening to music in an automobile while commuting to work or taking a trip. Lifestyle always dominates the criteria for creating aural space. When aural architects spend too much time in the laboratory, fascinated with their creations, they may forget the larger issue. Listeners choose a presentation system primarily based on its compatibility with their respective lifestyles, and only secondarily for its aural properties. As lifestyles evolve, and as virtual spaces evolve, aural architects adapt to a combination of both.

During the last decade, the automobile and suburban vehicles have changed from a utilitarian means for moving people, to actual living spaces—rooms on wheels. As a driver, you are a captive audience in your seat, which defines the sweet spot. As a commuter, you may spend more time in your automobile than in your living room, especially if you are commuting in a dense metropolitan area. When you buy an elegant vehicle, you do not notice the incremental cost for an expensive sound system because it is built into the sale at the time of purchase. Many vehicles are sufficiently large that they provide the designer with the possibility of creating a controlled sound field. A presentation system can be designed and optimized while the vehicle is being engineered, and the designer can take responsibility for system integrity. Sound-absorbing treatment of the interior is already present to reduce the noise produced by tires, engine, and traffic. And, unlike living rooms, there are few, if any, possibilities for adding furniture to automobiles.

The automobile manufacturer, by controlling the properties of the interior, can treat aural design as a complete system—positioning the seats, orienting the windows, selecting the presentation format, mounting the loudspeakers, designing the acoustics, and adding signal processing. Being small and noisy, compared with a living room, however, the automobile is still a acoustically hostile and constrained environment. For designers of virtual-space systems, the automobile is the last place they would choose to replicate the experience of a concert hall. Yet because the economic rewards are potentially large, the effort continues to justify investing in such systems.

From the myriad problems that need to be solved, two stand out. First, being a very small volume, on the order of 3 cubic meters (110 cubic feet), the automobile acoustic space produces strong resonances at lower frequencies that are perceptible and difficult to suppress. Roger Shively (1998) describes major transversal resonance modes as having amplitude peaks of 12 decibels in the region around 150 Hz. Between 300 Hz and 1,000 Hz, there are numerous weaker resonances that correspond to other modes. By emphasizing some frequencies over others, these resonances color the sound. Second, rather than being equidistant from both loudspeakers, each passenger is generally much closer to one than to the other. Creating either a single large sweet spot for all seats or many small individual sweet spots, one for each seat, requires complex signal processing. Ignoring this issue results in listeners hearing the sound from just one speaker.

Attitudes toward coloration range from total indifference (social) to sophisticated interventions (technical). Although only partially effective, equalization filters can reduce the extremes of frequency response anomalies. Shively (2000) recommended that each loudspeaker have its own multiband equalizer. Alternatively, by manipulating the geometry of the environment, such as the angle of the windshield, a designer can reduce the severity of resonances, thus improving timbre and spatial perception (Shively and House, 1996). Research is likely to continue for the immediate future, but there are currently no ideal solutions available, only incremental improvements. Most listeners, however, become readily habituated to coloration, which has existed since the introduction of radio into the automobile.

Because of the proximity of the surfaces and loudspeakers to the listeners, the clarity of the direct sound is also corrupted by secondary direct sound from multiple loudspeakers and by sonic reflections from nearby acoustic boundaries, both of which arrive within a few milliseconds after the first direct sound. In contrast to a typical living room, where there are no nearby acoustic boundaries, it is easier to create a sweet spot in the center of the space. Theoretically, competing reflections could be cancelled by using the kind of signal-processing algorithms found in transversal binaural systems.

Despite the lack of a comprehensive technical solution, many expensive automobiles already include surround-sound systems using multiple loudspeakers and advanced signal processing. These systems generally have at least seven channels, for example, three in the front, two on the sides, and two in the rear (Nind, 2001). For larger vehicles with an extra row of seats, Neal House (2001) proposes another pair of side loudspeakers. When technology produces inexpensive distributed loudspeaker arrays, no doubt, they will become widely accepted for one simple reason. As the number of sources increases while injecting constant sound power into the space, the loudness of each source becomes correspondingly lower. Each loudspeaker contributes a smaller percentage of the total. The sound source closest to a listener no longer dominates because its loudness has been greatly reduced; regardless of proximity to one source, the listener hears multiple sources, which produces an enveloping sound field. This approach has already been demonstrated in larger spaces, and a variant with signal processing will eventually appear in automobiles.

Audio engineering research, closely linked to marketing, is chiefly concerned with determining the preferences of real listeners for systems that can be built now and with designing systems that are pleasing to people who purchase automobiles. Changing listener expectations is a perfectly viable solution to an otherwise intractable technical problem. Elderly consumers who formed their expectations from a lifetime of concert hall experiences have been replaced by younger consumers with expectations from the contradictory virtual spaces of multimedia home theater systems. Each generation has its own spatial sensibilities, and the next generation will have grown up with the aural architecture of automobile spaces and portable headphones, yet another set of expectations.

Just as early humans adapted their religious ceremonies to caves, modern humans are adapting their musical entertainment to the automobiles. In both cases, the space existed before its acoustic properties were recognized. This implies that social adaptation to the automobile is just another example of a response to an acoustic accident. The automobile environment is an acoustic "accident" that arose from our desire to transport ourselves in private spaces.

Loudspeaker Arrays for Large Virtual Spaces

With enough loudspeakers, sufficient signal-processing power, and the freedom to customize a space, could you reproduce any musical space in your living room? The answer is yes. During the last few decades, scientists and mathematicians have proved that an audio reproduction system *could* replicate the sound in *any* volume of space within any other space. Such a system, using today's technology, is not science fiction. For example, you could duplicate the sound within a 100-cubic-meter (3,550-cubic-foot) space of your favorite concert hall, perhaps rows L through M and seats 122 to 145, in your living room. There would be a large sweet spot; every seat in your living room would match the listening experience of the corresponding seat in the concert hall.

By understanding the properties of an ideal system, we acquire a reference for other systems, especially those with primitive compromises based on commercial pragmatics, cultural values, hidden agendas, artistic variability, and elusive definition of taste. However mathematically complex the details, the implications of choosing one audio reproduction system over another are important for artists using such systems. The acoustic science of virtual musical spaces has proven to be as arcane—and as relevant—as that of traditional musical spaces.

Either from a lack of scientific interest or from insufficient education, most artists do not understand the properties of audio reproduction systems used for creating virtual spaces. If available systems outside of the laboratory were closer to the ideal, there would be no need for artists to understand their properties. However, existing systems are far from ideal. For this reason, let us first consider some of the key scientific ideas in an ideal system, and then discuss the consequences of using primitive compromises.

Water waves on the surface of a pond are analogous to sound waves in the acoustic space of a concert hall. If we drop a pebble into the water, a wave radiates from the point of impact, like a sound wave radiating from a source. If we then place a floating cork on the water surface, it bobs up and down as a wave traverses it. The cork's up-down motion is like an eardrum moving in response to the *pressure* of the sound wave. The critical idea is the distinction between sound pressure at a single point in space, the sonic parameter that we hear, and sound waves, the movement of sound energy throughout the space. We do not hear sound waves; we hear sound pressure moving our eardrums.

Replicating a sonic experience throughout a space requires that the sound pressure at every point in the reproduction space match the sound pressure at each corresponding point in the reference space. Mathematically, the only way to achieve such a point-by-point match in pressure is to duplicate all the waves traversing the space. These myriad sound waves create a unique sound pressure at every point. In a binaural sound presentation system, as discussed earlier, only the pressure at the right and left ears is replicated—at two points in space—which is far simpler than replicating all of the waves that exist throughout the space. A binaural recording can ignore the spatial distribution of sound waves. If, however, we wish to reproduce the spatial experience over a wide area, to create a large sweet spot, we must reconstruct all the sound waves.

Extending our water wave analogy, let us place a few dozen corks in a circle defining the acoustic area to be reproduced and attach an electronic sensor that detects the up-down motion of each cork—the equivalent of placing a circular configuration of microphones in a concert hall. The signals from these corks have enough information to fully determine all the waves within the circle. In another pond, the reproduction space, we place motorized paddles that are able to produce up-down motion of the water, like loudspeakers vibrating air. By connecting each cork to its corresponding paddle, like connecting each loudspeaker to its corresponding microphone, we can mathematically prove that the waves within the circle of the *reproduction* pond will match those within the circle of the *concert hall* pond. Because the waves match at all points, and because waves give rise to cork movement or sound pressure, listeners will hear the same sound in either of the two spaces.

As a feasibility study, Marvin Camras (1968) first demonstrated the wave-field approach with a series of microphones along the periphery of a rectangular room and a matching series of loudspeakers outdoors: "When one was within the boundaries of the listening space, he had the sensation of being inside the room, instead of outdoors. After crossing the borderline of the rectangle, one felt that he had passed through a door leading to the outside." The experience of the room was reproduced outdoors and was described as being dramatic. This was a clear early example of a portable space.

Where a single microphone connected to a single loudspeaker duplicates the sound pressure at a single point, a wall of microphones and a corresponding wall of loud-speakers reproduce a two-dimensional sound wave traversing the surface, literally a wall of sound. Listening near the virtual wall of microphones is identical to listening near the wall of loudspeakers. This technique is called "holophony"—the aural parallel of holography—because in both cases the waves, whether light or sound, are being replicated over a two-dimensional surface. Conceptually, a wall of microphones is a two-dimensional *spatial* sound-wave sensor, a metamicrophone, and a wall of loud-speakers is a two-dimensional *spatial* sound-wave generator, a metaloudspeaker. For example, the left and right loudspeakers of a traditional stereo pair could be replaced by single metaloudspeaker array over the entire front surface. When six holophonic acoustic surfaces are combined, the fully enclosed metaspace becomes a perfect virtual space.

Ignoring the pragmatic issues, numerous scientists have proposed acoustic holophony as a means for creating a new virtual space rather than recording from an existing space—an electroacoustic virtual concert hall. Augustine Berkhout (1988) suggested that the aural architecture of space could use acoustic holophony rather than bricks and mortar. And Kazuho Ono, Setsu Komiyama, and Katsumi Nakabayashi (1996) suggested that a signal-processing system with an array of 300 loudspeaker channels would be sufficient to create such a virtual space. In this approach, an algorithm produces the same acoustic processing that would have taken place in a concert hall. Aural architecture thus becomes the design of algorithms as surrogates for the geometries of walls and surfaces. There is no aural difference between these two forms of sound presentation.

Yet architects, like engineers, live in the world of compromises. Using holophony or wave-field synthesis to construct sound fields involves numerous trade-offs in the density of sensors, the upper frequency limit, and spatial accuracy. How many microphones and loudspeakers are actually needed? Using the mathematical transform of spatial spherical harmonics as an analytic tool, Darren B. Ward and Thushara D. Abhayapala (2001) showed that high frequencies require a denser array of microphones. To put the issue into perspective, the wavelength of a 15,000 Hz sound is about 2 centimeters (1 inch), and at two samples per wavelength, there would need to be 100 microphones per meter, or 1,000 microphones in a circle with a perimeter of 10 meters (35 feet). As the volume of the space increases, the enclosing surface area also increases such that the number of sensors grows very rapidly, indeed. An ideal system, one that is accurate over the entire region, is hardly practical, even if theoretically possible. When the density of speakers does not correspond to the ideal, the system introduces spatial errors. Although such errors are measurable, they may be inconsequential in an artistic environment; indeed, they may be imperceptible to even the most sensitive listeners.

Motivated by the need to find practical alternatives to this cumbersome method for recording and then reproducing spatial acoustics, scientists have invented alternative approaches to sensing the spatially distributed sound waves. One such technique, called "ambisonics," is mathematically equivalent to holophony but avoids the need for an array of microphones spread over a wide area. Consider a visual analogy. We can predict a visual experience anywhere within a space if we know the visual properties of every surface element. Alternatively, if we captured every detail of light arriving at a point, perhaps with a circular stereoscopic camera, we could also predict the visual experience any place in the space. In one case, we know the details of the light entering a region from the periphery, and in the other case, we know the details of the light arriving at a point from every direction. The information is the same in both cases.

Returning to our water wave analogy, we can see that the circle of corks captures a complete description of the waves entering the circle. The same information, albeit in a different format, is also embedded in the tiny details of a wave at a single point at the center of the circle. Such details include the wave's pressure, direction, curvature, change in curvature, and so on. If, instead of thinking of a cork as a single object moving up and down, we think of the pond's surface as gelatin with detailed surface undulations, we could observe in these undulations the microscopic differences in movement resulting from the wave's *shape* and use them to reconstruct the spatial waves. Applying the ambisonic technique, we could replace the circle of corks with a single metacork at the center of the circle, one that senses the wave shape.

In the world of sound waves, the metacork would be a special multichannel microphone whose outputs contain information about the detailed shape of the sound wave. Ordinary microphones sense only sound pressure or sound speed. The complexity of an ambisonic microphone, as determined by the number of internal channels, needs to increase rapidly when attempting to capture fine details. Michael A. Gerzon (1973) showed that an ambisonic microphone with a large number of channels is equivalent to the circle of microphones. But a single microphone, however complex and expensive, is more practical for recording than a distributed array. Rozenn Nicol and Marc Emerit (1998) showed that the ambisonic technique is a special case of holophony, and David Malham (1999) treated wave-field synthesis, holophony, and ambisonics as three equivalent systems that shared mathematical consistency, not just ad hoc artistic equivalence. Science has thus established the intellectual foundation for implementing an ideal virtual-space system.

Although mathematically pure, such systems do not necessarily match the needs of the musical arts. Uniformly spaced loudspeakers allocate spatial accuracy uniformly around a circle. With a small number of loudspeakers, spatial errors, which are relatively large, are the same in the front as they are in the back. Yet traditional music is spatially asymmetrical—spatial details are important in front where musicians sit on stage, and the remaining space contains diffused and enveloping reverberation, which is highly tolerant of spatial errors. Just as traditional composers wrote music with spatial assumptions about the concert halls, modern popular musicians and their audio engineers make assumptions about their listeners' acoustic space. Those assumptions do not include a large number of speakers in a uniform circular array.

However much advocates of ambisonic recording promote its elegant properties and practical advantages, it has never acquired mainstream acceptance, although some music has already been recorded ambisonically. At least for now, ambisonic and other holophonic techniques serve only to demonstrate that our ability to achieve a technically elegant solution is less relevant than other artistic, historical, and cultural forces. Technology creates new choices, but society selects from among the many choices, often ignoring questions of spatial accuracy and fidelity. Home listening environments rarely tolerate a prescribed geometric pattern for a large number of loudspeakers; laboratory elegance does not translate to the home. The properties of an audio reproduction system must therefore include a high tolerance for gross errors in the listeners' configuration, as well as for the degrading acoustics of the listening space itself. A sound system installed by a consumer having no formal training and no inclination to study a 300-page manual needs to be highly tolerant of gross deviations from its optimal properties. In this sense, the aural architecture of virtual spaces, notwithstanding advanced technology in scientific laboratories, is still an adaptive reaction to the properties of residential living spaces and consumer sound systems.

Electroacoustic Support for Live Auditoriums

Among those who have strong opinions about spatial acoustics of musical spaces, discussions often degenerate into passionate debates about the importance of *natural* acoustics, with *artificial* acoustics being denigrated by implication. In such debates, *artificial* becomes a synonymous with "electronically processed sound," and *natural*, a synonymous with "sound waves interacting with surfaces in the musical space." For classical music aficionados attending live performances, "artificial acoustics" means that they are hearing an electroacoustic intervention. But even in the "natural acoustics" of a concert hall without electronics, listeners hear the acoustic interventions of sound-dispersing statues, sound-reflecting ceiling panels, sound-diffusing walls, and sound-absorbing panels. There is only one relevant question. Does any particular intervention benefit the aural experience of a musical space? Debates about natural versus artificial are thus spurious and misleading.

Man-made musical spaces are the result of human intervention, and electroacoustics is simply an aural technology of the twentieth century. At the inception of electroacoustics in the late nineteenth century, microphones and loudspeakers were used to improve intelligibility in public spaces. And with the addition of a delay line, it was possible to emulate the sonic reflection from a hanging ceiling or a performance stage wall. By the end of the twentieth century, complex signal processing had already appeared in theaters, public auditoriums, opera houses, and concert halls, as well as in the sound systems found in homes and automobiles. In each case, electroacoustic intervention altered our spatial experience.

There are several reasons for using electroacoustics in performance spaces. First, some auditoriums have acoustic defects that cannot be corrected by a simple physical modification of the space. Second, modern auditoriums are usually intended for multiple uses with conflicting requirements for their acoustic properties. Electroacoustics allows for instantaneous and flexible manipulation of aural architecture. Finally, electroacoustics can expand the acoustic arenas in such remote locations as under a long balcony. Signal processing and the physics of sound each have unique constraints and freedoms. Although, when even slightly misused or misadjusted, an electroacoustic intervention degrades sounds into an unpleasant aural experience, when it is implemented properly, listeners are no more aware of an electroacoustic intervention than they are of an acoustic one.

The appropriate electroacoustic intervention in a performance space depends on seven factors: whether the space is public or private, large or small, expensive or inexpensive, flexible or inflexible; whether the performance is music or oratory, live or prerecorded; and whether listeners are one, few, or many. By habit and convention, we associate a set of factors with a specific space: concert hall, opera house, theater, or home theater room. For example, a concert hall auditorium is an expensive, large, public space, usually with microphones, whereas a home theater room is a (relatively) inexpensive, small, private space, usually without microphones. For aural architects, two factors dominate the acoustic design: the size of the audience in the sweet spot, and whether a live performance produces feedback when microphones and loudspeakers share the same space.

Using a common conceptual framework, we can view any electroacoustic system as a means for sensing sound in one part of a space and injecting it into another. Every system has one or more microphones somewhere, and one or more loudspeakers somewhere else. Microphones and loudspeakers may be far apart in time and space, as in prerecorded music, or they may be in the same space at the same time, as in amplifying an actor's weak voice. In each case, electroacoustics enlarges the acoustic arena, and modifies the perceived size of the aural space.

When placed at remote locations, loudspeakers and microphones are acoustically isolated from each other with only a unidirectional electronic connection. When sharing a space, however, they form a feedback system; the sound leaving the loudspeakers enters the microphones where it is amplified and reemerges from the loudspeakers. With enough amplification, the system oscillates, producing a loud and unpleasant squealing, growling, or ringing sounds. With somewhat less amplification, the music is still spectrally colored, like the sound of a barrel. Every form of electroacoustic intervention must control feedback. Do the benefits of an electroacoustic intervention manifest themselves without also producing oscillation, coloration, or ringing?

Sound reinforcement (amplification), the oldest form of electroacoustic intervention, supplements insufficient loudness, allowing a listener seated under an overhanging balcony in the rear of a concert hall auditorium to hear a singer whose delicate voice would otherwise fail to fill the auditorium's large acoustic space. The need for vocal amplification has been known since the early Greeks, who, some contend, inserted a miniature megaphone into the mouth opening of their theater masks (Lewcock and Rijn, 2001). Two millennia later, Lee De Forest (1921) received his patent on the triode vacuum tube for amplifying electric signals, and George D. Edwards (1926) and others invented sound reinforcement systems for auditoriums. Electroacoustics as sound reinforcement became an architectural tool, providing an easy way to enlarge an acoustic arena.

The speed of sound itself, however, poses a significant problem for amplification. The sound produced by a loudspeaker on the back wall of a concert hall, for example, would normally arrive before the direct sound from the performance stage because the loudspeaker is closer to listeners seated in the rear of the auditorium. Since the earliest sound determines the aurally perceived location, listeners would aurally perceive the source as being behind them. Delaying the loudspeaker sound, however, would reverse the order of arrival and switch the perceived location back to its actual location. Helmut Haas (1951) demonstrated that if a strong but delayed echo arrives within about 20 milliseconds after the direct sound, listeners experience an increase in the sound's loudness but perceive no change in its location. This illusion is useful and compelling, and is equivalent to a reflection from a virtual wall.

Although the importance of high-quality audio delay was readily apparent to earlier acousticians, they lacked the technical means to implement it. The search for an audio delay technology began immediately after the birth of electronics (McCutchen, 1927), and by 1928, scientists at the Bell Telephone Laboratories had created a damped spring as a means of producing signal delays (Wegel, 1932). For the next half century, improving the quality of audio delay was a major research goal. Indeed, audio delay was *the* primary problem of sound reinforcement—one not truly solved until the digital audio revolution. The first commercial application of digitized audio technology was the prosaic delay line (Blesser and Lee, 1971), which replaced the garden hose as a means for creating long audio delays in sound reinforcement.

Could electroacoustic intervention also supplement inadequate reverberation in spaces with too much sound absorption? By mounting a multiplicity of loudspeakers with audio delays over the performance stage, R. Vermeulen (1958) attempted to increase the reverberation time of the La Scala Opera House in Milan. At the end of his experiment, he concluded that his system provided "evidence that electro-acoustics has come of age when well-known musicians are willing to accept the assistance of loudspeakers, not to produce greater loudness but to improve the quality of their live concerts" (Vermeulen, 1958). Yet Vermeulen's system was basically a failure, lasting only three years: though sufficient to increase the reverberation time, the injected energy also produced unwelcome feedback and coloration.

After many years, scientists and engineers realized that the reverberation problem should be considered from an energy perspective. Inadequate reverberation time means that the sound-absorbing surfaces are removing energy too rapidly. As a discrete sound absorber, every listener in the audience removes a fixed *percentage* of the incident sound energy. A reverberation enhancement system, one that increases the reverberation time, needs to replace that lost energy. Halving the rate of energy loss doubles the reverberation time. If a distributed array of microphones senses the average energy in the space, and a distributed array of loudspeakers then uniformly reinjects energy, the system cancels the effect of excessive absorption. To avoid emphasizing one frequency and one part of the space over others, the signal-processing algorithm and the location of microphones and loudspeakers must average and randomize that energy in time and space.

Such systems had already been developed as artificial reverberators for use in recording studios. The mathematics for reverberators based on energy has been known for some time (Gerzon, 1976), and practical algorithms have been used to create studio-quality reverberators (Jot and Chaigne, 1991; Jot, 1992). Such devices can solve the problem of inadequate reverberation in real-life spaces (Poletti, 1996, 1999). Signals from a large number of microphone channels are simultaneously reverberated to feed an equal number of loudspeaker channels. At every frequency, the total energy injected by the loudspeakers is proportional to the total energy at the microphone.

Since the statistics of time-space averaging are highly complex, let us consider the following simplified analogy. Let us liken the air in a concert hall auditorium to the water in a pan, the frequencies of sound energy to dyes, with each frequency having its own color, and blotter paper at the bottom of the pan to sound-absorbing material. Injected at the front of the pan, a teaspoon of red dye, representing, say, a trumpet note, rapidly spreads throughout the water, mimicking enveloping reverberation. The blotter at the bottom of the pan, mimicking sound absorption, gradually absorbs the red dye at a rate proportional to its concentration. Eventually, the water in the pan is again clear (the reverberation has died away). The duration of the clearing process is like the reverberation time. To extend it (by electroacoustic enhancement), we add optical sensors (an array of microphones) that together measure the total quantity of red dye present in the water (reverberant energy), and periodically, dozens of tiny droppers (an array of loudspeakers) add tiny amounts of additional red dye throughout the pan. Assume a 1-second reverberation time. During a given interval, if the blotter removes 1 percent of the dye and the droppers add 0.2 percent of the dye, the rate of clearing decreases from 1 to 0.8 percent, and the reverberation time is now 1.25 seconds, compared with the original unassisted time of 1 second.

With only one dropper, the injected red dye would be concentrated at one point in the space. But with a large number of droppers each distributing tiny amounts of dye over the full area of the pan, a uniform cloud of red dye is created, rather than a concentration at one location. The principle is clear: multiple loudspeakers uniformly distributed throughout an auditorium uniformly bathe the space with added reverberance. The performance space becomes one large sweet spot.

Unbeknownst to the audience, but appreciated by musicians, such systems have already been installed in many performance spaces. For example, in order to increase the reverberation time from 1.5 seconds to over 2 seconds, a version of Poletti's design was installed in the Prague Congress Center, a multipurpose auditorium used for discussions as well as symphonic concerts (Noack, 2002). The Church of the Living Word in North Hills, California, uses a similar system in order to increase a paltry 0.9-second reverberation time to something adequate for concert music (Holbrook, 2002). Other installations include the Roda Theater in Berkeley, California, the Hayden Planetarium in New York, the Vernon and District Performing Arts Centre in British Columbia, and many others.

Although Poletti's design is based on mathematics and a firm scientific theory, empirical engineering approaches have also solved the problem. David Griesinger (1991) modified readily available standard studio reverberators, designed to the highest standards, to serve as the foundation for his enhancement system. Because his system contains optimized time-varying characteristics to randomize periodic patterns, thereby reducing coloration, Griesinger's approach requires a smaller number of channels to achieve a comparable effect. A limited number of microphones, on the order of five to ten, drives a group of independent reverberators, perhaps as many as thirty-two, whose outputs are then combined in a matrix to provide a large number of loudspeaker signals. The largest installation using Griesinger's system had over 300 loudspeakers, which was more than adequate to disperse the injected energy throughout a given space. Versions of this system have been installed in the Elgin Theater in Toronto, the Deutsches Staatsoper in Berlin, the Hummingbird Center in Toronto, the Circle Theatre in Indianapolis, the Morbisch Festspiele in Austria, and many others (Griesinger, 2000).

Designers and installers of these electroacoustic enhancement systems function as true aural architects. They usually do their work long after a space has been built, and mostly in the form of remedies for discovered acoustic defects. Ideally, the original aural design process should simultaneously include both the geometry of physical surfaces and algorithms for electronic support. Without electroacoustic intervention, a fixed space cannot be optimized for lectures in the morning, Wagnerian opera in the

afternoon, and Gregorian chants in the evening: the requirements for these three performances are contradictory. Electroacoustics thus provides a new dimension to aural
architecture: instantaneous spatial changes by adjusting acoustic parameters, either
manually or through presets. It can allow an aural architect to create temporary musical spaces to match the enduring legacy of our diverse musical heritage. It can allow
a conductor to use aural space as a musical element rather than adapting music and
musicians to an immutable acoustic structure. Although these prospects may seem to
smack of science-fiction fantasy, science and technology are currently available to
achieve them provided only society wishes to invest the necessary resources.

Art and Engineering Converge in Enveloping Sound

At the beginning of the twenty-first century, those adventuresome audio enthusiasts
who had already installed surround-sound systems in their homes were enjoying music
presented with five discrete channels and one low-frequency effects channel, the so-
called 5.1 surround system. There were, in fact, many alternative formats and configurations available, designed in various development laboratories by some of the brightest minds in the audio engineering world and representing a range of artistic and
scientific choices. Why, then, did this particular format of audio reproduction become
dominant, and what determines which format will gain acceptance in the future? Part
of the answer is found by examining the behavior and motivation of those innovators
with the means to support risky experiments in mass markets.

In the late 1970s, a small but rapidly growing audio industry proposed the quadraphonic format for surround sound. It was an ill-conceived concept using a matrix of
four channels, which proved to be a disaster, artistically, technically, and economically. It would be the audio industry's last large-scale autonomous innovation in aural
architecture without borrowing from the technical infrastructure of larger industries.
In the 1980s, digitized audio grew as a derivative extension of computer hardware and
software. And in the 1990s, the audio industry absorbed and modified cinematic concepts of aural space rather than inventing its own approach. Like the proverbial elephant, large industries go where they want, and the smaller audio industry follows as
a borrower of digital technology. Sound transducers are the notable exception. For the
aural architecture of consumer sound systems, audio engineers became the chefs who
blended ingredients from the cinema and computer industries to create a surround-
sound system for consumers.

Cinema Pioneers Advance Spatial Sound

Even with its historical preoccupation with the visual modality, the cinematic industry
also has had its share of aural architects. Yet as we review how this industry unwittingly contributed to aural architecture, it will become clear just how much the audio

component of cinematic space was an artifact of other artistic, economic, and technical choices. In some cases, decision makers acted without recognizing the implications of their choices. The result can therefore be viewed as a social accident, which then became part of our inherited legacy of aural space.

With its large economic resources and professional staffs, and with the avid support of the public, the cinematic industry has dominated experiments with surround sound. That industry had both the means and the organization to engage in massive experiments, whereas audio engineers and composers, with a few conspicuous exceptions, had neither. Moreover, as a concentrated vertical organization, the cinematic industry controlled the entire delivery chain from artistic content to production facilities, delivery mechanisms, and presentation venues.

Surround sound acquired sponsors and standards when the cinema industry finally began a massive refurbishing of theater audio, motivated in part by the success of George Lucas's *Star Wars* with its swooping rear-channel special effects. Lucas sponsored Tomlinson Holman to develop a new audio presentation system, which became a licensed standard called THX ("TH" for his initials, "X" for experiment). In order to use the THX symbol, theaters now have to demonstrate that their installation, including their acoustic treatment, meets strict requirements. A small group of influential individuals, functioning as passionate artists, sophisticated engineers, and pragmatic executives, made the decision to sponsor and then launch the initial THX surround-sound standard.

After the audio industry created the compact disc (CD) for high-quality stereo audio, and after the cinema industry enjoyed the success of videotape in the home, the next step was the digital versatile disc (DVD), an inexpensive means for distributing quality audio and video. The DVD format includes stereo audio and four additional channels to implement 5.1 surround sound. Thus home theater was born as viewers attempted to replicate their cinema experience at home by purchasing equipment configured to match this standard. It specified a layout for the five loudspeakers that approximated the elongated rectangle of the cinema theater. Those wishing to take full advantage of this new cinema experience had to acquire a surround-sound system for their homes.

Although a reproduction system with five channels is just barely sufficient to create the spatial impression of a diffuse sound field (Hiyama, Komiyana, and Hamasaki, 2002), it was a dramatic advance over stereo, which had endured for three decades without competition. But three years after the introduction of 5.1 surround sound as part of DVD format, the audio-only version is still not in wide distribution; surround sound has spread throughout the culture only as an adjunct to video. According to the Consumer Electronics Association, the DVD has been the fastest growing consumer product ever, faster than the CD; its penetration was expected to exceed 175 million households by 2004 (Owsinski, 2001). In just four years after its initial release, a half billion DVD discs have already been shipped. The DVD's success has brought surround sound into mainstream consciousness, for the most part as an accidental consequence of video, which traditionally dominates audio whenever they appear together.

There may be social explanations for the dominance of surround home theater over surround sound. Whereas stereo audio is portable, allowing the listener to be engaged in other activities while listening, surround sound, like video, is not. Any surround system requires listeners to sit within a small stationary region enclosed by the array of loudspeakers. Configuring a room for surround sound requires as much effort as arranging a space for home theater. In addition, a generation that grew up with television, movies, computers, and video games has an elevated need for "eye candy," something to occupy vision while listening. Thus musical space has become only one of the many spaces in a total "audio-video" environment.[2]

To appreciate the degree to which the audio engineering industry overlooked the aural architecture of virtual space, at least in comparison to the cinema industry, we need only review the history of audio systems. The first systems for broadcasting, recording, and cinema all used a single channel connected to a single loudspeaker, which amounted to listening to an aural environment through a window. Even at the beginning, the inadequacies of a monophonic presentation were apparent. Already in 1881, soon after the telephone became available, an experimental system using two channels was used to transmit a performance from the Grand Opera in Paris to a gallery at the Palais de l'Industrie (Hertz, 1981). In 1911, a patent was filed for two-channel recording (Offenhauser, 1958). A practical application of stereo would wait for Alan Dower Blumlein (1933) to invent a means of recording the second channel on the grooves of records using both left-right and up-down modulation. Shortly thereafter, the cinema industry added a second optical track to provide its version of stereo, and somewhat later broadcasting invented side-channel modulation of an FM carrier for stereo.

The addition of a second channel allowed listeners to aurally localize the sound source along the lateral dimension. Horizontal localization allowed concert listeners to aurally perceive the instruments on stage at their correct locations, and cinema audiences to align dialogue with its visual location. But even though two channels were far better than one for aural localization, a major deficiency in stereo was also readily apparent: weak auditory image stability at locations between the two loudspeakers. The reason is straightforward. Whereas a real sound source emits a *single* sound wave and the source's location can be sensed by the arrival time and differential intensity at the listener's right and left ears, a pair of loudspeakers emits *two* sound waves. Creating the illusion of a sound location between the loudspeakers depends on varying the amplitudes from each; for example, increasing the loudness from the left moves the auditory image in that direction. But the illusion works only when the listener is roughly equidistant from both loudspeakers. When the listener is too close to the left loudspeaker, it dominates, and all sounds appear to be located at that point, at the left.

Suppose you were sitting on the left side of a cinema theater watching an actor in the middle of the screen. With only a left and right loudspeaker, you would perceive the actor's voice as originating from the left loudspeaker, a spatial inconsistency. Indeed, the actor's aural location is accurate only when the actor is directly in front of the left or the right loudspeaker, and nowhere else. Under that condition, there is only one sound source. As cinema screens grew in width, perceptual holes between widely spaced loudspeakers became more significant. A solution was needed. In the 1950s, the Todd-AO format used five groups of loudspeakers behind the screen but typically only one was active at any given time; a control track determined which loudspeaker emitted sound. Sony's SDDS format also allocates five of eight channels to the front screen. There would always be a sound source roughly aligned with any given actor regardless of the actor's location. This solved the contradiction between the auditory and visual modalities.

The importance of hearing as support for vision is well known. Aural localization automatically determines where we should position our gaze, without which scanning would be required; it makes visual target acquisition more rapid and accurate (Perrott, 1993). For example, reacting to the sudden onset of an actor's voice in a film, we can immediately locate the actor, even when there are many on screen, without having to observe the actor's lips. Consistent and integrated sensory experiences produce less cognitive load, which is more than a matter of artistic discretion. In addition, film directors have additional artistic freedom when they can rely on sonic rather than visual events for focusing attention on an action.

Without diminishing the artistic importance of aural localization errors, when you remain in one place and listen to an audio reproduction of pure music that lacks a video counterpart, such errors are, in effect, not contradictions because you have no visual reference points. Instability in an auditory image is apparent only when you move about a space, and then aurally experience a source location that moves with you. Even though, as a connoisseur, you may be uncomfortable perceiving a musical instrument at the wrong location, you have no reliable means of knowing its actual location. In contrast, contradictory locations between vision and hearing are disconcerting, more apparent, and do not require special sensitivity.

An appreciation for improved auditory image stability in cinema only belatedly transferred to reproduced music. The cinema industry improved stability by using a center channel almost a half century earlier than the audio industry. Yet the knowledge and technology were available to both industries. Paul Wilbur Klipsch (1958) proposed a simple phantom center channel as the sum of the left and right channels, which would have improved stereo, but the idea was ignored. Similarly, the failed experiment with the quadraphonic format not only ignored the problem of auditory image fragility, but exacerbated it (Bauer, 1971). In his study of cinema sound from the perspective of an audio engineer, John Mosely ironically observed that, whereas cinema engineers pursued spatial issues, aural localization, and ambient surround, audio engineers pursued signal quality, frequency response, and dynamic range. A semipermeable wall separated the two groups, and to some extent, it still does. The first aural architects of virtual aural space came from cinema, not audio.

In addition to addressing auditory image stability, the cinema industry also brought radical concepts of musical space to the larger public by using surround-sound technology. The film *Fantasia*, a brilliant combination of classical music and animated cartoons conceived by Disney and the legendary conductor Leopold Stokowski, was released in 1940 with Fantasound (Klapholz, 1991). In its most sophisticated embodiment at the New York Broadway Theater, Fantasound contained three channels, left, right, and center. For the few theaters that could afford a fully configured sound system, an additional 65 loudspeakers were distributed throughout the audience area, so that listeners were fully immersed in a sea of sound (Culhane, 1983).

Fantasia was a milestone. To achieve the goal of moving individual sound sources freely and independently throughout the theater space, William E. Garity and J. N. A. Hawkins (1941) designed their own methods and equipment. The requirements of surround sound, in the context of 1930s technology, placed unique demands on the recording, mixing, and reproduction processes. Up to 33 microphone signals were recorded on nine master tracks, then mixed down to three optical sound tracks, and finally, expanded to feed dozens of loudspeakers throughout the theater. By allocating one of the four optical tracks to control the variable-gain amplifiers in real time, audio engineers could share a small number of sound tracks among a large number of loudspeakers, at one moment providing direct sound, and at another providing ambient. Also, the newly invented constant-power panning potentiometer (pan pot) allowed a single source to be smoothly cross-faded between neighboring loudspeakers, thereby creating the illusion of motion. It was a massive and audacious undertaking, like none before.

Not only did the production of *Fantasia* legitimize surround sound in the minds of the public, but it also illustrated that a broad dissemination of advanced sound presentation systems required a harmonious partnership among four groups of contributors: financial supporters, artistic visionaries, audio engineers, and talented musicians. If any one of these four groups of contributors had dominated the others, or if one had been missing, the film experiment in an integrated audio-video surround system would certainly have failed.

Everyone on this project shared a common vision—a marriage of artistic creativity and economic pragmatism. Garity and Hawkins (1941) emphasized that "the public has to hear the difference and then be thrilled by it, if our efforts towards improvement of sound-picture quality are to be reflected at the box office. Improvements perceptible only through A-B comparisons have little box-office value...Simulation of live entertainment is not our objective. Motion picture entertainment can evolve far beyond the inherent limitations of live entertainment." They succeeded in achieving their dream, and a half century later, *Fantasia* had become a surround-sound icon that was recently rereleased for a second time.

In the late 1970s, three decades after *Fantasia*, the cinema industry was only just then upgrading the sound system of the average theater, many of which still dated from the 1930s. Even with advances in signal processing, economic constraints still limited film sound to two optical or magnetic tracks. The bottleneck was in the delivery of sound, not in the creation or presentation of sound. But with the eventual arrival of digitized audio, the cinema industry gained the flexibility it needed to deliver five sound channels and one effect channel, all of which fit within the audio budget of film tracks. Thus 5.1 surround sound was developed as a practical system for the average theater installation, and it was configured for the typical long, rectangular shape of theaters, which arises from the size of the screen and corresponding viewing angle. Cinema audio applications forced the audio industry to upgrade sound systems for cinema theaters to support a 5.1 format in a long, narrow space.

By the late 1990s, consumers could easily view and listen to cinema at home using the same format found in many commercial theaters. The professional infrastructure for recording, mixing, producing, and delivering sound in this format was already in place as part of film production. No additional investment was required. At the beginning of the twenty-first century, for a typical listener, surround-sound virtual space is a 5.1 surround-sound system in the living room. Any musical artist wishing to distribute surround music to a wide audience assumes the spatial properties of a 5.1-format sound presentation system, and accepts the acoustics of a typical home living room, which must be taken as a given, like the acoustics of cathedrals and concert halls. And like their predecessors from earlier centuries, contemporary composers and musicians once again find themselves in a reactive role. In this case, aural architecture of virtual space was the child of cinematic space, just as centuries earlier, performance space was the child of religious space.

Society could have invested in perfecting any of the many competing virtual-space sound systems. Technically, the 5.1 format could have been a 10.2 or even a 30.7.2 format, and transversal binaural could acquire special headphones with adaptive signal processing and head tracking. The choice often has little to do with available technology or spatial accuracy. The critical issues are rooted in social, economic, and artistic factors: portability, size of the sweet spot, public or private listening, artistic expectations, economic allocation of resources, and other factors that manifest a culture's values, traditions, and history. These all determine whether a given sound system will be acceptable, desirable, and marketable. Once a format becomes widely distributed throughout a culture, its properties become a culturally accepted norm.

Because surround-sound technology happens to have been a recent development, we simply know more about its history. There is every reason to believe that this pattern has been repeated throughout the millennia in all cultures. Like all historical accidents, the story of surround sound could have easily had a different outcome. The aural architecture of a virtual-space sound system using the 5.1 format was not inevitable.

The Mixing Engineer as Conductor and Aural Architect

The roles played by acoustic architects, audio engineers, conductors, and arrangers are apparent and widely recognized, but hidden from public view in their recording studios are the audio mixing engineers, skilled professionals who simultaneously function as musicians, composers, conductors, arrangers, and aural architects. They create music and musical space, and their creations are then distributed to a broad range of listeners. Apart from a small group of professionals connected to audio and music, few people even know that such persons exist or what they do. The audio mixing engineer, who melds the components of music, and the design engineer, who created the studio equipment, are both important parts of the family of aural architects.

Musical compositions rarely contain any notation about spatial acoustics. Even when conductors, composers, producers, sponsors, and all those with a professional stake in music influence the audio mixing engineers' concepts of musical space, the final sound is *their* artistic conception. They may think of themselves as being only skilled audio experts practicing a craft, trusted service providers, but they have also acquired many of the traditional responsibilities assumed by aural architects. They continuously exercise that responsibility by selecting processing equipment, and by choosing how that equipment is used. The most successful audio mixing engineers are wooed, honored, and financially rewarded, mostly without public recognition.

By choosing the parameters of spatial synthesizers and artificial reverberators, and by routing the various musical sources to their respective channels, audio mixing engineers create spatial experiences. They have no formal rules. Choices that are currently in vogue, however compelling and believable, will eventually disappear, just as past choices are to be found only in the dusty archives of music libraries. Like all manifestations of aural architecture, musical space is influenced by professional fads, emotional undercurrents, innovative technology, and marketing fantasies, as well as by the artistic desire to explore new art forms. But audio mixing engineers, unlike traditional architects, create a new spatial experience during *every* recording session, using a time scale of hours not decades, and at minimal cost.

With the introduction of surround sound, audio mixing engineers were forced to adapt the old rules and paradigms that had been used for mixing a stereo production. Answers were needed for new questions. Tomlinson Holman (2001), the father of THX cinema surround, pointed out that the most important decision in creating surround sound was the choice between the two primary listener perspectives: the *in-audience perspective*, where the listener sits in the best seat in the house, sonic activity is located at the front, and surround creates reverberant ambience; and the *on-stage perspective*, where the listener sits in the midst of the musicians, encircled by active sound sources. Both choices have precedents in traditional performances. Seats on stage have occasionally been used for overflow seating in live performances, and recreational musicians also enjoy listening to music from within their ensemble.

In a survey of listeners' preferences, quoted by Holman (2001), the in-audience perspective was preferred 3 to 1, but for the few individuals who preferred the on-stage perspective there was a 2-to-1 youth bias. Several factors explain these preferences. First, young listeners have less experience with classical concerts, and are therefore less committed to traditional artistic arrangements. Second, the on-stage perspective is still too new to be a refined art. For those old enough to remember early stereophonic recording, in order to emphasize the difference between monophonic and stereophonic presentation, recordings featured bouncing ping-pong balls and trains roaring through a crossing. Many surround-sound recordings using the on-stage format resemble the early period of the stereophonic format—lacking artistic subtlety. Finally, during the last few years, surround-sound systems appropriate for on-stage presentation have become widely available.

Aside from the many possible explanations for the dominance of in-audience perspective, the current generation of surround systems, the 5.1 format, is not particularly well suited to creating an encircling experience. With other presentation systems, such as binaural headphones and ambisonics, the audio mixing engineer can place sound sources anywhere in the horizontal circle. But with the 5.1 format, there is a strong bias toward the front, where three of the five loudspeakers are placed at 30-degree angles and the two side (rear) loudspeakers are placed at 120 degrees from each other and at 90 degrees from the front. The surround circle is thus discontinuous, with a tightly spaced frontal region and sparsely spaced rear and side regions.

A phantom sound source location between two physical loudspeakers can be created by using the technique called "panning," where a balanced mixture of sound from two loudspeakers creates the aural impression of a sound source location between them. After years of mixing stereophonic recordings with phantom sound source locations between the two front loudspeakers, and after years of psychophysical research on aurally perceived location, the rules for locating a phantom sound source between two real ones have been well established. And these rules determine the maximal angle between neighboring loudspeakers. Widely spaced loudspeakers do not readily support the perception of a phantom auditory image located between them. In other words, the 5.1 format supports continuous locations in the front, but only two discrete locations on the side. Should the on-stage perspective gain artistic prominence, with a need for uniform aural localization around the circle, at least eight loudspeakers would be needed in the sound presentation system.

There are two other constraints imposed by the sound presentation system. First, elevation, the third spatial dimension, cannot be used if the presentation system has no means of creating a sound field pointing upward or downward. Binaural systems and those surround systems with loudspeakers mounted above the listener can, however, take advantage of all three spatial dimensions. Although seldom used, height is a legitimate location in classical music. Second, sound presentation formats have different constraints on the distance between the musical source and the listener. Binaural systems can aurally place sound sources as close as the listener's head. A musician could aurally sit on the listener's shoulder or in front of the listener's nose. In contrast, conventional surround systems cannot create the aural experience of a sound source any closer than the loudspeaker themselves.

Just as real performance spaces have geometric and acoustic properties that limit the artistic freedom of composers, conductors, and musicians, so every virtual space system also has properties that limit the artistic freedom of the mixing engineer. The two sets of properties produce different biases. In the case of the 5.1 format, however, its properties are actually quite close to those of a real concert hall auditorium, which implies an in-audience location for listeners, a front location for musicians, and reverberant ambience from the side.

Except when musicians are located at the side loudspeakers, mixing engineers face the old issue of creating enveloping reverberation. The means for creating reverberation has been a central topic since the early days of monophonic recording. With the monophonic format, reverberant ambience emanates from a single point in space. The stereophonic format expands the quality of reverberant ambience by having two loudspeakers spread across the front to create a lateral and diffuse reverberant sound field, but it does so from only one direction. Even with sparsely spaced loudspeakers, the 5.1 format expands the possibilities for high-quality reverberant ambience even further. Unlike aural localization of sounds sources, the aural perception of ambient reverberation better tolerates widely spaced loudspeakers.

From the perspective of "natural" reverberant ambience, the audio engineer who designs an artificial reverberator, and the audio mixing engineer who adjusts its parameters, together replace the acoustic architect who builds concert hall auditoriums. When used in the traditional way, the reverberator produces reverberant properties equivalent to those of a real space. For a monophonic format, a reverberator produces one output; for the stereophonic format, two outputs; and for the 5.1 format, five. Other than the number of outputs, a 5.1 presentation format does not constitute radical change.

From about 1930 to 1980, an ideal reverberator was viewed as an economical alternative to building a real performance space. Could such a device adequately mimic the sound of a real space? Could it adequately reproduce the acoustic effects of its walls and ceiling and those of the materials on its surfaces? If so, then it could replace the acoustics of a real concert hall auditorium.

At some point during the evolution of spatial simulators, physical *naturalness* became an unnecessary constraint, being replaced by artistic *meaningfulness*. During the last two decades, unbeknownst to most listeners, reverberators have produced creative acoustics with no pretense of imitating real spaces. Parameter presets for these imaginary spaces have names such as "Small Foley," "Buckram," "Auto Park," "Jazz Hall," "Plated Gate," "Beefy Hall," and so on (Lexicon, 2000). Alternatively, parameter presets for reverberators that, at least crudely, approximate the acoustics of real spaces have names such as "Konzerthaus Mozart," "Kings College Chapel," and "Concertgebouw" (Owsinski, 2002). These names of real or imaginary structures, when attached to the parameter presets of artificial reverberators, are only semantic tags for certain kinds of spatial sound. Does a particular musical rendition sound better in "Beefy Hall" or in "Boston Symphony Hall"? The answer depends on the artistic sensibilities of the audio mixing engineer. For listeners with a passion for classical music, the ambivalence between artistic and natural spatial acoustics remains, and some artificial reverberators are still being developed to implement the acoustics surrounding a particular seat in a specific concert hall (Reilly and McGrath, 1995).

Other than centuries of tradition, why should enclosed physical spaces be the exclusive models for the aural architecture of musical spaces? The enclosing walls of concert hall auditoriums, for example, may simply have originated from the need to shelter the audience from the weather, to isolate the space from extraneous sounds, and to provide a means for controlling access. Their spatial acoustics are then an artifact, not of artistic requirements, but of these needs, and our musical traditions have adapted to them. Even now, because musical spaces involve compromises between nonartistic factors, they often have significant acoustic defects.

As it turns out, studying enclosed physical spaces as models for musical spaces has been marginalized as an academic distraction, and the preponderance of recorded music simply ignores "real" spaces. Consider the following example where the mixing engineer combines the acoustics of a small chamber, with elevated clarity, presence, and intimacy, for the solo violinist, and the acoustics of a large concert hall, with diffuse and enveloping reverberation, for the orchestra in the background. In the final mixing, two musical spaces are actually overlaid: a chamber and a concert hall. The result may be delightful, artistically exciting, and aesthetically consistent, but conceptually the space is unreal, contradictory, like an Escher picture. It exists only in the listeners' minds. Since the creation of a virtual musical space is now more an art form than the scientific simulation of a physical process, the traditional requirement of a single musical space vanishes as irrelevant.

Other constraints inherited from the legacy of concert halls are also vanishing with the advent of electronically supported aural architecture. Any parameter can be changed at any time. In real spaces, acoustic parameters are tightly coupled and mutually dependent on each other. With an artificial reverberator, they may each be controlled by a separate knob. For example, the reverberation attack may be fast or slow; the sustain may be long or short; the decay may be linear, nonlinear, or abrupt. Sonic reflection density may increase rapidly or slowly; resonance density may be high or low, and as a special effect, the spectrum of a reverberant sound may expand during the decay. The reverberation may have strong sonic reflections corresponding to real walls, or it may mimic the acoustics of caves composed of multiple chambers. Multiple reverberation processes may be cascaded to form a series of coupled spaces. There is no limit to the ways that a reverberator can be used.

When Laurens Hammond (1939) first included a spring reverberator inside his portable reed organ, he was in fact merging an artificial space with a vibrating reed, both contained within a single housing. This early fusion of spatial attributes with musical instruments continued during the remainder of the century. Today, spatiality is just another musical attribute. A listener can sit inside a virtual violin, such that its virtual body *is* the musical space. Similarly, as discussed earlier, there is no requirement that a virtual resonant space for a concert provide seating for a large audience. Virtual spaces have a new set of rules.

Once a spatial parameter is connected to a knob, button, or key, a reverberator becomes effectively indistinguishable from a musical instrument, played in real time by a musician. Musicians can change acoustic parameters other than pitch. For example, a pianist can change the decay rate of vibrating strings by pressing a pedal; an organist can change an organ stop by selecting a particular rank of pipes. In both cases, acoustic parameters are part of the musical composition, and in both cases, the pedal or stop can be used to change an acoustic parameter of the space—aural architecture as an extension of the musical instrument. From this perspective, musicians, and not the audio mixing engineer, should control the parameters of artificial reverberators in the recording studio. As a part of a musical composition, space can then be made to grow in size, walls can move in and out, surfaces can morph from hard to soft, and the sound source can approach and recede from listeners. If we accept that a virtual space is subservient to artistic meaning, it need not remain static.

Advances in the art, technology, and techniques of the audio mixing studio have produced artistic and social paradoxes. More than a few popular music groups have gained fame and fortune by creating music with audio mixing and processing equipment in a recording studio, and then distributing their performances only as recorded music. But when these groups attempt to replicate the unique sound of a studio recording in a live concert, their music, lacking the creative manipulations of the audio mixing engineer, sounds weak and ineffectual. Because they cannot create the sound of their studio recordings in a live performance, singers must now silently lip-sync to their prerecorded, preprocessed, and premixed music played through loudspeakers. In a true role reversal, rather than recorded music re-creating the experience of live performances, live performances re-create the experience of recorded music. Life imitates art.

Although virtual spaces evolved from physical spaces, the two types of space diverged rapidly, to become distant cousins. There are four principal reasons why this occurred. First, we live in physical spaces, not in virtual ones (at least not the overwhelming majority of us). Second, we inherited the traditions of auditory awareness of physical spaces over the span of thousands of years. Third, the acoustic properties of physical spaces are highly constrained, whereas those of virtual spaces are almost entirely unconstrained. Finally, our artistic grasp of virtual space is, at best, rudimentary. The aural architecture of virtual musical spaces is a pure art form. But because of the arcane nature of skills required to manipulate spatial experiences, the art is limited to a small group of specialists with a combination of artistic sensibility and technical skills.

The music critic, Justin Davidson (2005) observed the paradox of what he calls "iPodspace": "Today, the environments that music occupies have gotten either very small or very large: the aural isolation of headphones or the anonymity of the stadium. Live, unamplified music still exists in the cloistered precincts of the concert hall.... But for much of the world, music has become either a solitary experience or a form of mass ritual. Yet the history of music is inseparable from the history of places where people gathered." Virtual spaces for music are no longer related to social spaces for people. Just as visual architects design very different kinds of space, aural architects design different kinds of musical spaces.

6   Scientific Perspectives on Spatial Acoustics

Not everything that can be counted counts, and not everything that counts can be counted.
—Albert Einstein[1]

Scientific knowledge of spatial acoustics has advanced sufficiently during the last half century that aural architects now have the ability to implement specific acoustic properties in both real and virtual spaces. Compared with the previous century, our knowledge of physical and perceptual acoustics is well advanced. On the other hand, scientific and engineering research has already picked much of the low-hanging fruit on the acoustic tree, leaving behind many difficult, intractable, and philosophic problems. In addition, social and cultural forces now play a larger role in the problems that remain. Progress becomes elusive when scientists struggle just to formulate questions that can be answered meaningfully using scientific methods.

Because detailed knowledge about spatial acoustics is relatively recent, our culture does not have a long tradition of designing aural spaces. In contrast, our knowledge of visual space was already advanced by the sixteenth century. Painters already understood the rules of light, color, reflectivity, perspective, and shadows. There is still no established counterpart for aural painters. Can our newly acquired scientific knowledge about acoustics finally change this historic pattern? Can aural architects, functioning as artists, use scientific tools to create aural spaces?

Surprisingly, the answers are ambiguous. Although we understand far more about the physical behavior of sound in a space than ever, our knowledge is still far from complete. For some questions, there is no methodology for acquiring answers. And for other questions, answers require massive computational resources. Sound is actually more complex than light for three simple reasons. First, light waves moves instantaneously, whereas sound waves move relatively slowly. Second, the highest frequency of visible light is less than 2 times as great as the lowest, whereas the highest frequency of audible sound is 1,000 times greater than the lowest. Third, relative to the size of object and surface variations, the wavelength of light waves covers an extremely narrow range, whereas the wavelength of sound waves covers a wide range, large at low frequencies and small at high frequencies. For these reasons, it is easier to simulate, record, and analyze a visual space than an aural one. Turn on a light in a room, and the visual impression is static and immediate. Film is remarkably accurate at capturing and retaining visual details, and the human visual system is almost as simple as an array of light sensors: the neurological equivalent of film.

Unlike the simplified problems of spatial acoustics to be found in introductory textbooks, such as how to catalog the reflections of a sound in a rectilinear box, questions about real spaces seldom have straightforward answers. Textbook examples create the illusion that simplifications are only for the benefit of students. Hidden from the students is the fact that only these simplified problems have clean, elegant, and compact solutions. The science of real acoustics is a messy subject. Constraints on assumptions, questions, methods, and philosophy all limit the scope of scientific conclusions. Unfortunately, these limitations are impossible to evaluate without training in physical acoustics, statistical mathematics, and perceptual psychology. Such subjects are abstruse. Discussions about the theoretical foundation of statistical acoustics are currently intelligible to at most a few hundred specialists. Assumptions buried under elegant conclusions undermine their applicability. When real spaces do not match required assumptions, aural architects find themselves in the same situation as their early counterparts—lacking predictive tools. However, when spatial questions are consistent with assumptions, the resulting insights are useful and robust. The following discussions attempt to provide an introductory feeling for the difficulties involved in using acoustic science to analyze and predict the experience of aural space.

When decomposed into their component parts, questions about auditory spatial awareness concern three related disciplines: *physical science*, which represents physical acoustics with mathematical equations; *perceptual psychology*, which describes perceptual acoustics with subjective measurements; and *cultural anthropology*, which understands cultural acoustics in phenomenological terms. Formal science is best at exploring physical acoustics, modest at perceptual acoustics, and weakest at cultural acoustics. Interesting questions result from melding these three aspects of acoustics together, but the enlarged scope then produces additional complexity. A narrow question within the scope of one aspect may not contribute insight into phenomena that are a combination of all three.

To appreciate the complexity of the aural experience of a concert hall, let us consider how these three aspects of acoustics contribute to auditory spatial awareness of a particular hall. First, there is a direct relationship between physical spatial properties, such as geometry and materials, and the resulting acoustic parameters, such as reverberation time. Within this realm, scientific and mathematical questions lead to predictive models with numeric answers. Second, there is the relationship between perceptual attributes, such as enveloping reverberance, and its physical parameters, such as measures of spatial diffusion. Knowledge in this area is progressing, but still incomplete.

We know that some changes in some parameters are clearly audible, whereas others are not. For example, a small change in reverberation time is aurally perceptible, whereas a doubling of the sonic reflection density from 10,000 to 20,000 per second is not. Finally, there is a relationship between auditory perceptual parameters, such as spectral balance, and human experience of those parameters, such as aural warmth and ambience. These bear directly on the bases of auditory art, emotion and affect, all of which derive their meaning from the specific culture. These cultural concepts are part of the phenomenology of aural space. Although not intractable, phenomenology as an intellectual formalism remains controversial and subject to extensive debates among scholars. From our perspective, phenomenological questions are interesting because aural architecture is not only the physics of an acoustic space, but also the sociocultural aspects of what we bring to the experience of a space.

As long as we recognize the biases and limitations of these three aspects of acoustics, and as long as we preserve our humility and skepticism, we can extract valuable insights from each. When a problem is formulated in terms of physical acoustics, the results may provide an exact answer, but its relevance to describing an aural experience is often tenuous. When a problem is formulated in terms of cultural acoustics, the result may be compelling, but the physical basis for the aural experience is equally tenuous. The famous quantum physicist and Nobelist Max Planck summed up the dilemma when he said: "Science cannot solve the ultimate mystery of Nature. And it is because in the last analysis we ourselves are part of the mystery we are trying to solve" (Barrow and Tipler, 1986).

Evaluating the Aural Experience of Concert Halls

In his first book on concert hall acoustics, the prolific scientist and acoustic architect Leo L. Beranek (1962) began with a quotation from the conductor Eugene Ormandy, "But I don't want to hear a pin drop, I want to hear the orchestra!" This explosive remark, with his arms waving for emphasis, was in response to the manager of a world-famous concert hall who had just said that his concert hall had "perfect acoustics because everywhere in it one could hear the sound of a pin dropped on stage." By rejecting the engineering concept of an acoustic space as a transparent audio channel, Ormandy was interested only in how the space enhanced the experience of his orchestra. He was expressing dissatisfaction with the manager's criteria for judging quality.

In an ideal world, using the social values of the various acoustic stakeholders, aural architects first ascertain the desired aural experience using common language. They then convert aural requirements into a specialized perceptual vocabulary—experiential dimensions—a bridge between art and science. To use an automobile analogy, a desire to rapidly overtake other vehicles (sociocultural value) maps to responsiveness (perceptual attribute), which maps to acceleration (measurable physical parameter), which maps to engine architecture (physical design). From the engineering perspective, with sufficient knowledge, designing a concert hall would be like designing an automobile.

The Elusiveness of Quality Preferences

Based on the assumption that the mystery of concert hall acoustics should yield to the scientific method, Beranek's multidecade investigation (1962) took him to over sixty of the most famous concert halls in twenty nations on five continents. He interviewed and questioned world-renowned conductors, performers, music critics, and experienced listeners, most notably, those who could compare their experience of different spaces. Among those interviewed were musical luminaries such as Erich Leinsdorf, Leopold Stokowski, Isaac Stern, Bruno Walter, Leonard Bernstein, Herbert von Karajan, and Charles Munch. The consensus among such connoisseurs was that some concert halls are loved and admired by all, and some are sufficiently disliked that they are relegated to artistic purgatory. Beranek conclusively showed that musical experts, at least within a shared time period, evaluated the quality of concert halls consistently.

Using a lifetime of professional experience, Beranek sorted concert hall quality into 18 distinct concepts, each of which was represented by a word or phrase that served as a label for a perceptual experience. Some of the more prominent labels included "intimacy," "fullness of tone," "clarity," "warmth," "brilliance," "balance," "blend," "ensemble," and "texture." Each concept had a working description, but one that mixed natural language, acoustic measurements, and auditory perceptual attributes. Although Beranek readily admitted that the concepts in his initial catalog were interdependent and ambiguous, he had at least laid a foundation to solve the problem of defining quality.

A few examples serve to illustrate the relationship between Beranek's terms and the listening experience. *Clarity* and *definition* are the degree to which individual sounds stand apart, retaining their distinctness: horizontal clarity applies to successive notes, whereas vertical definition applies to simultaneous notes. *Intimacy* is the aural sense of being in proximity to the performers, as if the space were small. Based on intuition, Beranek further expressed intimacy as being related to the delay between the direct sound and the onset of the first sonic reflections: small delays produce aural intimacy. *Warmth* is the sensation produced when low frequencies dominate high frequencies, which in the extreme produces a boomy quality. *Blend* is a measure of the way that sounds from different instruments mix together.

In addition to providing an encyclopedic compilation of invaluable data, Beranek's study revealed one particular pattern. Of the 47 concert halls analyzed, the best 25 halls had less than a 30-millisecond delay between the direct sound and the first sonic reflection, and the worst 8 halls had delays between 40 and 70 milliseconds. Later research confirmed that this delay parameter was indeed critical and needed to be kept within a narrow range. This insight led to the design of acoustic hanging clouds, shown in figure 6.1, whose surface generates early sonic reflections within the critical time window, and whose acoustic success gave rise to a mood of optimism.

With their scientific training in acoustics and a passionate interest in classical music (a delightful combination of work and pleasure), acoustic researchers turned their intellectual focus toward analyzing other acoustic parameters of concert halls. Unfortunately, understanding the perceptual basis for hall acoustics proved far more difficult than initial expectations. For the next four decades, researchers searched for the perceptual dimensions of musical spaces, adding to, removing from, and modifying Beranek's initial catalog. They devised both measurements and equations that connected perceptual attributes to physical metrics of the sound field. They performed psychoacoustic experiments to demonstrate the relationship between perception and acoustic parameters. And they simulated how spatial geometry influences a sound field. On the one hand, these committed researchers added, and continue to add, valuable insights into the aural experience of concert halls. We understand much more now than we did forty years ago. On the other hand, the results were riddled with contradictions, ambiguities, assumptions, and confusions about the nature of the problem. Why was it that evolving insights about concert hall acoustics did not converge into a clear picture?

We begin with the problem of language: What do particular words actually mean? Representing experience using words is by no means straightforward. What, for example, is the opposite of "soft"? Is it "hard" or "loud" or "brittle"? There is no natural vocabulary for sound. The adjectives we use to describe music and acoustics are borrowed from the vocabularies of sight, touch, and taste: "a bright sound," "a dark melody," "a transparent loudspeaker," "a warm room," "the sweet spot" (Marks, 1982). Theoretical linguists still struggle with how adjectival meaning is to be represented. Words that have fairly close synonyms do not always have the same antonyms (Murphy and Andrew, 1993). Yet the meaning of adjectives to describe auditory impressions is fluid even in Japanese and Chinese, languages that have a similar vocabulary for sound attributes (Iwamiya and Zhan, 1997): listeners in these two cultures seldom agree on description of sounds.

Could the techniques used by social scientists to overcome problems of language also be applied to the acoustic study of concert halls? Many have tried. Recently, A. G. Sotiropoulou, R. J. Hawkes, and D. B. Fleming (1995) selected some 80 volunteers to evaluate their auditory perceptions of live classical music in three concert halls. Listeners were given a questionnaire containing 63 polar adjective pairs—semantic dimensions—and they were asked to assign a numeric value to each pair. To ensure that the list was broad and comprehensive, researchers first created a much larger list using a thesaurus, literature, and other sources. Then, in a preliminary phase, subjects sorted the list into three categories: meaningful, ambiguous, and irrelevant. Only the most relevant pairs were then used in the next phase of the experiment. Assuming that the list of adjective pairs was meaningful and consistent, Sotiropoulou and

Figure 6.1
Acoustic hanging clouds in the Holy Spirit Lutheran Church produce early sonic reflections but without decreasing the acoustic volume. Courtesy Chris Brooks, Orpheus Acoustics. www .orpheus-acoustics.com

colleagues then performed a factor analysis in order to remove redundancy from the ratings. For example, answers along the dimension "voluminous-thin" strongly correlated with answers along the dimensions "sonorous-thin" and "mighty-small." Using statistical methods, they collapsed the data into four *independent* auditory perceptual dimensions, which they called "body," "clarity," "proximity," and "tonal quality." Initially, the category "body" had more than 23 adjective pairs. Their findings, though interesting, were neither conclusive nor readily extensible. At best, such studies suggest that listeners of a particular culture experience spatial sound as a composite of independent dimensions, but the identity of those dimensions remains controversial.

The consensus among those acoustic experts who also have auditory perceptual training is that the spatial dimensions must somehow subsume aural experiential concepts such as intimacy, reverberance, spaciousness, envelopment, and ambience. Experts have an intuition about the appropriate labels for spatial auditory dimensions, yet surprisingly common labels, such as "intimacy," often do not survive formal procedures. No one knows whether these procedures are defective or whether intuition is simply wrong. Although the details vary, some research paradigms produced results that were more consistent (Wilkens, 1977) than others (Hawkes and Douglas, 1971). In addition, Trevor J. Cox and Bridget M. Shield (1999) noted that results depended on the choice of listeners, with less consistency among nonmusicians.

One way to avoid the problems of language is to provide individuals with a defined vocabulary of reference prototypes. Consider the following instructions: "Here is a sample of a sound that is called 'hot,' and here is another called 'cold.' Using these two anchors as references, evaluate the following samples along the dimension of 'hot-cold,' 10 for 'hottest' and 1 for 'coldest'." This approach produces consistent results, and with training, any perceptual parameter can be mapped into such a vocabulary. The words are simply defined by the experimenter's proffered reference samples. We learn little about experience because the words are only aliases for the sound attributes being manipulated. Experimenters could be manipulating reverberation time, spectral balance, reflection density, or any one of a hundred metrics. Moreover, the physical dimension could be replaced with any preselected word pairs, such as "big-small," "white-black," or "fast-slow." In one sense, the experimenters are teaching their subjects to detect an attribute, and to then convert it into a rank, a number, a word, or a button push. However much these experiments tell us about the ability to discriminate between auditory attributes, they tell us nothing about artistic relevance of these attributes.

Before we explore other approaches to studying the perception of spatial acoustics, we should note that evaluating concert halls belongs to a broader research question: how to convert subjective quality judgments into reliable and predictive metrics of preference. Preferences in concert hall acoustics are similar to preferences in food, houses, automobiles, and literature. In struggling with the problem of subjective preferences, market researchers, functioning as applied social scientists, have already created methods for formulating questions and acquiring answers about human judgments.

In his compendium of how individuals make unconscious judgments, Malcolm Gladwell (2005) revealed robust patterns and consistent conclusions. First, with extensive experience, experts are remarkably adept at sensing critically important dimensions in complex situations. Second, nonexperts often perform almost as well as experts, although their abilities to make judgments are prone to gross errors and devoid of natural language. In studying loudspeaker preferences, Sean E. Olive (2003) provided concrete support for Gladwell's conclusions. Third, when first describing an experience in natural language, subjective preferences then degenerate into random judgments. Counterintuitively, introspection can actually undermine insight when verbalized (Schooler, Ohlsson, and Brooks, 1993). These results may very well apply to many aspects of auditory perception.

The extensive research into food preferences is but one example of converting multidimensional attributes into a single sensory experience (Stone and Sidel, 1993). Which attributes of an apple pie most contribute to making it pleasing? Note that the ability to taste the difference between two pies (perceptual differences) is only one aspect of ascertaining the recipe for the most desirable pie (preference judgment). If there is no perceptual difference, then there can be no preference. But the converse is not true: perceptual differences do not imply preferences. Even if you can taste a subtle difference between two pies, you might not prefer either. There is already an extensive literature on statistical and methodological techniques for handling the complexity of preferences. The literature includes psychometrics (Kline, 2000), which convert aural experience and auditory ability into numeric values, and multidimensional analysis (Borg and Groenen, 1997), which separates aural experience into its components.

Many researchers still assume that aural experience can be broken into separable perceptual dimensions, essentially polarized ranges, such as "large-small," "close-far," "intimate-aloof," "boring-exciting," "spicy-bland," and so on. This assumption is an extension from physical phenomena and mathematical constructs, such as "east-west" and "up-down." With food, dimensions such as sweet, sour, salty, and bitter have biological counterparts that match the words. If such an approach is to be useful for analyzing spatial experience, we must first demonstrate that separable auditory perceptual dimensions exist, assign them names and labels, and show that the combination of these dimensions predicts both aural experience and preferences.

Multidimensional scaling techniques avoid the problems of assuming knowledge about the dimensions. Subjects are presented with pairs of samples and asked to estimate the degree to which they are similar or dissimilar. Nothing else is required of them. For example, they are given a list of cities scattered around the globe and told to let similarity be the geographic distance between pairs. Close cities are similar; distant cities are dissimilar. After the application of statistics, the results show that there are two primary dimensions, which we call "A" and "B." The method does not assign meaning to the dimensions; that is a task for the researcher. But in this illustrative example, the researcher notices that dimension A corresponds to latitude and dimension B to longitude. Further analysis of the data shows that the earth is round.

The same approach has been used in spatial acoustics, examining the similarity of simulated spaces. The resulting dimensions still appear as A, B, C, and so on, but unfortunately, the meaning of these dimensions, when expressed in natural language, is not yet obvious (Berg and Rumsey, 2001). By sidestepping the burden of finding auditory perceptual dimensions, Manfred Schroeder, Dieter Gottlob, and K. F. Siebrasse (1975) and Michael Barron (1988) demonstrated a correlation between subjective preferences and physical attributes. Auditory perceptual representations are more elusive than physical or preferential metrics.[2] Researchers had hoped to find such auditory perceptual dimensions as reverberance, clarity, envelopment, and so on, but failed. Among the three representations of spatial attributes, physical, perceptual, and preferential, there has been progress made on selected pairs but little success on the triple combination. As a concession to the difficulty in describing auditory perception, Petra Muckel, Franz-Josef Ensel, and Brigitte Schulte-Fortkamp (1999) studied how subjects described sounds by their spontaneous associations with images, memories, and feelings, all of which emphasizes those emotional components that arise from personal history.

Because subjective experience is frequently dominated by culture, it is distinctly possible that every subculture shows major differences when making perceptual and preferential judgments. For example, Barron (1988) showed that listeners divided into one of two groups, preferring either "intimacy" or "reverberance." In every time period, musicians, acoustic architects, and ordinary listeners are simply different auditory subcultures, each with a unique vocabulary, experience, and perceptions. The problem may lie with an imperfect language or with crude research paradigms, rather than with variations in perception. Even in a relatively uniform culture, language differences result in perceptual differences. It is simply not obvious that there should be consistency in auditory perceptual and preferential judgments. And if there are, in fact, wide variations, any attempt to find consistent results must fail.

Currently, we know more about the subjective dimensions of food and automobiles than about those of concert halls, either because aural architecture of concert halls is a more difficult problem or because society has not invested sufficient resources in studying aural architecture. Perhaps, both are true. Nevertheless, many musical spaces have consistent and well-earned reputations for quality, and scientific analysis can sometimes explain why those spaces are preferred (KnicKrehm, 2004; Griesinger, 2004). Although people make judgments and scientists make measurements, linking culture and science is a more difficult problem than pursuing either by itself. As we discussed earlier, preferences depend on cultural attributes, and science has nothing to say about shifting attitudes.

Struggling with Subjective Measurements

Experimental techniques for measuring subjective judgments often ignore or de-emphasize a hidden bias that dominates experimental results. How should subjects be selected? Should they be randomly selected from a telephone book? Or should they be selected from a college course? From a professional auditory society? From a group of volunteers trained and compensated for their time? The answer to this question dominates the results. Experimenters who select their subjects from a specific subculture are therefore measuring the training and experience of that group of subjects, rather than the baseline auditory perception of our species.

By training subjects as part of the experimental paradigm or by selecting subjects who have been self-trained through personal and professional activities, auditory experimenters attempt to create a uniform population. They do so to serve two independent functions: to enhance auditory sensitivity to a target attribute (creating a new auditory subculture) and to provide a consistent cognitive link to such an attribute (producing consistent responses in experiments). Most experimenters assume that similarity in culture and training ensures that subjects are responding to the same attribute (F. Rumsey, pers. commun.). Proceeding from this assumption, Francis Rumsey and colleagues (2005) showed that both naive and experienced subjects have similar sensitivities to audio that had been degraded in bandwidth and spatial precision. In the restricted task of evaluating the quality of audio encoders, David and Mark Moulton (1998) demonstrated that the results were dependent on the listeners' backgrounds and, more important, on how much they practiced.

Auditory researchers often select professional colleagues as subjects in their studies of concert halls. Members of this subculture have a heightened auditory perceptual sensitivity, a common working vocabulary, an understanding of research paradigms, the incentive to devote the necessary time, and experience with the class of stimuli being tested. The argument for using this specific subculture is efficiency. These subjects have already been trained, which obviates the need to train naive subjects in practice sessions lasting hundreds of hours. But experimental results are then only indicative of the behavior of a trained population.

Another obvious choice for subjects in the aural evaluation of concert halls is the subculture of musicians. Yet, even though they have extensive experience and an interest in spatial acoustics, their biases are different from those of the audience and acoustic architects. Musicians experience a concert hall from the perspective of the performance stage, whose acoustics are often unrelated to those of the auditorium. Musicians are preoccupied with hearing each other in order to play together in synchrony and with consistency. In contrast, the audience is passively listening to a composite experience in another acoustically coupled space. Reverberation and spatial ambience play different roles for the two groups. Other than a few isolated studies, aural architects designing concert halls ignore the special needs of musicians, who are left to adapt to the acoustics of their space as best they can. There are some exceptions. Dennis Noson and colleagues (2000) evaluated the role of early sonic reflection for singers, and Shin-ichi Sato Yoichi Ando, and Saho Ota (2000) studied the acoustic preferences of cellists.[3]

Exploring the aural experience of spatial acoustics in the general population, which precludes the use of specialized training that would implicitly create a subculture, makes researchers dependent on ordinary language. But absent a *common* vocabulary for describing sounds in general and spatial acoustics in particular, communicating experience is problematic. Acoustic defects in concert halls give rise to an imprecise language. An aural experience might be described as "sterile," "flat," "brittle," "harsh," "muddy," or "boomy." But what do such words really mean? Are they dependent on the culture, experience, education, personality, or the mood of the listener? Only in extreme cases do words have precise meanings. Unsophisticated listeners use a fluid and ill-defined vocabulary for expressing aural experiences.

If determining a useful perceptual vocabulary is so difficult, perhaps we can circumvent the problem. Studies by experts who make assumptions about the perceptual relevance of the acoustic parameter being tested based on years of experience and who focus on the preferred value for that parameter, a limited goal that avoids descriptive words, have produced less ambiguous results. For example, by assuming that reverberation time was important, Walter Kuhl (1954) found that the preferred reverberation time depended on musical style. Using 28 short segments of music from many different genres, Kuhl analyzed 13,000 judgments about listener preferences on one acoustic attribute, reverberation time. For Mozart's *Jupiter Symphony* and Stravinsky's *Le Sacre du Printemps*, listeners preferred 1.5 seconds, whereas for Brahms's *Fourth Symphony*, representative of Romantic music, they preferred 2.1 seconds.

Since reverberation time is only one measure of a space, the same procedure might be repeated for an infinite number of other measures. By the 1980s, there were dozens of equations to measure specific acoustic parameters of spatial reverberation, including rise time, early-decay time, clarity, center of mass, and others. Peter Lehman and Henning Wilkens (1980) then examined subjects' subjective evaluation of these parameters; Trevor J. Cox, William J. Davies, and Y. W. Yam (1993) determined subjects' sensitivity to the single parameter of early sonic reflections. When Kuhl performed his evaluation, only a few parameters were considered meaningful, and reverberation time was at the top of the list. Now there are literally hundreds of plausible candidates for other acoustic parameters. Someone must decide what is worth knowing. With researchers generating thousands of questions, the burden now shifts to selecting the appropriate question rather than just finding the answer or, as is often the case, selecting a question to match paradigms that can provide numerical answers.

Lest you think that science is the only means for acquiring insight, consider an alternative formulation from the early part of the twentieth century. Before the birth of psychology as a behavioral discipline, there were intellectuals who functioned as psychologists using philosophical constructs that relied on introspection as a form of inquiry. Violet Paget (1932), a famous intellectual and literary figure, studied the psychology of music using an open questionnaire that asked volunteers to write down their personal experiences of listening to music. She then analyzed the result for patterns and principles. By embracing emotional variability and personal attitudes as having a decisive role in listening, she was overtly rejecting the presumption that there was constancy between physical sound and personal experience.

In discussing spatial ambience, one individual described the experience of entering the Cathedral of Santa Maria del Fiore in Florence: "I had instantly a very great sense of what I might call immersion: an utter change of mode of being into as different an element as water or the change from complete silence to voluminous sound. This effect lasts but a very short time, leaving indifference unless one is inclined for exploration of form." When entering the Abbey at Poets' Corner, the individual described how the "exploration of form" became the feeling of an enclosure, or an experience of being immersed in an ambience. "I had a fine example of how both architecture and music can affect one massively as by immersion in another element. It is a massive change, a momentary rebirth. . . . The effect is great and delightful. One sees passively and hears passively and is perfectly happy (and often deeply moved). After a few minutes (or seconds?) one begins to 'look' and to 'listen,' and the state is broken, the charm is gone or a quite different one takes its place. One stands up to the sounds and the sights instead of being plunged into them" (Paget, 1932). Although this description of the acoustic space is from a single subject, it provides a class of insight that cannot be acquired from formal methods.

Ultimately, all sensory research struggles with the problem of how an individual perceiver communicates internal experience to an external observer. Descriptive analysis, one of the oldest and most widely used research methods, is elegantly simple, yet it lacks an intellectual foundation. Everyone has access to natural language. Although such data can be structured, descriptive analysis resembles introspective psychology, a process of validating the intuitive insights of individuals in a selected group.

A century later after Paget's informal questionnaires, the rules of formal science replaced the rules of introspective phenomenology. Science now limits the domain of inquiry only to questions amenable to numeric answers that are statistically valid and repeatable. Within these constraints, experimental science is more useful, powerful, and reliable than introspection. Nevertheless, these numeric metrics are only shadows of the phenomena, not the phenomena themselves. Understanding and measuring are neither equivalent nor interchangeable concepts. Both are required.

Even though we now understand more about concert hall acoustics than we did in 1900, the early comments of R. J. Hawkes and H. Douglas (1971) still hold: "Assessing concert hall acoustics reveals disagreement as to the number of different dimensions of acoustic experience, the terms used to describe these, and their physical correlates.... These results show that the subjective assessment of acoustics is made on more than one dimension, and that although there are general overall similarities, these dimensions do vary with type of music and concert hall, [as well as with the background of the listener]."

Because listeners seldom have the opportunity to hear the same music in two different halls without intervening weeks or months, to compare concert hall acoustics, they must rely on memory, whether short- or long-term. Moreover, the acoustics of a rear-balcony location are very different from those of a front-row orchestra location, and sampling different locations by walking through the hall during a concert is socially unacceptable. Thus, generally speaking, you cannot compare the acoustics of two musical spaces in the same way that you can compare two apples sitting on a tabletop. For these reasons, almost all research involves contrived cases using spatial simulators, which allow us to create pairs of controlled cases with known parameters. That minimizes the problem of memory, but spatial simulations are only crude approximations of real acoustic spaces.

Finally, we arrive at a simple but rather inelegant proposal: define high quality acoustics for a specific auditory subculture as the absence of objectionable acoustic defects. A theoretical acoustic defect is not a defect if no one in that subculture objects to it. Unlike acoustic quality, acoustic defects are readily apparent, and when they are explored, perceptual attributes become apparent. A strong sonic reflection perceived as an echo is an acoustic defect. The absence of low-frequency energy is readily apparent. Excessively long reverberation time creates aural soup. The lack of acoustic spaciousness is obvious when listening to music outdoors. Extreme cases are instructive. Beranek's list of concert hall attributes can also be transformed into an inventory of potential acoustic defects. For evaluating the experience of an acoustic space, focusing on its acoustic defects may be more productive than optimizing quality.

From a different perspective, acoustic defects address the philosophical problem regarding the nature of acoustic quality. Listeners with natural ability and extensive auditory training can learn to perceive even the most irrelevant acoustic attributes of a space. In contrast, the general population has much less refined abilities. Hence, acoustic quality is strictly dependent on the choice of the listener population, ranging from millions of average listeners to the world's expert listeners. Acoustic defects can be considered unpleasant acoustic attributes that are perceived either by a large group of listeners or by a small group of influential sponsors, hence justifying the effort to identify, understand, and remove them.

Although no single approach will lead to *the* answer for spatial quality, the composite of flawed methods provides a slowly converging picture. Because the flaws and biases of each method are different, they form a series of checks and balances. A strong perceptual phenomenon is likely to manifest itself regardless of the research method or choice of subculture, and a weak phenomenon, however real and statistically significant, may not be worth studying. From this perspective, consensus gradually emerges.

Two Manifestations of an Acoustic Defect

To illustrate how we might evaluate acoustic defects, let us consider nonuniform frequency response. This acoustic defect manifests itself as changes in the loudness or timbre of any note that is selectively amplified or attenuated by the acoustics of a space. With a nonuniform response, musicians hear some tones and overtones as being emphasized or suppressed; engineers and scientists observe these as changes in amplitude at different frequencies. It is the same phenomenon. Composers, musicians, and conductors use intensity as a major component of their art, and any spatial attribute that overrides their intention can be considered an acoustic defect. This much is straightforward.

When an audience is seated in parallel rows on a planar surface, floor interference patterns produce a marked dip in the middle frequencies. Georg von Békésy (1933) first observed that "the attenuation was greatest at 800 Hz [where] the resulting frequency distortion was rather large and could be easily perceived." The physical phenomenon is straightforward. Consider a direct sound of a 500 Hz tone, having a period of 2 milliseconds, when combined with its sonic reflection delayed by 1, 3, or 5 milliseconds. The direct and reflected sounds are out of phase, and if they have the same amplitude, when summed, they completely cancel each other. Neither the direct sound nor the sonic reflection is audible under such conditions; the listener hears nothing at this cancellation frequency.

Normally, we think of sonic reflections originating from a larger number of irregular surfaces, walls, and ceiling, but the surface of the audience also produces a sonic reflection. Very shallow angles, such as when sound grazes the audience, result in small delay differences. Theodor J. Schultz and B. G. Watters (1964) first confirmed the phenomenon, the "seat-dip effect," and observed that it was very minor when a listener was standing or sitting in the first row of the balcony. Using a computer model of the audience environment, Joe LoVetri, Doru Mardare, and Gilbert A. Soulodre (1996) arrived at similar results. The effect was stronger in those rows more distant from the stage, although the physical process is actually more complex because people and their seats form a regular grid rather than a flat surface. The analysis of the physical process must take into account scattering and interference (Takahashi, 1997). In an effort to reduce the level of the effect to a value below audibility, Davies and Cox (2000) suggested making "changes to the profile of the seats and floor [by] the introduction of a pit under the seat." However effective a technical solution, it is simply not practical in a real concert hall.

The aural coloration produced by audience seating is actually a special case of a much larger topic: short-term aural coloration produced by sonic reflections. Some experts have noticed that they could perceive spectral anomalies at particular distances to surface walls. Sitting or standing changes the spectrum. Tor Halmrast (2000) enlarged the scope of this topic when he suggested that inadequate spectral balance is a potential explanation for some defects that do not otherwise appear when using conventional methods. But this leads us back to Schroeder's offhand comment (1975) that all interior surfaces should produce low-loss, high-scatter, truly random reverberation, which completely circumvents these issues. A massive deployment of acoustic diffusers, albeit visually ugly and unappealing, would remove this kind of acoustic defect found in many spaces. Strictly speaking, other than Schroeder's ideal, any regular surface is an acoustic defect, including seats in uniform rows.

The fact that virtually all of the listening public is unaware of seat-dip defect explains the lack of interest in correcting it. Yet experts argue that an accumulation of such acoustic defects degrades the aural experience even if listeners are unaware of the specific acoustic attributes. Scientists can measure this acoustic defect and prove that it is real. They can construct laboratory experiments showing that it is audible. They can train listeners to perceive the effect. They can suggest research paradigms that illustrate the subtlety of the effect. However, until the acoustic defect becomes objectionable to a significant percentage of the audience, it cannot be classified as a problem in need of a solution.

The problem of nonuniform frequency response produces a stronger manifestation when listening to music in a small space, which has numerous strong resonances and antiresonances that are orders of magnitude more perceptible than the seat-dip effect in a concert hall. For all the majestic elegance and relative spectral uniformity of concert halls and cathedrals, we spend more time listening to music in small, prosaic environments, such as living rooms, automobiles, jazz clubs, student dormitories, recording studios, and rehearsal rooms. Nevertheless, an inappropriate acoustic volume is readily apparent even to unsophisticated listeners, who perceive a bathroom as a highly defective acoustic space.

The physics of sound waves in small spaces are well understood, especially when the walls are parallel. Consider a sound wave at 115 Hz, which has a wavelength of about 3 meters (10 feet). If two parallel walls are also spaced 3 meters apart, exactly one wavelength of that sound fits between the walls, like the undulating wave of a taut wire anchored at each end. Sound at such frequencies produces a standing wave whose amplitude depends on location. At some locations, the signal is strongly amplified (antinodes); and at other locations, the signal is strongly attenuated (nodes). Other frequencies are not affected as much when the distance between the walls is not an integral number of wavelengths. In simple words, some frequencies show dramatic differences in amplitude depending on the location of the walls.

For a small room at low frequencies, the transmission gain from the sound source to the listener (frequency response) resembles a craggy mountain range, rather than a prairie. Peaks in the response originate from resonances, and valleys originate from antiresonances. Because the density of resonances increases with the square of frequencies, eventually the density is sufficient to produce statistical uniformity, which is perceptually equivalent to a flat response. Large spaces have statistical uniformity over the entire frequency range, whereas smaller spaces have individual low-frequency resonances that dominate. The transition threshold between statistical uniformity and discrete resonances, called the "Schroeder frequency" (Schroeder, 1996), is simply much higher in a small space. A large percentage of the music spectrum exists where discrete resonances dominate. In a small room of only 25 cubic meters (900 cubic feet), the threshold is 400 Hz, and in a small automobile of 2.5 cubic meters (90 cubic feet), the threshold is 1,200 Hz (Kuttruff, 1998). Below 1,000 Hz, the major musical fundamentals exist at frequencies where a small space produces isolated resonances.

Are discrete spatial resonances an acoustic defect? At extremes, the resonance effect is not the least bit subtle. Listening to concert music in a bathroom does not at all sound like listening to music in a concert hall. Training is not needed to appreciate the difference, and few listeners would choose a tiny space for music. As size increases, however, the effect becomes weaker. And for many listeners, even those who have invested in a high-quality sound reproduction system, there is little awareness of spatial resonances as an acoustic defect. Nevertheless, a few listeners find spatial resonances sufficiently objectionable that they are willing to invest substantial resources to designing a room just for listening.

There are several ways that you, as a listener, could mitigate the consequences of spatial resonances. First, you could simply use a very large room. Second, you could acoustically treat the surfaces and obliquely orient the walls (as in recording studios) to reduce the influence of standing waves. And third, you could use filter electronics, however primitive, to partially equalize the effect of resonances. In fact, the critical step for any listener is to first recognize that a problem exists. Yet few do.

Defects in frequency response are only illustrative examples. Several principles emerge from this discussion, principles that apply to any of a dozen acoustic parameters, not just frequency response. Earlier, we divided the acoustics of a space into three aspects: physical acoustics, perceptual acoustics, and cultural acoustics. When discussing an acoustic defect we must first choose which one or more of its three aspects we wish to address.

Scientists can measure frequency errors in terms of amplitude deviations over a selected bandwidth. That produces a set of numbers. The relationship between numbers and perception, however, is grossly nonlinear. Reducing an error from 20 to 10 percent or from 0.02 to 0.01 percent is an improvement by a factor of 2. The former may be experienced as a major improvement in quality, whereas the latter may appear as unchanged, already being at the highest quality. Perceptual metrics are only indirectly related to physical measures. Better numbers do not necessarily imply better experiences. A statistically significant result (not random) may still be socially irrelevant (nobody cares).

That everyone can hear the frequency response errors from the conical horn used for the early recordings of Pablo Casals is self-evident. But it does not follow that the brilliance of his performance is artistically degraded by those errors. We can think of musical aesthetics and audio quality as independent dimensions. And listeners have varying sensitivity to each. When carried to the absurd, listeners can train themselves to hear all kinds of acoustic defects, and eventually, such individuals become an extension of perceptual science rather than devotees of musical elegance. Some amateur audiophiles spend more time improving their sound system than listening to music.

To place the dilemma of quality into perspective, consider looking out of your bedroom window on a beautiful spring morning. Do you experience the unpleasantness of accumulated winter dirt on the windowpanes, or do you delight at the sight of flowers heralding the arrival of spring? For most of us, a modest amount of dirt is neither detected nor experienced as a problem. Ultimately, aural architects designing concert halls have to estimate how much acoustic dirt is objectionable to a specific auditory subculture. When do the acoustics need to be improved? The answer is simply when acoustic dirt noticeably degrades the musical experience for those who support concert halls. And that depends on art and cultural values, which transcend science and engineering.

The Sensation of Enveloping Spaciousness

The auditory perception of an acoustic space, like the visual recognition of a face, can be decomposed into separate attributes, some strong, some weak. A strong acoustic attribute is readily apparent without training. Conversely, a weak attribute, however real, measurable, and detectable by subjects in controlled experiments, is too subtle for most listeners to observe. In my opinion, frequency response consistency, except in extreme cases, is a weak acoustic attribute, whereas enveloping reverberation arriving from all directions is a strong one. Sorting acoustic attributes into weak and strong is central to all discussions about the physical basis for aural architecture.

The dominant auditory perception of a large space is *spaciousness*, a scientific label for experiencing enclosed spaces. Aural spaciousness is the sensation of being inside the music rather than separate from it, analogous to swimming underwater rather than being sprayed with a garden hose. In the process of analyzing the attributes of spaciousness, researchers eventually realized that it comprised two distinct, unrelated phenomena: *apparent source width* and *listener envelopment* or enveloping reverberation (Bradley and Soulodre, 1995a). Spatial acoustics can enlarge the aural perception of the source size, converting it from a tiny point to a massive object—sound source broadening. The same acoustics can also envelop listeners in a sound field without influencing the perceived size, location, or orientation of a sound source. Together, the two phenomena create the experience of spaciousness.

What is the relationship between the physical process of reverberation and the perceptual experience of aural spaciousness? Which physical attributes of a sound field produce source broadening, and which produce listener envelopment? The answer is critically important for the design of concert halls, surround-sound reproduction, and aural architecture in general. After many decades of scientific research, often with contradictory and ambiguous results, researchers currently believe that sonic reflections arriving laterally dominate the perception of space: sonic energy from the sides is more important than sonic energy from the front, back, or top. In addition, energy arriving shortly after the direct sound contributes to source broadening, whereas energy arriving slightly later contributes to listener envelopment.

These insights immediately led to a search for an equation that would convert the parameters of the sound field into a measure of the perceptual experience. Equations have predictive power. Creating equations is easy; creating equations that relate to perception is more difficult; and creating *the* equation that predicts spatial perception in all cases remains, at least for the moment, elusive. There are many candidates for computing aural spaciousness, but none does so reliably in a wide range of cases.

The history of this research illuminates many of the fundamental difficulties in perceptual science. Consider, for example, the competing choices for determining the boundary between early and late reverberation or between apparent source width and envelopment, which ranges from 80 milliseconds (Bradley and Soulodre, 1995b) to 160 milliseconds (Griesinger, 1999). Rather than a single number, the boundary between early and late might be represented by a gradual transition—yet another equation.

Similarly, there are also many approaches to measuring the perception of lateral sonic reflections. One approach compares the sounds in the two ears of a dummy head, approximating what a human would hear. An equation computing the metric *interaural cross-correlation coefficient* measures the degree to which signals to the two ears are similar. Highly dissimilar sounds from the two lateral directions produce a stronger sensation of spaciousness. Another approach uses a microphone with a figure-eight sensitivity to measure the *lateral fraction*, the percentage of sonic energy arriving from the sides. And still another approach measures *lateral gain*, the increase in loudness relative to what it would have been in open space. Different frequency regions make different contributions to aural spaciousness (Okano, Beranek, and Hidaka, 1998). Sonic energy at the frequency extremes contributes less than at the middle frequencies. Each investigation reveals yet additional factors that need to be considered in order to predict auditory perception.

An ideal equation, if it exists, and nobody has yet shown that it does, would incorporate the complex interaction among an ever-expanding set of temporal, spectral, and spatial measures. The task of matching numbers to auditory perception is tiresome because the equations are effectively attempting to duplicate the hidden signal processing of the auditory cortex. There is no reason to believe that cortical behavior is either simple enough or consistent enough among individuals across a variety of cultures to conform to a single equation. Often a small displacement in the location of a listener produces a twofold change in the measurement, but without a corresponding change in auditory perception (de Vries and Hulsebos, 2001). For example, moving forward in your seat dramatically changes some metrics but not your aural experience. With spatial averaging, the resulting measures of aural spaciousness become more consistent (de Vries, Hulsebos, and Baan, 2000), but not necessarily more accurate.

Rather than search for equations with exact metrics, scientists have created heuristic models: a careful composite of reasoned arguments and accumulated knowledge combined into a single concept. Although such models often omit details that would make them formally testable, a fundamental requirement of the scientific method, they are nevertheless useful because they contain a large corpus of knowledge. Indeed, heuristic models often provide more insight into the underlying phenomenon than testing equations.

David Griesinger (1999) proposed such a hierarchical model for explaining aural spaciousness. In his view, the early part of the reverberation process fuses with the direct sound to form a unitary sonic event, which increases the aurally perceived size of the sound source. With the right balance of early reflections, for example, a violin sounds like a large instrument with mass and power. Masayuki Morimoto and Kazuhiro Iida (1993) described source broadening as "the width of the sound image fused temporally and spatially with the direct sound."

Like all loud sounds, this fused sonic event also produces a temporary insensitivity (momentary deafness) to the reverberation energy immediately following. This is the principle of masking—loud sounds make inaudible the following weak sounds. Not only do the early reflections contribute to source broadening, but also they inhibit the experience of enveloping reverberation for about the next 100 milliseconds. The boundary between early and late reverberation is related to auditory masking, a topic still being researched.

As this momentary deafness abates, the sensation of background reverberation increases. Because the reverberation is also decaying, however, it is readily masked by subsequent musical notes that are louder. In other words, there is only a small time window, between about 100 and 300 milliseconds, for creating the experience of listener envelopment: the continuous reverberation process is audible only during this time window. This window is squeezed between early sonic reflections of a given musical note and the next note after it. Previously, we defined the experience of this window as running reverberance. Although it does not demonstrate detailed equations for running reverberance, Griesinger's model is appealing because it is consistent with a much larger body of literature.

The shape of the reverberation decay controls the sensation of envelopment because the shape determines the amount of sonic energy in the critical window. Low sonic energy in the window produces weak aural envelopment; high energy produces strong envelopment. In an artificial reverberator, where the details of the reverberation process are under the direct control of an experimenter, it is quite easy to influence the relative sonic energy in this window. As confirmation of the heuristic model, William G. Gardner and David Griesinger (1994) conducted experiments where listeners adjusted different types of reverberation curves until they appeared equally loud. The result showed a match in the intensity around 200 milliseconds, the peak of the critical window. Regardless of the shape of the reverberation process, similar sonic energy in the window produced a similar sensation of reverberation intensity.

More than a century after reverberation was defined as the time for energy to decay to 0.001 of its original value (a physical metric), scientists discovered that *perceived* reverberation depended mostly on the sonic energy in an early part of the reverberation. Griesinger (1995) argued that, because it is a good measure of the energy in the time window, the slope during the initial 350 milliseconds should replace the traditional measure. The improved model corresponds, at least in general terms, to the experience of those who are experts in reverberator design, where manipulating the shape and delay of the reverberation process is straightforward. In addition, the model explains inconsistencies observed in those concert halls with identical reverberation times but varying degrees of envelopment. This new model has been an improvement over the earlier one because the physical acoustics of concert halls were now linked to their perceptual acoustics.

Having briefly considered the physical and perceptual basis of aural spaciousness, let us proceed to its cultural basis. Let us assume for the moment that the perceptual properties for source broadening and listener envelopment are completely understood. Applying those results to real concert halls is not obvious. Each region of a concert hall is likely to have a different degree of spaciousness, and an aural architect cannot manipulate the design for uniformity in the entire space. Is a particular degree of aural spaciousness, perhaps too much or too little, a defect? Every seat has its own pair of numeric ratings for the two components of spaciousness, and listeners could select their seat based on their personal preferences for spaciousness. Informal evidence, however, suggests that the average listener is either unaware of or indifferent to aural spaciousness when choosing a seat. The most expensive tickets, which indicate strong preferences, do not necessarily have good acoustics.

Ironically, the implications of aural spaciousness are clearer in the context of recorded music because the audio engineer and the sound mixer can create a relatively uniform experience for those listeners with a particular sound system configuration. There are adjustments on artificial reverberators that correspond to the two measures of spaciousness. In the recording studio, perceptual parameters are aligned with artistic criteria and musical taste. Thus, as the major attribute of the aural architecture of musical spaces, spaciousness moves from a static characteristic of a concert hall to an active component of a musical art form. In fact, each instrument in a recording studio can have, and often does have, its own individual source broadening and enveloping reverberance.

The final aural architects of a musical space are the listeners who purchase and install sound systems in their homes or automobiles. The ability of such systems to create aural spaciousness ranges from ultraprimitive to elegantly complex. It only requires a moment to appreciate the differences among various cases: a portable boom box on the beach (no spaciousness), a table radio in the kitchen (limited spaciousness), and a surround-sound home theater (full spaciousness).

Being able to hear aural spaciousness does not, however, imply that it is more important than other social and experiential aspects. When we select a seat in a concert hall, sight lines may be more important than acoustics. Domestic living spaces are used for many purposes other than listening to music. And more important, acquiring knowledge of home acoustics takes time and effort. Like average concertgoers, average listeners, in their private homes simply lack the interest and knowledge to be sophisticated aural architects.

Thus we may conclude that indifference toward aural architecture is only a reflection of cultural values. Visual architects are well aware that ordinary glass produces distorted images, but how many of us purchase plate glass with optical coating for our windows? A few people do; most do not. Can you see the difference? Absolutely. Can science provide physical and perceptual explanations? Yes. Does it matter? Probably not.

Scientific Models of Enclosed Spaces

Although scientific studies answer some questions about the acoustics of enclosed spaces in greater depth than needed, they also ignore important but intractable questions that are too difficult to answer or to formulate. Our limited understanding of auditory spatial awareness is largely the consequence of four interrelated factors. First, society has provided only a minimal investment for supporting research focused exclusively on the aural experience of spaces. There are but a few dozen researchers engaged in full-time study of spatial acoustics. Second, the current state of the art in computer modeling is not adequate to pose let alone to answer some of the more complex questions in this area of inquiry. Third, some questions about auditory spatial awareness will never be answered by science, even if presented in a scientific language, quite simply because they are fundamentally unanswerable within the rules of scientific inquiry. And fourth, there are many groups of unrelated questions about the subject that only modestly overlap. The answers depend on the way that the questions are formulated, the hidden assumptions in the particular paradigm being used, and the definition of what constitutes truth. Among the various disciplines involved in auditory spatial awareness, knowledge is neither well defined nor consistent.

Art, Science, and Engineering Have Divergent Goals
The motivation to study auditory spatial awareness depends on how the resulting knowledge will be used. Knowledge exists within a social and professional context. A few scientists and intellectuals have a passionate desire to understand an acoustic problem without regard for practical application of that understanding. Most others see their efforts as contributing to specific activities: architects to designing office buildings, audio engineers to recording music, musicians to incorporating space as a new musical frontier, and acousticians to upgrading a church's acoustics to support a larger congregation.

For those without an interest or background in science and engineering, understanding the physical nature of acoustics is as mysterious and inaccessible as decoding ancient hieroglyphics. For those not endowed with an artistic sensibility, and for those without an appreciation for the aural nuances of a concert hall, the attributes of a musical space are equally mysterious and inaccessible. Only a few individuals have a strong background in both areas. Surprisingly, most professional architects have little background in acoustics. The following discussion attempts to bridge the gap between art and science by interpreting the science of acoustics in terms of its artistic implications.

The science of aural space is, at one and the same time, extremely useful in providing predictive insights and useless in establishing an extensible model of spatial acoustics. How can these polar opposites be reconciled? Consider the analogy of a sand castle on an ocean beach. The perceptual attributes of the castle depend on the properties and relationship of each grain of sand. To fully understand the castle, we might have to examine every grain, its cohesion with neighboring grains, the concentration of water on its surface, its mass and density, and so on. Then, if the grains are reasonably similar, we might combine their statistical properties into concepts such as form, curvature, and shape. From the artist's macroscopic view, sand castles are works of art; from the physicist's microscopic view, they are assembled grains of sand. The relationship between the two views is not always obvious.

The analogy becomes relevant when we consider a concert hall as a sand castle. Every part of every surface of every internal acoustic object corresponds to an acoustic grain of sand, which then contributes to each sonic reflection and spatial resonance. For a few specific geometries, mostly degenerate, converting a microscopic to macroscopic viewpoint is simple. If the concert hall were a perfect rectangular space with three sets of parallel walls, the classic shoebox, the analysis is straightforward because every sonic reflection and resonance for all possible sound sources and listener locations can be readily computed (Kuttruff, 1973). Every college course on acoustics begins with this example simply because the mathematics is elegant, compact, accurate, and predictive. In contrast, almost every other geometry is too complex for a closed-form mathematical approach, and real concert halls are orders of magnitude more complex than a rectangular box.

Even if a scientist could determine every sonic reflection and spatial resonance in a geometrically complex space, perhaps with an empirical simulation using the world's fastest computer, how would we use the result? Listeners do not hear individual sonic reflections and spatial resonances, especially when the numbers of these approach millions. Rather, they perceive the acoustic attributes of the ensemble, represented by statistical parameters, such as sonic reflection density and spatial resonance density. Similarly, a chorus of singers is a statistical ensemble of voices. Our discussions have continually referred to one particular acoustic parameter: reverberation time, when the average signal level drops to one-thousandth of its initial value. Because it is an average—a statistical—operation, it is immune to small variations. Thus, even though they modify the detailed sonic reflection pattern, small changes in the temperature of the air or the movement of the audience do not influence the reverberation time. Thus, too, averaging of the *appropriate* acoustic parameters serves to remove irrelevant details.

Ultimately, all knowledge struggles with the dilemma of discarding irrelevant details while preserving relevant information. But relevance, itself, is a judgment that depends on the researcher's assumptions. One researcher's data is another's noise. The relevance of knowledge about aspects of spatial acoustics depends on how it is to be used. To illustrate how the choice of application changes the definition of knowledge, let us consider four examples of applied spatial acoustics. Although these examples share some common knowledge, from the perspective of their respective goals, the definition of knowledge changes.

In our first example, an acoustic architect is evaluating the aural properties of a proposed concert hall design before construction in order to choose critical parameters such as area, volume, geometry, and materials. Because of the cost and the consequences of acoustic defects, the designer needs to auralize multiple spaces with sufficient accuracy to detect any major problems. In this case, knowledge is the ability to construct simulated spaces that are sufficiently similar to the real space. The simulations need not be perfect, and they need not attain a level of accuracy such that no listener could tell the difference between the simulations and the actual space. Simulations are simply a form of insurance.

In our second example, a contemporary composer is creating a spatial experience as part of a music composition. However interesting it may be to simulate a real space, creating imaginary and contradictory spatial experiences is even more interesting. Aural space as art still requires knowledge, like a painter understanding pigments, but such knowledge is only in support of the art. By understanding the relationship between acoustic parameters and spatial perception, a composer acquires a wider repertoire of auditory artistic elements to evoke emotions and associations. Accuracy is unnecessary, and even irrelevant.

In our third example, an audio engineer is designing a spatial synthesizer to be used as a source of artificial reverberation in a recording studio. This is a hybrid of the previous two examples. Because the device is an artistic tool that serves as a surrogate of a real acoustic space, it has elements of both art and science. In addition, such a device is also used to enhance the aural impact of musicians and singers as a resonant space extension of their instruments and voices. Artificial space then becomes a special effect, a voice and an instrument changer.

In our fourth and last example, a neurobiologist is modeling the ability of the auditory cortex to discard spatial reflections when listening to speech in a reverberant environment. The goal might be to invent a diagnostic tool to measure brain lesions using a simple spatial test. Given this goal, the simulation can be crude, stylized, and highly simplified because the context is unrelated to any particular space. In fact, simple spaces, such as a rectangular room, are often used. The synthesizer is only a probing tool, like tomography, to visualize and model brain lesions.

Although the architect, composer, engineer, and medical researcher may or may not share technology and scientific knowledge, they always share questions about perception. They are all interested in the listeners' experience. Even with this common interest, however, they diverge in evaluating results and, more important, in their goals. The acoustic architect is concerned with audience comfort or discomfort; the composer is concerned with making an artistic statement about space to listeners; the audio engineer is concerned with providing listeners a useful, commercially viable tool whose sales will justify the cost of research and development; and the neurobiologist is concerned with patients' well-being. With different goals, they each have a method for acquiring results, a unique criterion for ascertaining legitimate truths, a different attitude toward risk of failure, and particular training appropriate only to the stated goal. Because they all speak their own private language, sharing knowledge across disciplinary boundaries is often difficult, if not impossible.

Between the extremes of auditory artists and acoustic scientists, we find people of varying sensibilities and attitudes who combine knowledge and skills in both art and science, including scientists functioning as artists and vice versa. Understanding such "hybrids" can be particularly challenging because we often do not know which viewpoint they are using to support a given assertion. If the goal is art, why should a simulation require an accurate model of a specific seat in an existing space? If the goal is science, explanations cannot use an artistic language, and all assumptions must still be carefully tested regardless of their artistic relevance. When the goal is art and science, dual criteria must be explicitly applied in a consistent manner. Inconsistency destroys their value to both art and science. The most glaring examples of the divergence between art and science are usually signaled by linguistic ambiguities. Consider the unstated distinction between an acoustic attribute that is perceptible (a matter of science) and one that is desirable (a matter of art). I have rarely encountered individuals who were sufficiently knowledgeable to take advantage of the overlap while preserving a clear sense of the differences. Hybrid artist-scientists can exemplify the best and worst aspects of marrying art and science.

Acoustic Measurements of Enclosed Spaces

Spanning more than a century, researchers from many nations have created a wealth of information about the aural architecture of spaces used for music, mostly concert halls. Although there are many methods for characterizing their acoustics, to one degree or another, every approach is a compromise and a simplification of an intractable problem. The appearance of being relatively straightforward is only an illusion.

As every engineering student knows, an acoustic or electrical system can be characterized by its response to a pulse, such as a click or spark. When a click is produced on stage, with a microphone placed at a choice seat, we can measure the way that the acoustic space changes the click, the *impulse response*, which completely defines the properties of sound transmission from stage to listener. Alternatively, when the space is excited by a sequence of pure tones, we can measure the amplitude and phase of each tone at the microphone to arrive at an equally complete definition, the *frequency response*. We can mathematically convert an impulse response to a frequency response, and vice versa. Either type of response provides a complete characterization of a sound transmission system. A concert hall, in theory, is such a system.

A careful examination of the mathematical assumptions for interpreting these two types of responses reveals a hidden trap when the theory is applied to real acoustic spaces. Although the theory is without problems, actual measurements produce believable, yet erroneous, results. The physical world simply does not fit the theoretical assumptions, which acousticians have chosen mostly to forget or ignore. To produce valid results, the system being measured must have two properties: *linearity* and *time invariance*. In a linear system, doubling the input must double the output; in a time-invariant system, all measurements must be identically replicable over time. Of the two requirements, the second is the subtlest and the most problematic.

For a system to be time invariant, physical changes are not permitted either during or between measurements. As the acoustic medium that carries sound, air is a critical part of the system. Although experimenters can keep objects in the environment stationary, they cannot control air, whose temperature and humidity influence the speed of sound (Hardy, Telfair, Pielemeier, 1942). A sound that travels for 300 meters (1,000 feet), bouncing from surface to surface, arrives at the listener's ears at a time determined by the speed of sound in air. Depending on the temperature of the air in that particular path at that particular moment, a sound might take 1.013 seconds to reach a listener; a moment later, it might take 0.993 seconds, a difference of 20 milliseconds. Although such small variations in arrival time are imperceptible, they are more than sufficient to destroy the assumptions embedded in the mathematics.

Changes in temperature also destabilize frequency response measurement. Vern Oliver Knudsen (1946) described a controlled experiment that was designed to measure the speed of sound one quiet Sunday morning before sunrise when there was neither perceptible wind nor thermal heating. Using a parabolic loudspeaker and microphone about 30 meters (100 feet) apart, he measured the response to a 4,000 Hz tone. "The tone at the receiving end was anything but steady. It fluctuated violently over a range of more than 10 [decibels] (3:1 ratio), with short periods (a tenth of a second or less) and long periods (several seconds) all jumbled together" (Knudsen, 1946). The magnitude of the fluctuations increased with higher frequencies and longer path lengths. At ten times the distance, the response would have been totally random. Even minor variations in temperature destroy time invariance.

To illustrate the difficulty in removing thermal turbulence from an enclosed space, Knudsen cited a similar experiment that was performed in an enclosed space of 550 cubic meters (19,750 cubic feet) 10 meters (100 feet) underground. The temperature of this space, being isolated from all environmental influences, varied less than a few thousandths of a degree over a matter of hours. The air was homogeneous and without turbulence. In this environment, the measurements between the loudspeaker and microphone were stable and without fluctuations. However, by introducing a small fan or a heated wire, the measurements again became unstable and time varying.

If a small fan destroys stability, imagine the environment of a concert hall. Even disabling the air-conditioning does not produce uniform temperatures because the concert hall is not isolated from external weather, which in Boston changes by the hour. In addition, as every air-conditioning designer knows, people in the audience are a major source of heat, on the order of 100 watts per person. At the beginning of the last century, Wallace Clement Sabine devoted an entire chapter to the subject of nonuniform sound velocity, concluding that with "many such differences in temperature, though slight in amount, the total effect might be great" (Sabine, 1922). Temperature determines the density of air. And thermal gradients result in variable sonic refraction indices, much like the shimmering heat waves radiating from the road surface on a sunny day. We cannot photograph the details of a distant object in such an environment. Just as thermal gradients bend light waves, they also bend sound waves (Humphreys, 1940). Although we can measure the impulse response, the data in the late reverberation are mostly noise, having little to do with the actual space.

As we notice from the dates of the previous references, they are more than a half century old; they were published in obscure journals, now located in dusty archives. Without knowing about this work and without realizing the underlying flaw in their assumptions, modern engineers and scientists are still inventing creative methods to solve an intrinsic problem: measuring the unmeasurable.

Recording an impulse response in a concert hall requires that reverberation be loud enough to overcome background noise, especially at the end of the reverberation tail, when it has decayed to a low level. To ensure sufficient loudness throughout the decay, the initial excitation must be extremely intense. Although a loudspeaker cannot generate the required narrow, high-energy pulse, a spark gap can. But it, too, has problems because extremely intense sounds violate the other mathematical assumption: linearity. The radiation pattern of a spark is not uniform because its high pressure makes air nonlinear (Wright and Medendorp, 1967). That nonlinearity, in turn, violates a cardinal assumption in the sound-wave equation. In addition, a high-intensity pulse is accompanied by intense heat. Locally heated air produces a thermal gradient that creates an acoustic lens, which then refracts any sonic reflections traversing that region (Lafleur et al., 1987). In simple language, a narrow sound pulse of sufficient amplitude cannot be created, and therefore, the impulse response cannot be directly measured.

In another approach, scientists widened the excitation pulse without increasing its amplitude, a technique that increases sonic energy while preserving linearity. Barron (1984) suggested a half- or full-cycle sine-wave pulse of varying widths to cover the sound spectrum; Augustine Berkhout, Diemer de Vries, and Marinus M. Boone (1980) proposed a high-amplitude swept sine wave of about 200 milliseconds; and Schroeder (1979) recommended low-level continuous pseudorandom noise. For each of these signals, the observed impulse response can be mathematically transformed back into a more standard representation (Borish and Angell, 1983), although these alternative signals work only for low frequencies. An extended-duration high-frequency signal remains too fragile for thermal waves, a conclusion confirmed by U. Peter Svensson and Johan L. Nielsen (1999).

Methods are readily available to accurately characterize the acoustics of an enclosed space under one of three restrictive conditions: the space must be small; the impulse response must be limited to early sonic reflections, which travel for a short time over small distances; or the sound spectrum must be limited to low frequencies. Because concert halls violate all three conditions, accurately characterizing their acoustics is impossible. No method can measure the details of high-frequency reverberation extending over a long time in a large space.

The mathematical problem of thermal waves illustrates an important paradox of art and science: variability is good for art and bad for science. As a result of the turbulence produced by air-conditioning in a large enclosed space, Yasutaka Ueda and Yoichi Ando (1997) measured wild fluctuations in its steady-state impulse response of sufficient magnitude to be perceptible. As we will discuss later in a different context, these variations make an important artistic contribution. The very property that destroys the mathematical assumption in physical measurements enhances the listening experience. Acoustic variability is part of nature. And ironically, the aesthetic value of random modulation in artificial reverberation algorithms, equivalent to thermal waves, was an empirical discovery made without any awareness of the research from the early 1940s. Although audio engineers can hear the beneficial contribution of randomness, which removes the sterile and mechanical quality of static sound patterns, even today, few audio professionals are aware of the underlying explanation. Nevertheless, most designers of studio reverberators explicitly include random time modulation (Gardner, 1998). Any designer of an artificial reverberator or a spatial simulator who claims to have successfully incorporated the acoustic details of a real concert hall by using a measured impulse response is naive, dishonest, or a marketing zealot.

Rather than measure real concert halls with nature's intractable problems, scientists can make a conceptual model of a musical space, and then explore the properties of the model using computer simulations. Acoustic models are easier to measure than real acoustic spaces. Models can assume constant and uniform sound velocity in a linear medium. In fact, those who create computer models have an infinite number of choices: include, exclude, or modify any physical parameter. Models can evaluate acoustic metrics with mathematical measurement tools. Models can process sound, allowing listeners to aurally visualize the space as if it were a real one. Models are flexible and inexpensive. With modest programming effort, for example, the ceiling of a hypothetical space can be lowered, raised, or tilted. Surface materials can be changed, acoustic hanging clouds added, or volume expanded. Listeners can instantly switch among multiple versions of a simulated concert hall, a simple way of comparing spaces without employing unreliable auditory memory. Models are particularly well matched to what-if questions, and thousands of variants can exist within a single computer. Yet, even as it solves one set of problems, this approach creates others.

To appreciate the consequence of creating an accurate acoustic model of a concert hall, we can compare its computational requirements with the power of available computers. Computers are now very powerful, recently having calculated π to 1.24 trillion places (Kanada, 2002). At this writing, the record for the world's most powerful supercomputer was held by NEC's Earth Simulator, with a computational speed of 35,860 gigaflops (billion floating-point operations per second) and a cost of $350 million (Tristram, 2003). Such computers are appropriate for any number of difficult tasks, from predicting the world's weather to displaying molecular reactions at the subatomic level. By comparison, characterizing concert hall acoustics is a relatively easy problem. Since the behavior of sound waves in an enclosed space is a deterministic problem based on the repeated computation of a mathematical expression—the wave equation—computers should be able to simulate a concert hall. Computing power relative to the quantity of data ought not to be an issue, but it still is.

A quick calculation illustrates the computational burdens. Assume that the interior surface of a concert hall is fully characterized in a database. Since the sound wave equation is relatively simple, and since that equation can predict the behavior of sound from instant to instant, the computer can compute the complete acoustic response with arbitrarily high accuracy. The algorithm partitions the space into tiny cubes of air, each small enough that the sound pressure is uniform within that volume. Each cube is then represented as a single point in space. As long as the density of points is higher than twice the wavelength at the highest frequency, the sound in the space can be fully reconstructed. Since these computations must be performed twice as often as the highest frequency, which is a mathematical requirement, the time response of the concert hall can be determined at every seat.

Ignoring the details, which are tractable but complex, what is the computation load for a 20,000 Hz audio bandwidth in a typical concert hall? Each cube would need to be about 7 millimeters ($\frac{5}{16}$ inch) on a side with a volume of about $\frac{1}{3}$ cubic centimeter ($\frac{1}{50}$ cubic inch). In an enclosed space with a total volume of 20,000 cubic meters (710,000 cubic feet), there would be some 60 billion such cubes, each of which communicates with at least six neighboring cubes. To determine the reverberation over a 2-second interval, the program must iterate for 100,000 time samples. At a microscopic level, the wave equation must be evaluated at least 3×10^{16} times, a number greater than the number of seconds since our first primate ancestors climbed out of the trees to become human beings. Specifying the interior surface is equally daunting. Using the same spatial resolution and adding the absorption coefficient at many frequencies, there would be something on the order of 1 billion data points to describe the surface interior. That is the number of seconds in a century.

Even using a supercomputer, such a computation would take days, if not years, to converge. And for each variant of a concert hall, a full simulation must be again computed. Someday, computers of even greater power may reside within the average acoustics laboratory, but probably not soon. For the moment, while a brute-force computation of a concert hall is theoretically possible, that approach is neither likely nor practical.

The computational load drops dramatically if we limit the problems. When the bandwidth is reduced to 2,000 Hz, one-tenth of the previous bandwidth, the computational load decreases by a factor of 10,000. If the size of the space is reduced to one-tenth, the load drops another factor of 1,000. And if we are interested only in the first 100 milliseconds, the load drops by another factor of 10. With these simplifications, the number of iterations is merely 300 million, well within the range of a personal computer. Yet again, we see the same issue: high frequencies in large spaces extended over a long duration create unsolvable problems in both computation and measurement. Thus we have answered our initial question: concert halls cannot be fully and accurately characterized with any technology that is currently available.

Let us now return to the world of technical compromises and clever algorithms. By including one new assumption, which has been validated in many common situations, researchers have simplified the problem: separate reverberation into an early part and a late part. The late part is a statistical process whose details are irrelevant as long as a few dozen parameters are accurately represented. This leaves the early part, which is amenable to computation and measurement. The boundary between the two parts, although somewhat arbitrary, is approximately 150 milliseconds for most situations.

Leaving aside late reverberation, which is discussed in the next section, the task of characterizing the acoustics at a seat in a concert hall is reduced to the problem of ascertaining the amplitude and time arrival of early sonic reflections. When so formulated, the problem corresponds to a three-dimensional version of billiards, an analogy that illustrates the algorithmic issues. In our version of this game, a rectangular billiard table (without pockets) has two balls: the sound-source cue ball, and the listener target ball. The scientist-player strokes the cue ball in a direction such that the cue ball hits the target ball. This corresponds to one of the many directions taken by a spherically radiating sound wave. There are many possible choices for a direction that will produce impact. The shortest path, without bounces on any side rails, corresponds to the direct sound. Alternatively, there are four other directions where the cue ball will hit the target after bouncing once on each of the four side rails, each corresponding to a single sonic reflection. There are twelve directions for the cue ball to hit the target ball after two bounces, and so on.

The number of directions that will result in the cue ball hitting the target ball is very large. How to find all directions, each of which produces a reflection? The most common methods are ray tracing (Sekiguchi and Kimura, 1991; Giménez and Martin, 1988), image folding (Allen and Berkeley, 1979; Borish, 1984), and combinations of each (Vorländer, 1989; Heinz, 1993). These algorithms iteratively search for trajectories that will connect the listener to the sound source with multiple sonic reflections from the surface boundaries.

To make the billiard table more representative of a concert hall, we must replace the table's rectangular shape with an irregular polygon, scatter miscellaneous objects on the surface, construct the balcony side rail with materials of varying elasticity, use boundaries that include nooks, crannies, and curves, and add the third dimension of height. In other words, we must include all those details which are acoustically significant, and which determine the accuracy of the result. Without sufficient detail, the result is only weakly related to an actual concert hall.

Consider some of the choices. Should the statues on the side of the upper balcony be ignored, modeled as concrete cylinders, or faithfully represented down to the level of mouth, nose, and eyelids? Is the balcony equivalent to a solid ledge? Should a human being, as a sound-absorbing acoustic object, represent each seat? Only a smooth surface segment produces specular reflections, like a clear image of a mirrored surface. Typical concert hall objects, like sparkling glass crystals hanging from a chandelier, are complex surfaces that disperse sound in all directions (Dalenbäck, 1995, 1996). In addition, the degree to which sound bends around a straight edge depends on frequency: higher-frequency sounds are shadowed by an edge, but lower-frequency sounds flow around it like warm gelatin (Torres, Svensson, and Kleiner, 2001). Surface materials absorb sound differently at low, medium, and high frequencies (Savioja et al., 1999). For all these reasons, making a model is more art than science, and labor-intensive art at that, the need for and cost of labor growing exponentially with successive increases in the model's accuracy.

Rather than create a computer model of a concert hall, researchers can construct a scale model using wood, fabric, metal, and other materials: building a miniature version of the space. Making scale models is, especially in art and architecture, an ancient and refined tradition. Predating computers by half a century, Spandöck (1934) demonstrated how a scale model could be used to simulate the acoustics of a space. Scale models were used to validate the planned music center in Eindhoven, Holland (Braat-Eggen, van Luxemburg, and Booy, 1993; Boone and Braat-Eggen, 1994), and to evaluate a renovation to an existing space (Cocchi and Farina, 1990). When building such models, researchers must still specify the amount of detail. And the techniques and trade-offs for sculpting real surfaces with carpentry tools are, in some sense, as difficult as creating an abstract surface in a computer database.

Scale models pose an additional problem that computer models do not: the need to find material whose acoustic properties are the scaled equivalents of those of the material being modeled. If the model is scaled by a factor of 10:1, then 10,000 Hz in the model corresponds to 1,000 Hz in the modeled space. Just as size scales, so, too, does frequency. The reflective and absorptive properties of wood, fabric, plaster, and people at 1,000 Hz must be replaced with other materials having the same properties at 10,000 Hz. For example, the sound-absorbing properties of air, which depends on humidity, only partially scale. To achieve the required accuracy with a 10:1 scale, Barron (1983) showed that air had to have a relative humidity of 2 percent. With a 50:1 scale, there is no gas medium for representing a scaled version of air (Polack, 1989). As a general conclusion, Mendel Kleiner, Rafal J. Orlowski, and Jakub Kirszenstein (1993) concluded that scale models and computer simulations each have their own assets and liabilities: some metrics were consistent, others were not.

The burden of physical accuracy, be it from modeling or from measurements, is important only when there is a reward for an exact simulation of a real acoustic space. In other cases, crude approximations are more than acceptable. Research effort grows exponentially with increasing requirements for accuracy. In some cases, the rewards for high accuracy can be large, and in others, there are none. For example, given the painful consequences of *not* correcting serious acoustic defects and the unacceptably high cost of rebuilding the hall, there was every incentive to achieve accuracy in the acoustic modeling of Avery Fisher Hall at Lincoln Center (Bliven, 1976). In contrast, if the goal is only the artistic experience of a hypothetical space, then artistic and perceptual liberties are more than acceptable. A physical model of a real space is actually not needed.

Evaluating spatial simulations in terms of art or science is clear only at the polar extremes; intermediate cases are ambiguous and contradictory. For example, a simulation of a rectilinear shoebox with smooth parallel walls (Kendall et al., 1986) is too crude for predicting the aural experience of a real acoustic space, and it is artistically too sterile to be musically useful. Implementing crude physical models of degenerate geometries is both bad art and bad science. When evaluating a spatial simulation, we must be clear as to its purpose. And in many cases, there is actually little crossover between art and science.

Reverberation as a Random Process

Having separated reverberation into early and late parts, and having considered discrete sonic reflections only during the early part of reverberation, let us now turn our attention to the late part, which we will consider as a statistically random process. Can a statistical viewpoint provide insights that are not otherwise obtainable? The answer is yes. Sonic reflections in a large space arrive by the millions, each with minimal loudness. Although a single reflection is inaudible by itself, the sum is what we hear as reverberation. Just as a single grain of sand is invisible, millions of grains together form the surface of a sand castle. We can think of the castle's grains and reverberation's sonic reflections as the statistical combination of individually insignificant details.

To appreciate the subtlety of reverberation, we first need to explore the statistical properties of the decay process. Consider a clarinet note in a cathedral having an 8-second decay time. The sound seems to stay around forever. Now play a much longer version of that same note in an acoustically dead studio. It, too, seems to stay around forever. In each case, the pitch is the same, and the sonic energy at every overtone is the same. In fact, if we were to measure the sonic energy at every frequency, the two cases would be identical. Yet even when we listen with one ear, which removes binaural effects, the two cases sound altogether different. One is the sound of a clarinet note in a dead space and the other the sound of a clarinet note in a cathedral. What accounts for the difference? There is more to the statistical story.

Although there is a distinction between a continuous clarinet note and its reverberation, it took decades before I fully appreciated the simplicity of the explanation: reverberation destroys the temporal fine structure of sound. With a pure clarinet note, each spectral component (overtone) is well defined, relatively constant, and time aligned with every other spectral component. The composite waveform of the clarinet note is periodic or quasi-periodic. Once reverberated, however, every spectral component acquires a statistically random time relationship to every other component. The destruction of temporal ordering is even more apparent with a sequence of notes. The musical note B immediately followed by F produces the same reverberation as the note F followed by B. Time ordering is destroyed. Similarly, the attack, sustain, and decay of a given component, as well as its vibrato and tremolo, are all stirred into a aural soup and cooked until nothing remains of those temporal details (Polack, Alrutz, and Schroeder, 1984). Only the intensity of the *total* energy at each frequency remains unchanged. Reverberation randomizes time.

As a means of confirming this assertion, Heinrich Kuttruff (1991) created two periodic signals with the identical frequency spectrum, but one was composed of narrow impulses and the other was continuous. They indeed sounded very different, as they should, but when they reverberate, they became indistinguishable. His example is elegant in its simplicity.

With this insight, we can demonstrate an alternative means for creating reverberation without actually using a space. Any process that preserves the sonic energy of each spectral component while destroying the time structure is reverberation. Consider the following example. We first measure the sonic energy of each component of a clarinet note. We then configure several thousand oscillators, each with an intensity corresponding to one frequency component of the clarinet. Not only do we hear the sum of these oscillators as a reverberated clarinet note, but also the summed signal is mathematically identical to reverberation in a cathedral. Each oscillator corresponds to a single resonant mode of the cathedral.

Think of a cathedral as millions of bells (resonating oscillators), each with its own pitch (resonance frequency), and each with a slightly different decay rate (reverberation time). The clarinet sound rings (excites) only those bells with a pitch corresponding to the frequency content of the clarinet. In other words, you are actually hearing the bells of space, not the original clarinet sound. With hundreds of millions of bells at all possible pitches, a cathedral can faithfully reproduce the frequency content of any musical instrument.

A large enclosed space creates a statistical process that can be described as either millions of sonic reflections or millions of resonances. Because there are so many of each, we cannot describe each one individually, rather, we describe them statistically, that is, we describe their average properties. The statistical location of grains of sand distinguishes a flat beach from a rolling hill. But what statistical attributes distinguish a bathroom from a cathedral? And what statistics need to be faithfully replicated in artificial reverberators? Are there statistical measures that distinguish between boring and interesting sounds? Which statistics should be part of the art of a musical space?

Before delving into the answers to these questions, we need to review what scientists have learned about the statistics of spatial acoustics. The resonance density, the number of resonances at a given frequency, is proportional to the volume of the acoustic space and the square of frequency. Large spaces have more than enough resonances at all frequencies to adequately cover the complete pitch range. Conversely, small acoustic spaces have insufficient resonances at low frequencies to form a statistical cloud. Small spaces break the statistical assumption.

Suppose you were to hear the lowest note of an organ in a tiny church, where there might be only one or two excited resonances, as opposed to hundreds in a cathedral. The reverberation decay of this note has a kind of "wow-wow-wow" periodic variation in amplitude, the result of two similar frequencies beating against each other. If the organ note excites only one resonance in the church, and if that resonance is not at the same pitch, then reverberation would mistune the music, a degenerate but real possibility. Knudsen (1932) mentioned cases where the pitch of the decay had been shifted by as much as a semitone in a small space. You would first hear the true pitch of the organ followed by the shifted pitch of the spatial resonance.

The threshold frequency, below which resonances are discrete and above which they merge into statistical clusters, is called the "Schroeder frequency" (Schroeder, 1996). The distinction between these two cases depends on the number of resonances that are excited by a single frequency. Schroeder (1962a) used three as the minimum, but statistical convergence is improved if there are at least a dozen. If the resonance density is high enough, and if the acoustic space is even slightly irregular, each resonance is independent of every other resonance in time, phase, and frequency. The Schroeder frequency, which is inversely proportional to the square root of volume, is the transition threshold for the statistical model. For a large space, the threshold is so low that the fundamental frequencies of all musical instruments are well above it. But for a small space, the threshold could be as high as 300 Hz, which splits the range of common musical fundamentals.

Above the threshold, the sum of many resonances obeys the "law of large numbers," called the "central limit theorem." According to this theorem, the sum of a dozen unrelated but similar signals approaches a Gaussian process, the most famous statistical distribution, and because the amplitude of reverberation is the sum of two such processes, it follows what is called a "Rayleigh distribution." In other words, when this statistical assumption is valid, the properties of reverberation are not dependent on the details of the acoustic space, but only on a *few* statistical parameters. The statistical assumptions define reverberation.

The statistics of an acoustic space also define its frequency response, which determines the intensity of each spectral component. All spaces produce a frequency response that is far from flat, resembling a sequence of mountains and valleys. Schroeder (1954a, 1954b) showed that the frequency response of a mathematically ideal space has randomly spaced peaks whose separation is inversely proportional to the reverberation time. For example, in a space with reverberation time of 7 seconds, there is one peak per hertz, but in a dead space with a 700-millisecond reverberation time, there is only one peak every 10 Hz. In addition, as a Rayleigh distribution, the variations in amplitude are independent of all other parameters. On average, there is a 10-decibel (3:1) variation in response from one neighboring frequency to another.

Should we care about the lack of a uniform frequency response? The answer depends on each specific case. Even for a relatively pure musical tone, such as one made by a flute, there are still sufficient variations in pitch and amplitude to smear the sound spectrum over a large number of peaks and valleys. A listener hears only the *average* over the frequency response variations, and that average is uniform. If, however, an electronic music composition uses a long-duration computer-generated sine wave, it is too pure to produce averaging across peaks and valleys. Its loudness rises and falls with extremely small changes in pitch, a kind of acoustic roller coaster not under the control of the musician and dependent on the listener's seat location. Musical sounds must, or should, have enough spectral width to smear the frequency irregularities of the space. And when so smeared, irregularities are imperceptible.

On the other hand, there are exceptions where this conclusion does not hold. We have already mentioned inadequate resonance density at low frequencies in small spaces. The frequency irregularity is then perceptible as the pitch progresses through peaks and valleys. Another exception results when the reverberated energy is significantly lower than the direct sound and early sonic reflections. The assumption about frequency smearing is no longer valid because the peaks and valley may be widely separated relative to the smearing power of the music. Sitting close to the orchestra, especially with nearby surfaces producing strong reflections, is a typical case of spectral coloration. The theoretical basis for this conclusion has been confirmed in empirical tests (Schroeder and Kuttruff, 1962; Ebeling, 1982), which showed that reverberation dominates the direct sound only beyond the *reverberation circle* (Kuttruff, 1954; Chu, 1980), the distance from the musicians beyond which reverberated energy is much louder than the direct sound.

In a small musical space, such as for chamber music, the reverberation circle is larger than the space itself; everyone sits within the reverberation circle. Conversely, in a large space, listeners beyond the first few rows are well outside the circle; everyone sits outside the circle. But in a mixing studio, with the adjustable parameters of an artificial reverberator and controllable ratio of direct and reverberated energy, the tight linkage between the spatial size and the reverberation circle is broken; you can sit outside the reverberation circle of even a small space. The experience of spaciousness is stronger at locations beyond the reverberation circle, in seats where the statistical assumptions are valid.

Even as a fully random statistical process, reverberation has several parameters that provide for its unique personality. The rate of decay, reverberation time, is the most apparent parameter, varying from a small fraction of a second to sometimes more than 10 seconds. We also observe that the decay rate depends on frequency, with high frequencies decaying more rapidly than low frequencies. The shape of the decay is not always a pure exponential, especially when the space is multiple spaces that are acoustically coupled. Many churches, which give the visual impression of being a single space, are actually multiple acoustic spaces. In general, we need think about reverberation only in terms of these and other statistical parameters—the elegance of (relative) simplicity.

Powerful Statistics with Invalid Assumptions

When used with mathematics, science is very effective at revealing the nature of physical phenomena. The science of spatial acoustics is no exception. Insight from investigations should produce better concert halls with predicable acoustic properties. Sabine, as the father of statistical reverberation, demonstrated the value of this approach over a century ago when he designed Boston Symphony Hall using an equation for reverberation time based on a statistical model of a space.

To demonstrate the applicability of statistics to reverberation in concert halls, we need to ask several questions. What assumptions are required for statistical randomness? Does a concert hall meet those assumptions? Would an artificial reverberator sound like a concert hall if it replicated the same statistics as a concert hall? There are two approaches for obtaining answers: theoretical and empirical. Each has the unpleasant possibility of producing results that are potentially useful while being theoretically invalid. How do we know when we can trust the results?

Sabine's pioneering work illustrates the general problem of validity. Without realizing it, he made assumptions about statistical consistency that were not tested and, in fact, could not be tested. When his assumptions were invalid, his equations produced erroneous results. Yet his basic equations are still in use today for two reasons: they are simple, and they have some predictive value for some spaces under some conditions. When applying statistical acoustics to spaces, the problem of relating validity, utility, and assumptions remains with us today. In Sabine's first application of statistics, he was lucky that his assumptions were sufficiently valid to produce a good concert hall. It could have been otherwise. Mathematical elegance becomes ambiguous and confusing because most statistical assumptions are often either invalid or unverifiable in real spaces.

Our earlier explanations of reverberation described the process either as millions of sonic reflections (a time-domain view) or as millions of resonances (a frequency-domain view). There is no dichotomy: both views are correct. There were other dualities buried in the discussions. Reverberation can be considered either as a deterministic or as a random process, using detailed waveform or statistical parameters. Similarly, any sound wave can be analyzed as a signal having either sonic energy or amplitude. Each description has a utility that depends on the goals, questions, and underlying assumptions. These alternative views are not contradictory. Scientists and engineers often switch back and forth between them, but such freedom depends on assumptions that are fragile, often forgotten, and severely constraining.

Consider the concept of a resonance, which assumes that the process is deterministic, always producing the same response from a given stimulus. This assumption is the bedrock of signal-processing engineers. The presence of thermal waves, which make the system time-varying at high frequencies, breaks the assumption required for mathematical resonances. Terminating the reverberation process after 10 seconds also destroys the concept of a resonance. Moving the conductor's podium changes the excitation of resonances. All of these examples undermine assumptions required for a resonance, but they do not destroy a statistical model based on sonic energy.

When we view reverberation as a statistical *energy* concept rather than as a *deterministic* signal process, we describe it using a different type of mathematics with different rules. Those rules, more robust than the ones used for deterministic signals, remain valid for minor changes in the environment. But the rules for statistical energy require other assumptions to be valid. Violating them again breaks the validity of conclusions. Unfortunately, statistical assumptions are subtle and beyond the grasp of many scientists and engineers; only a handful of physicists actually understand the underlying theory of statistical acoustics.[4]

Some of the basic ideas of statistical acoustics are, at least partially, comprehensible without formal training. They are relevant to understanding musical space. Let us begin with *randomness*, a concept surprisingly difficult to define. It can mean that the underlying process has elements of physical indeterminacy, such as the sound of air escaping from a balloon, which at the detail of a waveform signal cannot be determined or predicted by observing the process. It can also mean that the details of a process are too complex to understand or describe. Thus the height of the students in a classroom could be determined by measuring every student, but we *choose* to view that property as if it were a random process described by such statistical parameters as average and variance. We expect height to have *statistical regularity*, a well-behaved distribution of values among independent samples. If, however, the students had been previously sorted by height, a fact that could be unknown or unknowable, the distribution would no longer be statistically regular. In the same way, if reverberation has statistical regularity, probabilistic methods will prove highly useful in handling the large amounts of data, but we need to be aware of the appropriate concept of randomness. Finally, particular sounds that are not in the least random according to mathematical metrics are perceived as random, and conversely, other sounds that are random according to strict mathematical definition are perceived as having a predictable patterned structure. The science of how we aurally perceive statistical parameters is still in its infancy.

In their study of the mathematics of statistical processes, Wilbur B. Davenport and William L. Root (1958) cautioned that "the belief in statistical regularity is an induction and is not subject to mathematical proof." Even though we can measure particular properties of a process, we cannot prove the nature of the underlying process. Unlike the noise from air escaping from a balloon, a tape recording of that same noise is completely knowable, reproducible, and predictable. In fact, after repeatedly listening to it, we can anticipate what comes next, and can recognize particular segments. Although a tape recording of balloon noise has statistical parameters that are all consistent with a random process, we cannot say that the same sound from a tape recorder is a random process. It is not. Any signal that repeats itself is no longer random. An artificial reverberator that implements a single impulse response of a concert hall is no longer a random process because, like the tape recording of the balloon noise, it repeats on each listening. In fact, listeners can recognize and label samples of noise after repeated exposure. Perceived aural randomness and measured statistical randomness are not the same.

Turning to one of the most arcane assumptions in spatial statistics, when can we assume statistical consistency and equilibrium throughout a given space? By analogy, let us consider the salt content of a lake. If salt is dissolved and in equilibrium, we could analyze any sample of water rather than examining the entire lake. But this conclusion requires a demonstration that the salt in the lake is at least close to equilibrium, which may or may not be true. If we substitute alcohol for salt, however, the water is less likely to be in equilibrium because alcohol evaporates from the surface, which would then have a lower concentration than at the bottom. Without statistical consistency, we would have to measure or calculate every part of the space over all time.

When is a reverberant space in equilibrium? Unfortunately, there are different definitions of *equilibrium*, and the choice depends on the nature of the application and the formulation of the problem. Generally, the strongest assumptions produce powerful conclusions, but the strongest assumptions are also least likely to be valid in real acoustic spaces. Let us explore the intellectual burdens that need to be faced when using statistics to answer questions about real acoustic spaces.

One assumption is that the probability of a sonic reflection originating at point A on the surface and arriving at point B is equally likely for *all* A and B. Although some geometric shapes have this property, concert halls do not. This property means that the likelihood of a sound wave traveling from under your seat to your neighbor's right shoe must be the same as that of a sound wave traveling from a point on the ceiling to a point on the side wall. Clearly, this is not true. This degenerate example illustrates the problem with mathematical assumptions—they are absolute. On the other hand, the assumption may be sufficiently true under most conditions to produce sufficiently valid results.

Jean-Dominique Polack (1993) argued that a reverberation process is purely statistical when sonic reflections at all points exceed ten. But his conclusion requires the introduction of the assumption of *ergodicity*, which, in its strongest formulation, assumes that the statistics for all points in the space at one moment in time are the same as for a single point over all moments in time. Think of the acoustic space of a concert hall as a thousand bags of a thousand marbles each. You could evaluate the statistics of its marbles by collecting one marble from each of the thousand different bags—observing samples of sound at a thousand different points in the space at one instant. Alternatively, you could evaluate the statistics of a thousand marbles collected from a single bag—observing samples of sound at one point but over an extended period of time. For the assumption of ergodicity to hold, both evaluations must produce the same results. Strict ergodic equilibrium implies a uniform distribution of sonic energy over space *and* time.

To be ergodic, an acoustic space must satisfy strict requirements on its geometry and surface reflectivity (Joyce, 1975). The geometry must be sufficiently irregular such that *every* sound wave will eventually, if one waits long enough, traverse *every* point in the space. Parallel surfaces violate this assumption because sound waves can continuously bounce between them without ever being distributed to other parts of the space. Sound must spread uniformly throughout the space at a rate that is much faster than the decay process. In other words, ergodic spaces must have long reverberation times because excessive sound absorption continuously changes the energy in some parts of the space. James B. Lee (1989) argued that concert halls are not ergodic because they do not satisfy these requirements. Not only is sound absorption significant, but it is also concentrated on only one surface, the audience.

The ergodic assumption can be replaced by a weaker one, the assumption of full mixing, namely, that sound is uniformly distributed throughout the space—fully mixed—without any memory of its history (Krylov, 1979), with *mixing time* being the time needed to achieve that condition. Because the assumption of temporal uniformity is replaced with the weak assumption of no memory, however, a fully mixed space is not necessarily ergodic. For example, when sound is fully mixed throughout the acoustic space of a concert hall, sound originating from either the performance stage or the upper balcony produces identical reverberation. After the mixing time has elapsed, there is no memory of the original location of the sound source. Without sufficient mixing, reverberation statistics depend on the location of the listener. Polack (1992) argued that mixing is actually equivalent to the more familiar concept of a diffuse sound field, which listeners aurally experience as being fully immersed within enveloping reverberation. In a fully mixed concert hall, the late reverberation would be identical at all seats.

Mixing time is purely dependent on the geometry of the acoustic space. Some shapes produce rapid mixing. In irregular spaces, mixing time is approximately proportional to the average distance between all pairs of points on the surface, the mean free path (Polack, pers. comm.), which is proportional to volume divided by surface area. Larger spaces therefore have longer mixing time. Similarly, structured geometries that are not ergodic will also exhibit much longer mixing times. Geometries that are equivalent to coupled spaces may, in fact, never become fully mixed (Jean-Dominique Polack, pers. comm.). Many cathedrals are better modeled as multiple spaces with connecting paths between them rather than as a single space (Shankland and Shankland, 1971), and the region under a balcony can also be viewed as a separate but coupled space. For these geometries, mixing times can become long.

Our simplified discussion leads to four conclusions. First, in varying degrees, all concert halls violate the required statistical assumptions of ergodicity and full mixing. That, in turn, weakens or invalidates the conclusions proceeding from them. Second, most statistical theories do not include a practical means for testing assumptions, even if they are valid or almost valid. Third, even when the assumptions are only weakly true, the statistical approach can still provide useful and predictive results for all but the most degenerate spaces. Statistical methods are just tools, and as such, they are far more relevant to spatial acoustics than deterministic mathematics. And fourth, statistical methods provide metrics that relate more directly to perceptual attributes. With spatial acoustics, we perceive statistics, not waveform details.

On a final note, having a deep understanding of statistical reverberation is not a prerequisite for designing a good concert hall or an elegant spatial simulator. Ad hoc methods, using artistic and scientific rules of thumb, often yield more productive results than highly constraining theoretical formalisms. Yet the combination of both art and science is more powerful than each by itself. Acoustic science, rather than providing unchallengeable conclusions, is also an art—the art of applying formalism to ambiguous, untested, or intractable situations. In the end, we have the "art of science" and the "science of art," two parallel intellectual systems that coexist and complement each other.

Aural Consequences of Spatial Statistics

Although only a few scientists have a strong grasp of the physics and mathematics of statistical acoustics, there are at least four good reasons why musicians, composers, and mixing engineers should have at least a basic appreciation for the way that reverberation statistics influence their art. First, many acoustic defects have surprisingly audible manifestations, like the "slap echo" in MIT's Kresge Auditorium produced by the curved rear wall. Composers, conductors, musicians, and mixing engineers need to avoid such defects that degrade their art. Second, statistical parameters determine how listeners perceive space, which in turn enables a contemporary composer to manipulate statistical parameters rather than spatial metrics. For example, instead of having to change the volume of a space as a physical parameter, a composer can, with a reverberator, electronically adjust variations in the decay envelope as a statistical parameter. Third, a designer of reverberators who desires to mimic the experience of a real space must duplicate the statistics of its reverberation. Artificial reverberators rarely model the physics of sound in an enclosed space, but they successfully duplicate the relevant statistical parameters. Fourth, because contradictory statistics degrade the illusion of a real space, statistical consistency is required for a compelling illusion. But, since spatial contradictions are also a useful artistic technique, the composer can intentionally create artistic exceptions by including spatial contradiction. For example, some statistics may create the perception of a large spatial volume, whereas others create the perception of a small space. And finally, all those who consider themselves to be educated on the subject of musical space should, at least to some extent, have an appreciation for the properties of statistical acoustics.

In addition to being a mathematical theory, statistics is also a language to describe nature, and a way of thinking about problems. Statistical parameters are a useful bridge between physics and perception. Whereas an acoustic physicist asks if a statistical conclusion can be *proven*, given a set of assumptions, a perceptual acoustician asks if a statistical parameter *empirically* matches the experience of an acoustic space. The metrics for high-quality reverberation become apparent when we examine the perception of statistical parameters that are used to describe sonic reflection and resonance distributions.

During the early part of the reverberation process, long before there is full mixing and statistical validity, reverberation should still be *perceived* as being smooth. This is an artistic, not a mathematical, criterion. Perceptual smoothness in the early part is actually more important than statistical validity in the late part. The early part is louder, and for that reason, less likely to be masked by continuous music. Ironically, statistical theories are relevant only when reverberation has already decayed to a low level. From an artistic perspective, statistical mathematics addresses the wrong question.

The lack of physical equilibrium and spatial consistency during the onset of reverberation is readily apparent from the perspective of a single listener in a given seat or from computations throughout the entire space—two views of the same process. Consider a click as the sound source. From the perspective of a single listener, just after the direct sound, the rate of sonic reflections begins to increase. As their density increases, there is a corresponding decrease in the energy of each sonic reflection; sonic energy is progressing from individual sonic reflections to continuous sound. Ideally, the listener should hear, not the individual reflections, but a continuous broadband decay of white noise because a click, like white noise, has the same sonic energy at all frequencies. From the perspective of the entire space, the initial click exists as a tiny spherical volume of increased pressure at one point in the space, perhaps no larger than 0.1 cubic meter (4 cubic feet), and although it eventually fills the entire space with a volume of 10,000 cubic meters (330,000 cubic feet), the interesting part of reverberation occurs long before complete filling. Early reverberation is simply a transition region between a single and infinite sonic reflections.

We hear acoustic defects in the spreading process as temporal flutter. If there are gaps or peaks of sonic energy, if the spreading has a rhythmic periodicity, or if the time to become dense is too long, we may perceive reverberation as harsh, fluttering, or as a coarse growl. To avoid these acoustic defects, Manfred Schroeder (1962b) casually suggested an ad hoc rule: sonic reflections should be arriving at the rate of a 1,000 per second within 100 milliseconds after the direct sound. A few decades later, David Griesinger (1989) recommended elevating the target value to 10,000 per second for transient sounds. Their "folk science" acquired the aura of truth. The absence of scientific verification, however, limits our understanding of the underlying perceptual process. An ad hoc rule using the metric of sonic reflection density for perceptual smoothness is adequate in many circumstances, but the issue is more complex.

Scientific disciplines other than acoustics have answered related questions that are directly applicable to reverberation. For decades, perceptual psychologists have studied the perception of auditory roughness when modeling the properties of the human auditory system. Although making a model of the human auditory system (pure science) serves a different goal than designing an acoustic space (applied science), with the appropriate translation, we can apply the results from these formal studies to our folk science. "Roughness" is the label used to describe the auditory perception of variations in the envelope of a continuous signal, and "temporal flutter" is our label for variations in the envelope of reverberation.

A simple example illustrates the similarity. If a 1,000 Hz tone is amplitude modulated at a 2 Hz rate, the sound still has a pitch of 1,000 Hz, but it also has a periodic loudness envelope of 2 Hz, as is the case in the musical effect *tremolo*, where intentional variation in intensity of a tone that adds texture to an otherwise static sound. In our example, the 1,000 Hz steady tone is the *carrier* and the 2 Hz amplitude envelope variations are the *modulation*.

There is no restriction on the choice of carrier or modulation. Either can be pure tones, wideband noise, narrowband noise having a pitch, or any other signal. We can think of the carrier and modulation as two channels or two components of a signal, which the auditory system treats separately. In the case of a reverberated click, the carrier is wideband noise that results from millions of reflected clicks summed into a single process. But unlike constant noise, there is a uniform decay as well as an amplitude envelope of random variations.

How sensitive is our auditory system to the magnitude and frequency of amplitude modulation? The answer to this question predicts how listeners will perceive variations in the envelope of reverberation. Using broadband noise as the carrier, Neal F. Viemeister (1979) determined that listeners' sensitivity to envelope variations was constant when the variations' frequency did not exceed 30 Hz, above which sensitivity gradually decreased until the variations became inaudible at 1,000 Hz. In simple terms, rapidly changing envelope variations are hard to detect, but slowly changing variations are readily apparent. The sensitivity to envelope variations depends on their amplitude and repetition rate. Variations within a millisecond interval are not detectable.

A related class of experiments (Grose, Hall, and Buss, 1999; Snell and Hu, 1999; Miller and Taylor, 1948; Harris, 1963) measured the ability of listeners to hear gaps in an otherwise continuous signal. A gap is an extreme form of amplitude modulation where the carrier is turned off for a short interval. These experiments showed a consistent pattern: gaps that were smaller than 1 millisecond are inaudible. Because the duration of the gap in these experiments and the period of the modulation frequency in previous experiments were the same, this suggests that the human auditory system has a single time window during which it integrates the sonic energy of the signal.

Expanding the model still further, Lutz Wiegrebe and Roy D. Patterson (1999) showed that the envelope of the carrier is actually a perceptible signal. The auditory system demodulates the envelope, as well as the carrier, to create two internal signals, each with its own set of perceptual rules. These experiments measured the properties of that second perceptual system: envelopes. Just as listeners can hear tones (carriers) from about 20 Hz to 20,000 Hz, so they can hear envelopes (modulations) below 1,000 Hz. The intensity and frequency of the envelope are two major parameters that contribute to audibility.

What attributes of the envelope do listeners hear, and how do they describe their perceptions? Periodic envelopes are perceived as having a weak pitch. E. M. Burns and Neal F. Viemeister (1981) demonstrated that subjects recognized melodies that were based on the sequence of frequencies used to modulate noise, and David A. Eddins (1993) showed similar results when the carrier was narrowband noise centered at various frequencies. When the envelopes were not pitchlike, subjects described the sound in terms of "roughness," "raucousness," or "harshness" (Terhardt, 1974), which parallels the descriptions of reverberation with insufficient reflection density. Moreover, Daniel Pressnitzer and Stephen McAdams (1999) found that signals with the same envelope spectrum could be made to have different degrees of roughness by changing the temporal shape of the envelope. Specifically, fast onset and slow decay sounded rougher than the slow onset and fast decay.

With these insights, we return to the question of when we perceive smooth early reverberation. When averaged over 1-millisecond intervals, reverberation energy should be constant. Schroeder's recommendation of one sonic reflection in an interval of this size was simply wrong. Conversely, Griesinger's recommendation of ten reflections was consistent with this conclusion because an average of ten independent sonic events in each interval is sufficient to reduce the energy variations in the sensitive region below 30 Hz. But that conclusion also has a critical, if unstated, assumption: sonic reflections should all have the same statistical distribution. A strong repeating sonic reflection, even if accompanied by a high density of smaller reflections, still sounds rough. For example, reverberation might sound harsh if some unusual geometry produced quasi-periodic energy bunching. The true measure is not reflection density, but the average energy of reflections over a short time interval.

This discussion illustrates the danger of relying exclusively on folk science. Yet for designers of artificial reverberation systems, there has been little interest in researching temporal flutter because folk science has proved to be adequate in *most* applications. Unknowingly, folk scientists used a crude approximation for the correct metric: reflection density instead of reflection energy. I have no doubt that their basic conclusion will survive additional research on temporal flutter to refine the model of the human auditory system.

To better understand flutter in the broader sense, let us simplify one of the model's details. However unmusical it may sound, a click is a scientist's most severe transient signal for evaluating a reverberation process. If a click produces smooth reverberation, so will less severe signals. This is generally the case. But when the click is replaced with a 100-millisecond raised-cosine tone burst at 500 Hz, essentially a pitchlike click that is representative of a short musical note, the reverberation tail will *always* manifest a disturbing flutter. In fact, two unrelated acoustic mechanisms can produce a similar perception of flutter. All types of reverberation processes, including the best concert halls and the most expensive artificial reverberators, produce perceptible envelope variability for this *specific* signal, a short tone burst of about 100-millisecond duration. Because this kind of flutter is a consequence of its spectrum, let us call it "spectral flutter," to distinguish it from the temporal flutter of a click. Spectral flutter is not an acoustic defect, even when experienced as an artistic defect.

Careful examination of a tone burst reveals why it can produce reverberation with a perceived flutter. Mathematically, reverberation must always have the same spectrum as the original signal, albeit randomized, and the tone burst's width determines the rate of the resulting envelope variations. Folk science assumed that random variations would *never* be perceived. In our example, the specific signal duration happens to produce slow variations that happen to match the high-sensitivity region of the auditory system, at about a 4 Hz rate. Longer and shorter bursts produce less audible envelope variations.

This brings us to an obvious question: why does music not manifest this kind of spectral flutter? There are three main reasons. First, we listen with two ears, and the independent flutter in each ear reduces its perceptibility. Second, music rarely contains a single frequency. Each overtone also has its own independent flutter, which again makes the combination less audible. And third, musical notes rarely have this same specific duration. For all these reasons, spectral flutter is rarely perceived during musical performances. On the other hand, it is easy to imagine an electronic composition using 100-millisecond tone bursts that would exacerbate this natural process. The flutter would not be apparent in an acoustically dead recording studio, where direct sound dominates sonic reflections, but it would manifest itself in any reverberant space. The very freedom to create any kind of musical tone is also the freedom to create unpleasant surprises.

As if the life of an acoustic scientist is not difficult enough, this kind of spectral flutter can also be created by acoustic defects. With more typical musical signals that last a long time, and which have a gradual onset, the reverberation can manifest a very unpleasant "wow-wow-wow" sound, which we referred to earlier with an organ note in a tiny church. If only two proximate resonances are excited, the reverberation will have a beat frequency that is equal to their frequency difference. If, for example, a sound excites resonances at 89 Hz and 91 Hz, then the envelope of the 90 Hz decay will have a 2 Hz periodic variation. Although this can be dismissed as a special case of inadequate resonance density, the phenomenon is universal in smaller acoustic spaces.

In the discussion of statistical reverberation, we observed that a cloud of resonances distributed randomly produces a perceptually smooth decay. But resonances also have random rates of decay as well as random frequencies. In less ideal situations, a few resonances in that cloud might have long decay times. What happens when the decay rates are not uniform? Consider a degenerate example. Assume hundreds of resonances have a decay rate of 2 seconds, but also assume that two of them to have a decay rate of 3 seconds. Although, at first, all of the resonances are contributing to a statistically random process, after a few seconds, all but two are quiet, and you hear just those two remaining resonances. This now becomes equivalent to the earlier example of the organ note in the small space with two resonances at 89 Hz and 91 Hz. Reverberation decay that begins smoothly can gradually acquire a periodic flutter during the decay process. Theoretically, all reverberation must end with four, three, two, and eventually one active resonance. No matter how many resonances are first excited, there must be one, and only one, having the longest decay time, by the end of the process. However, in a high-quality acoustic space, a large number of resonances remain active well beyond the point where the reverberation has decayed to inaudibility.

The phenomenon of aural coloration also serves to illustrate the disparity in decay rates. Consider again a transient input that excites all resonances, but now also assume that the resonance cloud near 500 Hz has a longer decay time than the other resonances. As the decay proceeds, the reverberation tail, which initially had a broadband spectrum, gradually converges toward this one frequency. There are still enough active resonances to preserve the statistically smooth envelope. Yet the decay acquires a dominant pitch, a spectral coloration at 500 Hz, regardless of the spectrum of the original sound. As a general rule, the spectral balance in the reverberation tail always shifts toward the resonances that remain active for the longest time.

Aural coloration is the precursor to spectral flutter. Both phenomena are the consequences of a nonuniform decay rate. Conceptually, coloration eventually becomes flutter. Such coloration is rarely heard in large complex spaces for three reasons: first, the reverberation tail has already decayed well below audibility when coloration and flutter manifest themselves; second, with a large number of excited resonances, small differences in their decay rates are imperceptible; and third, with a many overtones at different frequencies, musical instruments excite many different resonance clouds.

Just as unsophisticated listeners are unlikely to appreciate the exquisite sounds of a Stradivarius violin, average listeners may not appreciate the exquisite sound of a revered concert hall. For those who have acquired a refined taste for musical spaces, however, the previous discussions explain how acoustic defects can transform brilliant reverberation into merely high quality. Unfortunately, but understandably, neither the average aural architect nor the average listener appreciates spatial subtlety to this degree.

Visiting the Inside of Reverberators

When visiting cathedrals, concert halls, or opera houses, you can usually walk through the entrance, look, and listen. But you cannot visit the inside of an artificial reverberator or spatial synthesizer. You can see acoustic hanging clouds and wall surfaces in a real musical space, but you cannot see delay lines and feedback matrices in virtual-space software. Nevertheless, even with the abstruse and opaque technology of software spaces, you can acquire some basic insights with the tutoring of a knowledgeable guide.

Spatial software and virtual environments have become the twenty-first-century version of concert halls, yet another instance of the shift from the real to the virtual. They have done so with the support of both the general public and the professionals, enabled, but not chiefly motivated, by technology. Let us touch on four of the many reasons why.

First, concert halls are expensive, and a modest-sized city is unlikely to have more than a few. Society now devotes fewer resources to building performance spaces than to mimicking the experience of real spaces; acoustic architects simply have fewer opportunities to practice their craft. In contrast, a competent signal-processing engineer can design a new audio algorithm in a matter of weeks, and a single such algorithm allows dozens of musical spaces to be evaluated. Nothing else is required to become the aural architect of a virtual space but a personal computer, inexpensive audio software, a modest sound system, and a little free time.

Second, almost all recorded music uses reverberators to replace or augment natural acoustics. Every recording studio, from amateur to professional, has some kind of reverberation device—they are ubiquitous. The cost of such devices ranges from $10,000 for professional models to publicly available freeware programs (Wakefield, 2000) that can be downloaded in minutes. Over the last few decades, dozens of companies and hundreds of engineers have developed and distributed a wide range of reverberators, far more than the number of concert halls built during the same period.

Third, music aficionados spend more time listening to recorded music than attending concerts. Even with a subscription to a concert series, a devotee attends concerts for a few hours per month. In contrast, recorded music is everywhere, from records and radio, to cinema, television, and video games. A typical listener may be exposed to artificial reverberation for many hours per day. Some of us may spend 500 hours a year commuting in our automobiles, much of it reserved for listening to music.

Fourth, recording, mixing, and audio engineers have become expert listeners as they have learned to compare and contrast dozens of unique approaches to the design of reverberators. Many have thousands of hours of listening experience. These experts continuously raise the audio quality bar, and then pressure researchers and developers to improve their products. The marketplace, though an unforgiving critic, can and does recognize the value of audio quality. Over the last three decades, each new generation of audio algorithms has destroyed companies that have failed to match the advancing norms of audio quality, flexibility, and artistic relevance.

A discussion of musical space would not be complete without examining the electronic alternative to the concert hall. How do you become familiar with the process of creating a software space? When you open the cover of a reverberator, study its front panel, or examine its user manual, you learn very little about its underlying design. To appreciate the defects, elegances, and trade-offs in reverberator design, let us take a guided tour into the inner world of signal-processing architectures.

To begin, there are three broad approaches to spatial emulation of a performance space: first, simulate all sound waves in three dimensions throughout the entire space; second, duplicate only the sound pressure that appears at the listener's ears; and third, produce a perceptual experience that is equivalent to the first two cases. This hierarchy is in descending order of difficulty. An exact simulation (first approach) is the most intellectually elegant, but still at least a decade away from being realized in real time. Focusing on sound pressure at the ears (second approach) fails unless the designer controls the playback environment, an onerous burden for most listeners. That leaves perceptual equivalence (third approach) as the dominant design approach.

We first divide spatial emulation into two temporal segments: discrete early sonic reflections and statistical reverberation decay, each of which are too complex to faithfully duplicate. In both segments, the only question is how to achieve perceptual equivalence using electronic building blocks.

Regardless of the specifics, there is one universal for all spatial reverberators—their dependency on audio delay using digital storage. Delay holds the information of the internal signals in the same way that the air holds sound pressure throughout a space. Signals move through delay lines, like one-dimensional sound waves. But how then should a pool of delay lines be organized? There are many choices, and each one produces a design with its unique aural personality, just as a specific geometry of a concert hall gives it an aural personality. Acoustic architects organize bundles of air; reverberator architects organize groups of delay lines.

The topology of signals that enter and leave each delay element is called the "architectural wiring diagram." At one extreme, the topology might have one large delay that is equal to the longest reverberation time. At the other extreme, the delay pool might be broken into thousands of small delays embedded within a complex network of interconnected loops.

Basic Components, Topologies, and Their Properties

In the world of reverberators, the basic signal-processing components are delays, multipliers, and adders, which in turn serve as the foundation for more complex modules. In the following discussion, we will develop a reverberator design by moving from these basic components to the more complex interconnections between them. Just as there are thousands of concert halls, so there are thousands of topologies, each with its own unique aural personality arising from design trade-offs.

Let us first consider the *feedback-delay* module, which comprises a delay line with a feedback gain of less than unity. To observe the temporal properties of this module, a pulse is injected at the input. An identical pulse appears at the output, delayed by the size of the delay line, which then recirculates back to the input to appear at the output delayed yet again. For example, with a delay of 100 milliseconds and a feedback gain of 0.7, a sequence of output pulses appears 10 times per second with decreasing amplitude of 1.0, 0.7, 0.5, 0.35, and so on. The resonance density is proportional to delay, and the reflection density is inversely proportional to delay. Because only one parameter, delay time, controls both density measures, they work in opposition; increasing one decreases the other. In the example with a 100-millisecond delay, the resonance density is 0.1 resonance per Hz, and the sonic reflection density is 10 per second. Although this feedback-delay module is a basic component, by itself it would be called a "slap-echo generator." Both the resonance and reflection densities are far too sparse to produce a sense of space, and neither density is statistically random because each has a periodic pattern.

Cascading multiple feedback-delay modules is an attractive improvement because the reflection density then increases rapidly. A single pulse input into the first module becomes n outputs, which then feeds the second module to become n^2 outputs, the third, n^3, and so on. Each module can be considered a reflection multiplier, although each also has a periodic comb-shaped frequency response with periodic peaks and valleys. And when the peaks and valley of many modules line up, as they will at some frequency, the effect is unpleasant, a filter that sounds like the inside of a barrel. The sound of cascaded feedback-delay modules does not produce a sense of space.

We can separate the property of a nonuniform frequency response from the property of multiplying reflections by adding a feedforward path to a feedback-delay module. When the path is added, the new *all-pass filter* module passes all frequencies while still multiplying reflections. Because it allows all frequencies to pass, each module in a series chain receives the same spectrum. And each module multiples reflections without producing the spectral problem of aligned resonance peaks. Such a topology was, in fact, the first topology published by Manfred Schroeder and Benjamin Logan (1961).

Using these two module types, all-pass filter and feedback-delay, a designer selects a topology that specifies how a multiplicity of them should be interconnected. A design might use two dozen of these modules connected with any number of possible topologies. Modules can be cascaded in series, placed in parallel arrays, embedded one within another, interleaved in a cross-couple pattern, and so on. In one approach, called a "single large loop," alternating feedback-delay and all-pass filter modules are connected in a ring. In another approach, three all-pass filter modules in cascade are used as the input to a parallel array of eight feedback-delay modules. Because an all-pass filter module that is embedded in another retains its flat frequency response, any topology with an all-pass filter module can also use an embedded version (Gardner, 1992). And any loop can contain any number of other recirculating structures.

As we will see, with one exception, all topologies are combinations of local and global feedback—sonic energy recirculation. The sonic energy of an injected pulse spreads so that it eventually fills the major delay lines, with attenuators gradually removing energy at a rate determined by the desired reverberation time. Delay lines are the equivalent of the air, and attenuators are the equivalent of sound absorption. The topology determines the properties of the spreading (filling) and absorption processes (decay).

For a reverberator, sonic energy spreading and envelope decay are analogous to the acoustics of a real acoustic space, but this analogy is also seductively misleading. Delay lines only support one-dimensional signals of constant velocity, not three-dimensional sound waves. A sound wave is specified by a particular azimuth and elevation angle. A delay line has no such angular directions. The number of signal paths in a reverberator topology is tiny compared with the infinite number of paths in a space. With so many paths, filling a space with sound happens rapidly and uniformly. Each path is a consequence of thousands of sonic reflections from unique surface elements, each of which has its particular acoustic texture, reflectivity, diffusion, and diffraction. Reverberators are more analogous to regularly spaced mirrored walls. In addition, the information-holding capacity of delay lines is many orders of magnitude lower than that of air in a real space. Finally, reverberators produce a deterministic response, whereas a real acoustic space has statistical properties. In other words, crude reverberator topologies are inadequate to mimic the sound of a real concert hall.

In a short article in an obscure engineering journal, Michael A. Gerzon (1976) published the mathematical foundation for a broad class of reverberator topologies. It took several decades for the implications of his formulation to be appreciated and applied to designs (Jot, 1992). Although mathematical training is required to appreciate the elegance of his approach, the basic ideas can be readily illustrated.

Consider a reverberator composed of 8 delay lines each of a different length, and 8 sets of 8 coefficients. The input to *each* delay line is the sum of the signals from *all* 8 delay lines. There is a path from every output to every input, and there is a coefficient that determines the amount of signal that can pass along each of the 64 paths. The number of paths is the square of the number of delay lines. If the 64 coefficients are selected according to a specific mathematical rule, the resulting *unity-orthogonal matrix* of coefficients creates an *energy-preserving reverberator*. "Energy-preserving" means that the total sonic energy entering the eight delay elements is always equal to the total sonic energy leaving. The matrix of paths and coefficients scrambles the distribution of sonic energy among the delays but preserves the total energy at all times. The topology is thus a giant feedback loop that recirculates sonic energy in all the delay lines through the matrix and back again to all delay line inputs. On each iteration, the distribution of sonic energy among the paths changes. Although this topology has an infinite reverberation time because energy remains constant forever, we can add attenuators to each delay line that removes energy at a *constant* rate. This is the basic topology for Gerzon's reverberator. Traditionally and unproductively, signal-processing engineers analyzed such structures by following the signal through each path in order to acquire an exact representation of the system. With so many cross-coupled paths, however, an exact analysis is intractable. But by using total sonic energy as the relevant acoustic parameter, the topology reduces to a *single* module of eight delays and a *single* matrix of feedback paths—two elements in one loop—the essence of simplicity.

This topology has many desirable properties. If internal attenuators remove sonic energy at a rate proportional to the energy present, the reverberation time is provably identical for all frequencies—there is no coloration in the reverberation tail. The distribution of resonances also has a nonperiodic and almost random pattern. The temporal behavior is also ideal. Because of massive cross-coupling, there are 8 reflections on the first iteration, 64 after the second, 512 after the second, 4,096 after the third, and so on. The sonic reflection pattern becomes random and dense after only a few iterations.

The reverberator design is unfinished because the choice of delay line values creates a conflict. If the delay lines are long enough to achieve a high resonance density, perhaps with total memory of 1 second, the sonic reflection density takes too long to build because three iterations take 0.5 seconds. Conversely, if the delay lines are all short, the resonance density is too sparse to produce a sense of space. All recirculating topologies struggle with this trade-off. Long delays are hard to fill and do not produce many sonic reflections, yet long delay lines are required in order to achieve an adequate resonance density. The conflict is resolved by using longer delay lines but with all-pass filter modules included within each delay line (Dahl and Jot, 2000). The combination of a feedback-delay and an all-pass filter module is still energy-preserving, yet on each iteration, the all-pass filter modules dramatically increase the sonic reflection density. These all-pass filter modules must be slightly modified to compensate for their frequency-dependent delay with a matching frequency-dependent attenuation (Blesser, 2006).

The optimization process is now reduced to selecting a particular unity-orthogonal matrix from the infinite number of possibilities. The design criterion is a matrix that is maximally dispersive without any bias that favors one path over another, which can be achieved by making every coefficient have the same magnitude but with appropriate signs. In addition, because there are multiple sets of coefficients with this property, the matrix can randomly change from one set to another. The reverberator is now finished, and it will sound like a real space, though not like a particular seat in a particular space.

One of the elegant by-products of Gerzon's formulation is that it also provides a means to analyze other recirculating topologies, all of which are subsets of his concept. In other words, we represent a topology as a matrix and delay lines, and then examine the properties at the matrix to see whether it is energy preserving. If not, there will be spectral coloration in the tail because some resonance will have a longer reverberation time than others. Is the matrix dense, filled with similar coefficients, or is it sparse, mostly filled with zeros? If sparse, the topology will only weakly disperse sonic energy among the delay lines. For example, a topology with multiple parallel loops or one single large loop is equivalent to a matrix with one diagonal having nonzero coefficients. Some of the cross-coupled topologies have more complex matrices but still with only a few coefficients that are nonzero (Stautner and Pluckette, 1982). Some topologies, such as two feedback-delay modules in a loop (Sikorav, 1986), are not even energy preserving.

With Gerzon's mathematical formulation, reverberation topology acquired a scientific foundation rather than remaining just an art form. Yet even though the design of matrix topologies has rigorous mathematics, there is also a degree of artistry involved in selecting the delay and gain parameters. The challenge is relating parameter selection to perception.

To appreciate the role of art in the design process, consider how Manfred Schroeder, one of the most famous acoustic and mathematical scientists of the twentieth century, selected the delay values in the world's first electronic reverberator. In Logan and Schroeder's patent (1963), they stated that delay values should be *incommensurate* by which "it is meant that the delays are selected with values having no common divisor." As a junior scientist, but one who still recognized the insufficiency of this explanation, I had the opportunity to ask Schroeder what was beneath his opaque assertion. He explained that "we just picked numbers and subjectively listened to the results until we were *happy*," or as he stated in an earlier paper (Schroeder, 1962b), "until the results were subjectively indistinguishable from real rooms—not only for speech and music but even for such 'trying' signals as very short clicks and wideband Gaussian noise."

History shows, however, that the art of listening to spatial nuances has advanced as rapidly as the art of signal-processing design, and systems that were considered perfect in their day have been quickly relegated to the trash heap of history. It was a race between individuals who learned to hear acoustic defects in the spatial illusion, so-called golden ears, and those who improved the design by removing those defects, virtuoso designers. Perceiving subtle acoustic defects is a learned art. What, then, is the role of science? It can explain acoustic defects once perceived, and it can provide ways to avoid specific defects. Science cannot prove the negative conclusion: the absence of acoustic defects.

Before leaving the subject of mathematics and topologies, we must amend our conclusion with one important exception. Reverberators can be constructed without any recirculation, without loops, matrices, or all-pass filter modules. Consider one very long delay line of 10 seconds. Assume thousands of signal taps at every possible delay, each with its own coefficient. This is the *direct-form* topology with complete freedom to implement any reverberation, in fact, any type of filter. We need only match the coefficients at each delay to the corresponding sample of a prototype reverberation signal. If we ignore the pragmatic consequences of requiring almost a million coefficients to represent reverberation of a cathedral, given such a prototype, the direct-form topology can implement all spaces exactly with no approximations, no compromises, and no dependency on artistic judgment or mathematical rigor.

When James A. Moorer (1979) observed that the impulse response from concert halls around the world sounded remarkably similar to white noise, he implemented a direct-form reverberator using a pseudorandom number sequence as the reference prototype. "The results were astounding. Although the synthetic impulse response did not produce a sound that could be identified with a specific concert hall, the sound was clearly a very natural sounding response" (Moorer, 1979). No doubt he would have used a real concert hall had the data been available to him. But he simply demonstrated that, for the late reverberation, using data from a real space was not necessary.

If the direct-form solution is so perfect, why, then, is it not the design of choice? Ignoring its additional computational burdens, which are rapidly being solved (Gardner, 1995), the direct-form approach still has one major problem: it is simply not possible to acquire a reference reverberation impulse response to use as the basis for setting the coefficients. Earlier, in a different context, we showed that it was impossible to measure the high-frequency response for the reverberation tail of a large concert hall. And without that ability, there is no real-space reference to use in the direct-form topology. A compromise consists of measuring the early part of the reverberation and then concatenating a precomputed random reverberation tail, as Moorer did. But why bother faithfully reproducing a *particular* statistically random signal? Unity-orthogonal recirculators achieve the same result with less cost and more flexibility. Later, we will show that the direct-form topology has another serious problem.

This leads us to a hybrid solution. Use the direct-form topology for early reverberation, where perceptual details of the space matter, and use unity-orthogonal recirculators for the late reverberation, where statistics govern perception. In this respect, the problem of designing high-quality artificial reverberation is finished; all future development involves creating unique variation to suit particular tastes.

Increasing Resonance Density with Randomizers

To appreciate the intrinsic difference between acoustics and signal processing, let us begin by comparing the information capacity of real and artificial systems. An earlier discussion showed that there were some 60 billion independent cubes of air in a typical concert hall, each of which holds an independent sample of sound pressure. A typical reverberator might contain storage for only 4 million samples of an audio signal, which is four orders of magnitude less. Only technical ingenuity can compensate for the disparity.

Some properties of real spaces cannot readily be duplicated with reverberator topologies. For example, real spaces have a resonance density that is proportional to the square of frequency, whereas almost every reverberator topology has a constant resonance density at all frequencies. This disparity originates from the nature of velocity: three dimensions in a physical space and one dimension in a delay line. As a result, a real space has a Schroeder frequency, below which resonances are discrete and above which they are a statistical cloud. The constant density in a reverberator results in all frequency regions being either discrete or statistical. If the resonant density is dense enough for high frequencies, it cannot simultaneously be sparse enough for low frequencies. Conversely, if the resonator simulates a small acoustic space with a low density for low frequencies, it will be mismatched for high frequencies.

The following generalizations apply to topologies that can be represented as unity-orthogonal matrix recirculators (Jot and Chaigne, 1996), a fancy name that also includes the standard feedback topologies: a large single loop, multiple loops, embedded all-pass filter and so on. We are excluding typologies that use multidimensional "waveguide meshes" (van Duyne and Smith, 1995), which are still laboratory

curiosities for small acoustic objects and present intractable computational requirements (Campos and Howard, 2000). For the moment, we are also ignoring implementation of direct-form topologies, that do not have resonances.

The requirement for a high resonance density dominates the design topology because that requirement implies a very large total memory, either as a few very large delay lines or as many smaller lines. Because large delay lines take a long time to fill, they
are typically limited to about 0.1 seconds. Only the number of such delay lines then
determines the total amount of memory. To achieve a resonance density of 10 per
hertz, an ideal target for statistical resonances, the topology would need 100 delay lines
of this average size. It is now possible to implement this quantity although it is still a
very expensive solution. But historically, when the reverberator arts were just evolving,
the cost of both memory and computation made this requirement unacceptable. Even
now, price is still an issue when balancing quality and cost.

Assume for the moment that the total memory is limited to 1 second, which results
in a resonance density of 1 per hertz. Such a reverberator produces periodic envelope
flutter when excited by a pure tone, an unacceptable imitation of a statistically random
envelope.

Defects are the mother of invention. One ingenious approach includes internal modulators that continuously and randomly move the resonance frequencies hither and
yon. Although the mathematics of a moving resonance are invalid (by definition resonances must be static), the concept is still useful for understanding how a cloud of resonances can be replaced with a small number of diffused (moving) resonances. Moving
resonances have been redefined for a few special situations (O'Brien and Iglesias, 2001),
but I know of no definition that can be applied to reverberators. The use of random
modulators in reverberators is pure art without mathematical support.

The first commercial reverberator used randomly moving delays of signals from the
outputs of feedback-delay modules (Blesser and Bäder, 1980), and some topologies
changed the delays within a feedback loop (Moore, 1981). Most professional reverberators still use some kind of random modulation as a way to independently manipulate
the statistics of the reverberation envelope, although there are an infinite number of
ways to vary the internal topology. A designer is faced with two related issues: what
parameters to change, and how to change them.

Consider changing the delay in real time. Variable delays have a special problem because digital signals are quantized in discrete time units, samples of the signal without
intermediate values. For example, there might be samples at 1.001 and 1.002 seconds,
but the delay change must progress smoothly through 1.0011, 1.0012, 1.0013 seconds,
and so on. It cannot jump from 1.001 to 1.002 seconds without producing noise. Delay
changes must make smooth transitions between samples, and this requires some way
of computing the intermediate values by interpolating. There are many solutions to

this problem (Dattorro, 1997b; Laakso et al., 1996). Every delay line can be varied randomly in real time, but the choice depends on the quality of the interpolator.

Alternatively, any gain can be changed by slowly moving its value, which does not require interpolation. Typically, a randomizer gradually fades the gain of a path between two values, or reduces its gain to zero, moves the path, and then gradually increases the gain up to full value. Because recirculating topologies can be represented as a unity-orthogonal matrix, and because there are infinitely many such matrices, the randomizer can fade from one set of matrix values to another while still preserving the energy-conserving property during the transition (Blesser, 2006).

After years of experimentation, reverberator topologies with randomizers produce an extremely close approximation of the desired envelope statistics. Yet all forms of random modulation produce unwanted sidebands, spectral artifacts that do not exist in real spaces without air turbulence. Functioning as empirical artists, reverberator designers have found clever ways to take advantage of random modulation while keeping artifacts inaudible. Artifacts are nevertheless still present. But they are masked by the aural complexity of the music and by the insensitivity of the auditory system to low-level spurious components.

Time Variations as a Basis for Naturalness

Although manufacturers of professional audio equipment frequently introduce new products as soon as a technology provides advantages, innovative ideas often have subtle problems hidden within a marketing language. As an example, Yamaha (2002) described its product as "not using contrived algorithms...faithfully recreating the original reverberation." Yet they neglected to mention that it is theoretically impossible to characterize the details of a large concert hall because thermal waves produce sonic instability that randomizes the speed of sound. Curiously, some of those "contrived" algorithms, deprecated in Yamaha's literature, include randomization that, although intended to solve a technical problem, also creates the naturalness of thermal waves. This example illustrates the opacity of complex technology intended for artists and audio mixing engineers. On the other hand, true audio experts, rising above both scientific specifications and marketing technobabble, evaluate their audio tools directly by listening, as true musicians would evaluate their musical instruments. For the mixing engineer functioning as an aural architect, a reverberator is just an artistic tool. Although replicating natural acoustics is an abstract topic of interest to acoustic scientists, replicating natural acoustics is irrelevant for those listening to music.

The story of random modulation has an ironic twist that has only become apparent during the last few years. Historically, randomized parameters were invented to overcome technical and economic limitations—to correct an implementation problem, making the spatial illusion more compelling. Randomization was less expensive than increasing memory storage and computation power. With the recent introduction of direct-form reverberators, randomization was no longer needed. Or so it was thought. The direct-form topology can reproduce the equivalent of an arbitrarily high resonance density, a statistical cloud, even though it, in itself, has no resonances. And the reverberation envelope would then have perfect statistics, as Moorer (1979) proved.

Informally and subjectively, however, a few experts described the reverberation from the direct-form topology as having a somewhat mechanical and sterile quality. Even though one of my most trusted colleagues, a professional musician and a manager of reverberator development department, could not adequately describe the basis of his impression, he experienced an aurally unpleasant attribute in this topology. I have no doubt that he heard something. Even though we did not perform any scientific studies, I proceeded to reexamine the role of randomizers from an entirely new perspective. The following inferences are speculative, but compelling.

As previously discussed, the failure to find a scientific technique to measure the reverberation tail originates from natural phenomena in large enclosed musical spaces. Air has variability. Air is not a delay line. There are whirling masses of air moving in random directions (turbulence), shifting thermal layers changing the direction of sound wave propagation (refraction), and fidgeting listeners who change which sound waves are being absorbed (resonance). Nothing is static. These phenomena create a measurement problem, but they are also intrinsic to any natural listening environment. Even though sonic reflections are enclosed within an acoustic space, a large space means long travel paths for reflected sound waves. A sonic reflection arriving 1 second after the direct sound has already traveled a fifth of a mile through this ocean of turmoil. Listeners live in air, and they have expectations about its properties.

Demonstrating the existence of physical variability in the reverberation tail does not in itself prove that this variability is audible. In pathological cases, Sabine (1922) argued convincingly that this variability is perceptible and undesirable. But what can be said about normal cases? To explore this question, we turn to a particular subspecialty of psychophysical research, the perception of repeating random noise, a class of signals represented in reverberation.

Consider that a click in a large space produces a reverberation tail that is nothing more than exponentially decaying noise. Two clicks separated by a few seconds produce two such sequences of noise. In a real space, the details of these two responses are always different. By definition, two responses from a random process are never the same; or alternatively, no matter how long one listens to the sound of air escaping from a small hole, the sound never repeats. In contrast, if a reverberator uses a static topology without randomizers, every response is identical. Can a listener detect whether two noise samples are identical? That is the key question.

What does the scientific literature tell us about this sensitivity to repeating noise patterns? Begin by considering stimuli composed of a single segment of noise that periodically repeats. For durations of less than 50 milliseconds between repetitions, listeners hear a pitch; for durations between 50 and 250 milliseconds, a staccato-like motorboating; and for durations from 250 milliseconds to about 1 second, a smooth whooshing (Guttman and Julesz, 1963). No training is required to perceive these properties as attributes of the entire waveform. Over the years, however, researchers have noted in passing that a few listeners experienced a repeating noise segment as such without an awareness of perceptible attributes for longer durations, up to 4 seconds in one case (Julesz and Hirsh, 1972), and 10 seconds in another (Warren and Bashford, 1981). How, then, did they identify the noise segment as repeating?

Richard Warren and colleagues (2001) studied long segments of noise from the perspective of auditory memory—remembering the *particular* attributes of *particular* segments of sound. No matter how random, every sound has its aural personality: unique gaps and local pitch. In their study, both experts and nonexperts learned to recognize noise segments that were asynchronously embedded within a continuous random noise. They recognized the target segment regardless of when it appeared, often remembering it for as long as 10 to 20 seconds. In his study of what is being remembered, Christian Kaernbach (1993) concluded that "when the auditory system is presented with repeated white noise, it will enhance details of this noisy structure which we otherwise would not perceive.... White noise seems to be filled with a lot of such potential features.... The physical basis for perceiving such features does exist, so perception should be possible.... As soon as the feature pattern reappears [multiple exposures] the features are taken to be informative [recognized]." A segment of noise is perceived as noise only on first exposure. Christian Kaernbach, Erich Schröger, and Thomas C. Gunter (1998) found that, among those individuals with an aptitude for perceiving small segments of repeated noise, there were observable manifestations in their measured brain potentials.

Consider a visual analogue: a computer program generates a sequence of random faces by arbitrarily selecting a nose, mouth, chin, eyes, ears, hair, cheeks, and so one, from thousands of possibilities. A sequence of such faces forms a random progression of millions of unique images. Yet you would certainly notice a face that appeared twice within a short interval, and you would recognize your mother's face on first presentation. In this sense, every face is a random face when viewed for the first time after being selected from an ensemble of 6 billion other faces; but with familiarity, a particular face becomes readily recognizable.

The human auditory system has two processing and memory systems for stimuli: long-term and short-term (Cowan, 1984). Although permanent storage is usually fragile and unreliable, listeners who can permanently retain auditory details are said to have an "echoic memory," like an eidetic memory for vision. The duration of a memory trace varies among listeners. Audio professionals and skilled musicians, in particular, are known to have long auditory memories. Although I cannot prove it, I believe

that my colleague perceived the static properties of the direct-form reverberator because he could compare audio stimuli over a relatively long time. Only synthetic spaces
(without randomizers) produce exact duplicates of a sound. For him, the absence of an
expected attribute—variability—degraded the spatial illusion.

There is limited research that explores the artistic relevance of spatial variability, except for the indirect study of Ueda and Ando (1997) and their colleagues. Using measurement data from a large gymnasium, they modeled the time variation of the high
frequencies as being equivalent to random shifts in the arrival time of reflections.
Then, motivated by this result, Junko Atagi, Yoichi Ando, and Yasutaka Ueda (2000)
demonstrated that listeners, when given a choice, preferred a slowly moving sonic
reflection to a static one. Although a single sonic reflection is not reverberation, the
result confirms the speculation that variability is preferred. The experience of live art
includes the expectation of the unexpected. And from this perspective, all recorded
music lacks variability.

To summarize: with one or more randomizers and with multiple embedded all-pass
filter modules, those topologies that can be represented as a unity-orthogonal matrix
are currently the best method for matching the statistical reverberation of a real acoustic space. They are even better if they also include a direct-form implementation for the
early part of the reverberation. The design details are still an art form, but given the
effort already invested, many commercial products now achieve the highest level of
quality. The future may bring alternative methods, but the criteria for quality should
remain stable.

Even if we accept the fact that some experts can hear the subtlety of reverberation,
we are faced with the larger question: who are the listeners? To *prove* that an illusion
is equivalent to the real aural experience, we would need to show that no one exists
who could hear any differences. That would be the strictest definition. Or perhaps we
should exclude the top ten world's expert listeners, defining all others as the "listening
public." Or perhaps we should exclude all expert listeners, or all listeners who have
spent more than a thousand hours in real concert halls. Within the general population, auditory perceptual sensitivity is itself a probability distribution. On the one
hand, many of my friends cannot hear the difference between a professional reverberator and one created by a hobbyist as a school project. But, on the other, many of my
professional colleagues can hear the subtle nuances of an acoustic space. Obviously,
there is a social and economic side to acoustic quality, about which science has little
or nothing to say. Who actually appreciates that quality? Is it worth achieving, and if
so, at what cost?

Social and Political Contexts Influence Designs

The art of reverberator design is the simultaneous optimization of several incompatible
or competing acoustic parameters that are required to create the aural illusion of an

acoustic space. Each topology requires a different solution to the optimization task. When complete, the design should have a sonic reflection density that is large, random, and with uniform energy in all 1-millisecond windows. The resonance density should also be large and random. In addition, some parameters must remain free in order for the user to select the desired properties for the aural personality of the space. Decay time should be selectable in different frequency regions, perhaps ranging from 1 to 10 seconds in three or more frequency bands. Other subtler acoustic parameters include the rate of reverberation onset, the duration of the sustain region before decay begins, and the shape of the decay. In addition, the early part of reverberation requires some number of discrete sonic reflections defined by their delay, amplitude, and output channel. If the reverberator is a surround-sound spatial simulator, the relationship among the channels is yet another dimension.

Anyone wishing to master the art of reverberator design is likely to experience initial optimism followed by unremitting frustration. The first few experiments will show promising results. Then, as sensitivity to acoustic defects increases, the optimization process soon leads to depressing frustration. Even after a half century, the literature is remarkably devoid of examples and analyses of high-quality reverberator designs. To appreciate the reasons for a lack of public knowledge, we need to explore the personal, social, and economic context of the reverberator designer. As one of them, I speak from experience. Consider a few issues.

It can take years of experience for a designer to acquire deep insight into just one topology. This insight involves inventing, ad hoc, the mathematical rules for specifying the size of every delay and the value of every gain, a total of perhaps hundreds of parameters. With the wrong choice of acoustic parameters, the most beautiful topology becomes dreadful. Similarly, the art of testing a reverberator design to find defects and optimize performance requires an understanding of the relationship of sound to perception. Different sounds stimulate different aspects of the reverberator design. And finally, like a magician, the reverberator designer is actually in the business of creating an illusion.

The most serious professional reverberator designers work for, and are paid by, companies that have a very strong interest in protecting their intellectual property. Designs quickly become their crown jewels, to be protected at all costs. Moreover, virtuoso designers often have a stake in the company, earn their livelihood from their specialized skills, and thus fully support corporate secrecy.

Unlike some industries that rely on patents for the protection of intellectual property, the audio industry seldom patents its important inventions. Secrecy is the primary vehicle for protection. When patents exist, the published information is almost always insufficient to produce a quality result. Although patent law explicitly states that a patent is *invalid* if there is inadequate information to duplicate the invention, inventors frequently omit the process of tuning the design. Such omissions are creatively subtle ways of cheating a legal system that offers a monopoly on inventive ideas if, and only if, they are made available to the public. Although there are a large number of reverberator patents, I would guess that 90 percent of them are without enduring value. A student of reverberators is thus faced with the burden of finding a few elegant gems buried among a collection of useless or irrelevant inventions that never appeared outside the inventor's laboratory, so-called vanity patents.

To appreciate the influence of social context on reverberator design, consider the story of one particular design from the late 1980s, which originated as the flagship product of a world-renowned company specializing in reverberator technology. A start-up competitor, after failing to achieve a design of comparable quality, spent many months reverse engineering that product, legal, but perhaps unethical. In those days, with discrete electronics, reverse engineering was tractable. Shortly thereafter, the start-up company failed, but its reverberator design began to float around the industry. Even though details were still treated as a trade secret, the design eventually appeared in two articles (Gardner, 1998; Dattorro, 1997a). It was an old design, long obsolete in terms of acoustic quality, but it is still unique in that both the topology and a good set of working parameters are widely available.

Although almost anyone can find or create a reverberator topology and adjust a few of its parameters, only an expert designer can create a reverberator that faithfully simulates a real acoustic space. Over the years, a few virtuoso designers have created their versions of a true masterpiece, something to be appreciated but not readily duplicated. Artistic secrecy is not new. Scientists are still trying to duplicate the sound of violins made by Stradivari (Hill, Hill, and Hill, 1963) and Guarneri (Horace, 1977), whose skills were lost because they neither recorded their techniques nor passed them on to future generations. It is difficult, if not impossible, to derive a recipe by analyzing the object produced by a complex process. Will the reverberators of the late twentieth century follow this path? Perhaps.

On the other hand, there is a simpler reason why experts have not yet published the complete book on the art of reverberator design. How many people are interested in learning such an obscure art? I would guess very few. In an age of mass merchandizing and commodity art, the highest quality is often unrecognized, unappreciated, and more importantly, socially irrelevant. Aural architects, the customers of such acoustic tools, use them to create aural spaces, and they determine what degree of quality is adequate for their needs. The year 2000 may prove to have been the peak of discerning quality.

7 Spatial Innovators and Their Private Agendas

Political intelligence can be defined as the decision-making capacity that enables social animals to further their self-interest in situations that involve rivalry and questions of power and leadership.
—Christopher Boehm, 1997

We have already examined the relationship of aural architecture to culture, but we have not considered aural architects as people, functioning within their personal contexts. Beyond the few famous aural architects readily identified from surviving publications, there are thousands of unnamed contributors about whom recorded history reveals little. How and why did this army of anonymous designers and builders create aural architecture? Although answers to such questions tend to be culture specific, by examining the modern context, several overall patterns become clear. And because of the universality of human nature, we can assume that these patterns also existed in subcultures of other periods, where data are unavailable.

The classical Greeks, medieval Romans, Renaissance Europeans, and ancient Mayans all had their aural architects; they were found in royal courts, trade guilds, theater companies, and religious orders. Today, aural architects are found in some thirty artistic, professional, intellectual, sensory, and folk pursuits and disciplines.[1]

We can think of society as being a collection of subcultures, each providing a context for other subcultures. Subcultural values, which may deviate from those of the wider culture, have a strong influence on the behavior of individual members of a subculture. This chapter, containing numerous anecdotes and digressions, illustrates how professional disciplines provide a social and intellectual subculture for aural architects—and how private agendas derived from subcultural values are frequently more important than acoustic attributes when designing spaces. An analysis of a few representative examples illustrates larger patterns.

The disciplines involved in hearing space are surprisingly many and diverse; indeed, the term *aural architecture* was coined to designate the fusion of these disciplines in selecting, designing, and experiencing spaces by listening. Yet a search of the literature for such phrases as "aural architecture," "aural space," and "hearing architecture" only produced two citations (Sheridan and van Lengen, 2003; Rasmussen, 1959) that related to hearing space. This raises several questions. Why are those who share this interest only vaguely aware of related activities in other disciplines? How do the differences in the social structure of each discipline influence the way its members behave? How can social and intellectual diversity be fused to address aural architecture in intellectually productive ways?

The problems encountered when fusing knowledge from diverse disciplines into an interdisciplinary view transcend our specific topic. Indeed, interest in reconciling different approaches to similar questions has given rise to the new discipline of *interdisciplinarity*, itself an interdisciplinary field comprising anthropology, social psychology, epistemology, and philosophy. A half century after the formal debut of interdisciplinarity, there are now many examples of interdisciplinary activities in education, engineering, research, and politics.[2] Although the principles of interdisciplinarity are neither widely known nor much appreciated, they are central to understanding aural architecture.

Nonexperts need guidance in evaluating and connecting information from dozens of isolated disciplines. They need to know how to identify essential information without becoming overwhelmed with details; how to evaluate scholarly contributions from a range of disciplines without being expert; how to translate the private and specialized language of each discipline into a common language; and, finally, how to discover the hidden biases that undermine the scope and validity of assertions.

For those who have not yet engaged in an interdisciplinary activity, but who have refined expertise in a single discipline, attempting to answer cross-disciplinary questions requires intellectual humility toward both others' disciplines and their own. By reexamining a discipline's working assumptions, those within it begin to realize the limitations of conclusions that initially appeared to be stable truths. For example, the literature on the perception of sonic reflections, which comprises diverse experiments, viewpoints, assumptions, and conclusions, is neither consistent nor coherent (Litovsky et al., 1999). When feeling intellectually humble, scientists studying this topic occasionally concede that their best theories are, in fact, not always useful for explaining real experiences. The "flaws" within each discipline are the natural result of compromises, often hidden and unspoken, which must be made when working with otherwise intractable problems.

The concept of evidence varies dramatically among those disciplines involved in aural architecture. Some emphasize anecdotal wisdom, whereas others stress mathematical formalism, empirical data, abstract models, or speculative inferences. Thus, the pedagogy of teaching the blind to use echolocation is mostly based on experiential anecdotes, whereas statistical acoustics is based on mathematical proofs, although, when applied to the messy world of real spaces, it has its own ambiguities. Indeed, however fond they may be of objectivity, scholars, scientists, and engineers still use intuition to resolve the ambiguities and inconsistencies of difficult problems.

Each discipline has both a personality and recognizable rules for membership. Those within a given discipline share attitudes; by working together over decades, they form strong emotional and intellectual bonds. To understand a discipline as a social unit, between the larger culture on the one hand, and the individual on the other, let us first consider the concept of a *subculture*: a small group of individuals bound together by shared values and goals. The next chapter will argue that evolution optimized survival in subcultures. Concepts such as society and culture, however useful for understanding major differences in the aural architecture of diverse societies and time periods, are too broad to explain individual behavior. Individuals are strongly influenced by their respective subcultures. The anthropologist Pearl Katz (1999) provides a good working definition for *culture* (larger society) and *subculture* (coherent small group) as "implicit and explicit basic assumptions—beliefs, values, attitudes, and ideas—both about the ways of viewing the world, emotionally and cognitively, as well as how to behave in the world with other people and with objects and tools. Culture is shared and learned. It includes skills for communicating through verbal and non-verbal language and other symbols. It also includes social arrangements for passing on skills to new generations." Extending this definition, a professional *discipline* then becomes a subculture whose members share particular skills, and who are usually paid for making recognizable contributions to society using those skills.

Society is more than a collection of individuals; it is also an amalgam of diverse subcultures. Musicians belong to an artistic subculture, audio engineers to an auditory subculture, researchers to an intellectual subculture, and interior decorators to an aesthetic subculture. When we speak of an audio engineer or a contemporary composer, we are actually referring to a small group of individuals with similar skills, values, ideas, perspectives, education, and reward systems—a discipline. Spatial innovators, quite apart from being aural architects, are, first and foremost, members of their respective disciplines.

As an abstraction, a discipline suggests a metaperson with human goals, motivations, and characteristics (Katz and Kahn, 1978). Although this is not in fact the case, when individuals coalesce into a coherent social unit, their collective behavior does resemble that of a recognizable entity. However much individual members may be identified with specific contributions, they represent the supporting infrastructure of the entire discipline. For example, even though I was the first to commercialize digital reverberators, it can be said that the audio engineering discipline as a whole created digital reverberators—as electroacoustic aural architecture: if not I, then some other engineer would have produced an equivalent design within a few months.

We can analyze the properties of the disciplines that, as a whole, create aural architecture. To evaluate these diverse disciplines, we will focus on two important dimensions: intellectual framework and cultural values.

The *intellectual framework* of a given discipline defines how it thinks about knowledge and assertions, how it formulates problems; the framework defines its knowledge base, philosophical rules of inference and expertise, as well as the language it uses to describe knowledge. When disciplines with different intellectual frameworks interact, inconsistencies among their knowledge-belief systems produce seemingly unbreachable intellectual boundaries. It is also difficult to translate across disciplines, or to know when there is genuine agreement or disagreement. By studying intellectual frameworks across a variety of societies, ethnoepistemology reveals that these frameworks are expressions of cultural views, not absolutes (Kornblith, 1997). As subcultures, disciplines also believe in the legitimacy of their rules of evidence for establishing "truth," which varies from discipline to discipline.

The *cultural values* of a discipline are subtler, and reflect the social behavior of its individual members. A discipline's values embody the professional goals, the economic support structure, the criteria for career advancement, the allocation of decision-making authority, the training and education of the next generation, and the enforcement power of the leaders and peer community. Cultural boundaries are manifestations of differences in social and organizational rules, which are seldom written. When disciplines with different social values interact, inconsistencies reveal the existence of social boundaries, which may be as unbreachable as intellectual boundaries.

Disciplines are distinct because of differences in one or both of these dimensions. Frequently, cultural and intellectual boundaries only become apparent when individuals from different disciplines encounter one another. In some cases, two disciplines may have similar cultural values but different intellectual frameworks; in other cases, they have different values but similar frameworks. For example, academic departments, such as acoustics and anthropology, share the cultural values of higher education but their intellectual frameworks are mutually incomprehensible. In contrast, psychoacoustic researchers at a government laboratory differ from their counterparts at a commercial laboratory in cultural values but not in intellectual framework. More typically, disciplines may differ in both dimensions, as for example, audio mixing engineers and acoustic architects.

Social Values: Goals, Rewards, and Careers

When like-minded professionals or academics live and work among colleagues in a shared environment, they coalesce into a coherent social unit with unique properties, creating a recognizable discipline. Acousticians from different countries, for example, may have more in common with each other than they do with members of their respective national cultures: they belong to the family of acousticians. The same is true for audio engineers, traditional architects, modern composers, and perceptual scientists. The arcane language and education of a discipline bind its members together, while excluding outsiders. Aside from their specialized skills and knowledge, professional disciplines and sensory subcultures function like tribes or extended families—individuals bound together with shared identity, support, goals, and history.

As individuals, aural architects are shaped by their connections to the larger culture, other subcultures, and microcultures. Each social unit has strict rules for governing self-esteem, financial rewards, political power, and access to new recruits. Even if not apparent or articulated, such rule systems are very powerful. The larger society regulates the behavior of disciplines, and in turn, they regulate the behavior of their individual members. The means of control are varied, including such social currencies as public recognition, political power, professional banishment, or anything that manipulates an individual's sense of self-esteem. Like all innovators, those who contribute something to aural architecture are therefore exhibiting behaviors that are responses to the governance process.

Specific individuals often have disproportionate power to influence a discipline. Their power, which may originate from forceful personality, creative brilliance, or political influence, leads the discipline in a specific direction. There is no better example than audio engineer and entrepreneur Tomlinson Holman, creator of the current 5.1 surround format. With the support and sponsorship of the powerful cinema industry, Holman created a compromise that balanced social, political, economic, and aural properties. It may not have been an ideal compromise, but it became a global standard. Competing choices, many with better aural and technical properties, failed because of a lack of political support. In other cases, disciplines have a democratic or even a chaotic quality that does not allow a single individual to dominate decision making.

The aural architecture that we hear results from complex interactions between three social units: individuals, disciplines, and the larger culture. For example, the acoustic engineers who built radio studios in the 1940s were individuals, but they were also part of the early audio engineering discipline, which was embedded in the larger culture. Each of these three social units influenced studio design: the larger culture was hostile toward reverberation as noise, the radio subculture was interested in creating publicly acceptable programs using primitive audio technology, and individual audio engineers were interested in managing their careers and financial rewards. This view of aural architecture furthers the notion that cultural influences (social and political) are often more important than intentional design (science and engineering) in determining the final outcome.

As a subculture, a discipline maintains an integral relationship to the values of the larger culture; it does not exist in isolation. In studies that do not specifically address aural architecture, numerous scholars have identified the intertwined relationship of disciplines to the larger society. Lynn Nader (1996) makes the compelling case that scientists are also embedded within the larger culture, even if they primarily identify themselves with their own professional subculture. Sarah Franklin (1995) challenged the notion that intellectual objectivity removes scientists from these cultural forces, and Robert M. Young (1972) concluded that science is much more like the messy world of social and political intercourse than working scientists care to believe. Sociologist Andrew Webster (1991) views science as a social construct that strongly influences, not just the behavior, but also the ideas of scientists. The more they are tied to the priorities of society, the more science and technology lose their neutrality. Embedded as they are in the larger culture of society, with its underlying politics of resource competition, disciplines are then shaped by society's tensions and power struggles, the dynamics of which are rarely chronicled (Nader, 1996). These observations apply equally to aural architects, which makes the politics of aural architecture a significant part of our discussion.

More and more since the late twentieth century, knowledge has become intellectual property to be sold, traded, and licensed, functioning as a surrogate for money. Does an architect advocate a specific design solely from belief in its artistic value, or because it would also enhance the architect's ego, career, or bank account? Does a perceptual scientist publish results only because they make a major advance in our understanding of spatial awareness, or because the publication leads to additional funding and career advancement? The underlying motivation is often unclear or hidden. Because these questions can be asked of every contributor, we clearly see the problem of objectivity. Those who innovate, and those who evaluate those innovations, cannot be purely objective. Individuals have multiple agendas. Generally, only those agendas that are socially acceptable are publicly articulated.

Understanding how artistic and intellectual disciplines function requires the application of principles of sociology and anthropology. Although anthropologists originally studied primitive societies, there is now a significant application of their same techniques to professional groups in modern complex societies (Peirano, 1998). Anthropology studies have already focused on physicists (Traweek, 1988), biologists (Latour and Woolgar, 1979), musicians (Born, 1995), and surgeons (Katz, 1999). Such studies make it clear that disciplines, like tribes, make alliances of convenience, compete for natural resources, propagate their worldviews, educate future generations, share common goals, and battle for dominance.

Many academic groups are like peaceful tribes working harmoniously on subjects of mutual interest, but many others are like aggressive tribes battling for dominance. Consider the experience of Ken Wissoker (2000), editor in chief at Duke University Press, who described his experience with the negative aspects of academic tribalism. Because knowledge creation for its own sake, however admirable, has no external validation process, political infighting at academic institutions is well known, widespread, and highly corrosive. For this reason, academics from different disciplines collaborating on teaching, writing, or research projects do not always share knowledge. Indeed, regarding colleagues from other disciplines as "interlopers", scholars behave "as if they were engaged in a war of territory, as if interdisciplinary [activity] was a zero-sum game" (Wissoker, 2000).

Disciplines often foster loyalty and defensiveness, creating biases that override the publicly stated goals. Wissoker's comments are a clear articulation of such emotions in academic departments. Those who work in commercial and governmental institutions make similar observations about political infighting for status, power, and money. Just as Antony Jay (1994) observed that the social rules first articulated by Nicolò Machiavelli (1958) now apply to companies rather than countries, disciplines are analogous to fiefdoms in medieval Europe. From a similar perspective, Andrew Whiten and Richard W. Byrne (1997), using the concept of Machiavellian (emotional) intelligence to explain aspects of human behavior embedded in small groups, concluded that, when based on considerations such as self-interest and territorial defensiveness, emotional "irrationality" may actually enhance individual and group survival. Consider that innovation is often both a threat and an opportunity, disrupting the status quo, even while providing new opportunities for progress.

Emotions are the secret story of disciplines not found in textbooks: social considerations existing side by side with artistic and intellectual creativity. Once we accept the premise that emotions play a significant role in scholarship, it follows that there must be a subjective component to all intellectual assertions and conclusions. Awareness of subjectivity introduces humility into the otherwise widespread hubris of sweeping intellectual assertions. Subjectivity is an attribute of all disciplines. In his study of emotional intelligence, Daniel Goleman (1997) commented "that the thinking brain grew from the emotional [brain] reveals much about the relationship of thought to feeling; there was an emotional brain long before there was a rational one." Emotions are an evolutionary solution to surviving, especially in social groups.

Although social dynamics, which are based on human emotions, are neither logical nor objective in the conventional senses of these words, Ronald de Sousa (1997) showed that emotions have, in fact, their own rationality. As an expression of emotions, behavior is "rational" when considered from the perspective of an individual's personality, whose real goal may be nothing more than making the world align with a private model of what life should be.

Resources, Decisions, Knowledge, and Political Power

The sponsors of aural architecture, with their own goals, language, and perspectives, do not use the same criteria as those aural architects who actually design spaces. For musical spaces, community sponsors consider the acceptance and satisfaction of the listening public and the social recognition these will bring them. For academic spaces, government sponsors consider the political consequences of funding particular spaces. For religious spaces, church sponsors consider the symbolic effect on their congregations. Sponsors evaluate spatial designs according to their own particular goals; they typically exercise their financial power over the design process without making auditory spatial awareness a primary concern.

Aural architecture results from the dynamic tension between two classes of stakeholders: implementers (in their disciplines) and sponsors (in their supporting organizations). Counterbalancing the obvious financial power of the sponsors are the specialized skills, arcane knowledge, and often intimidating social status of the implementers. In some cases, the implementers overpower the sponsors; in other cases, the sponsors overpower the implementers. The best designs represent a harmonious response to the concerns of both.

The architectural literature overflows with discussions about the tension between the competing demands of aesthetics and functionality. At one extreme, traditional architects often view the spaces they design in their ateliers as static works of art, like paintings or sculptures, of sometimes exquisite grace and beauty. Indeed, books on traditional architecture are filled with pictures of spaces and buildings showing few, if any, people within them. At the other extreme, some nontraditional architects design spaces primarily to accommodate the needs of those who occupy or inhabit them. We might suspect that sponsors, valuing function over form, would force the architects to be more social artisans than creative artists. But this is far from always the case. To make a space practical, responsive, and comfortable, the architect must draw on psychology, sociology, anthropology, and an understanding of sensory perception, including, not least of all, aural perception.

The preoccupation with architecture as art, which often resulted in spaces that were impossible to occupy, let alone inhabit, gave rise to social design. Robert Sommer (1983) defines this as "working with people rather than for them; involving people in the planning and management of spaces around them; educating them to use the environment wisely." Sommer's ideal contrasts with the messy world of social and emotional politics within the architectural disciplines.

The intimidating power of traditional "creative" architects was clearly revealed by two social scientists who examined their attitudes. The prominent personality psychologist Donald W. MacKinnon (1963) studied 124 traditional architects. A sentence-ranking test was used to divide the architect subjects into three groups based on the relative importance of creativity to their architectural decisions. The most creative group stressed adjectives such as "inventive," "determined," "independent," and "individualistic." The least creative group stressed adjectives such as "responsible," "reliable," and "dependable." MacKinnon (1963) concluded that a creative architect "thinks of himself as imaginative; unquestionably committed to creative endeavors; unceasingly striving for creative solutions to the difficult problems he repeatedly sets for himself; satisfied only with solutions which are original and meet his own standards of architectural excellence; aesthetically sensitive; an independent spirit free from crippling restraints and impoverishing inhibitions." Creative architects did not view themselves as particularly social, as if people were an impediment to aesthetic creativity. Spaces were sculpture, built of bricks and glass on a grand scale large enough to walk through.

Two decades later, as an extension and confirmation of MacKinnon's earlier work, Dana Cuff (1989) interviewed seven architects concerning their views of their profession. "Architecture is a form of poetry," he concluded. "Architecture is made for architects, for themselves." As one architect explained: "Buildings have a life of their own. People, to exaggerate the point, are at the behest of buildings." And with a total disdain for sponsors, another commented: "Clients vary a great deal, but most cannot understand three-dimensional space. They cannot visualize the final result" (Cuff, 1989). These architects, at least at this time in their culture, wanted to be insulated from social forces. Nor is this attitude unique to architects. Composers, audio engineers, computer scientists, and academic researchers have expressed similar attitudes. Because of their arcane expertise and high social status, such professionals often feel that they have the right to impose their views on the larger culture through their designs.

In contrast to the disproportionate power of a few famous architects who view themselves as autonomous artists, most architects are controlled by the financial power of their sponsors. By selectively allocating money, sponsors dominate a project's goals, values, and designs. For an architect providing a service, pleasing the sponsor then becomes a means for acquiring rewards, which include public accolades, professional power, and an enhanced material lifestyle. Although one type of reward is not necessarily better than another, in modern industrial cultures, wealth is a common measure of rewards, not just because of its purchasing power, but also because it bears directly on our sense of personal autonomy and self-esteem (Stanley and Danko, 1998). Government grants, financial rewards, and commercial marketing have replaced the older traditions of royal patronage and religious subsidies, but they exercise much the same influence on the behavior of those involved.

Like traditional architecture, aural architecture can be understood by considering the details of the governance process, which determines which ideas become immortalized in society's spaces, and which do not. We observe the nature of power and its influence on decision making by examining the choices made as the money is channeled from funding source to final destination. In other words, each decision maker attaches constraints to economic resources as they are passed along, as examples of commercial and governmental money channels will illustrate.

In most cases, research that provides an intellectual foundation for aural architecture, such as physics research into the statistics of sound waves in enclosed spaces or cognitive research into the perception of space using the neurological models of the auditory cortex, is supported by government grants. Let us follow the progression of taxpayer funds through governmental money channels, from the federal government, to granting agencies, to research institutions, to research groups, and finally to individual scientists. At the first stage of the process, government officials decide how to allocate money among various agencies. Each agency receives a portion of the total, which is then sorted among a large number of research institutions in many different fields that cover a wide range of topics. Some projects receive funding; others do not. Research support, especially for big science, raises larger political questions that go beyond the specific research (Fuller, 2000). Scientists with interests in specific aspects of aural architecture are usually constrained in what they research because many powerful decision makers, using criteria that are only partly visible, determine what research will be supported.

Reviewing the international literature on aural architecture, we find that some countries actively support research in its disciplines, whereas others do not. One of the best examples of governmental support in this regard is the Institute for Music Acoustic Research and Coordination (IRCAM) in France, the world's largest scientific research center dedicated exclusively to innovative music using advanced technology. Founded by Pierre Boulez in 1970, it now has some 90 visiting and resident scientists pursuing a wide range of artistic and intellectual activities from the perspective of the musical arts, including the aural architecture of virtual spaces.

An ethnographic analysis of IRCAM by the anthropologist Georgina Born (1995) showed a tight coupling between the institution's internal projects and its public image. By the 1990s, IRCAM was placing increased emphasis on promoting its scientific, technological, and artistic projects to the public. When the Ministry of Culture argued that, as an institutional beneficiary of public funds, IRCAM should be evaluated by the public rules, creating a sympathetic view of the institution became as important as, or perhaps more important than, the actual work being done by the artists and scientists it sponsored. Centralized control of activities provided an efficient means for supporting IRCAM's public relations goals. Private and individual projects conducted behind the locked doors of offices, studios, and laboratories replaced the artistic and intellectual democracy of open collaboration. Accompanying this cultural shift, the rise of computer technology and artificial intelligence made computers and computer analysis central to psychology, cognitive science, and acoustics research. For example, research into spatial acoustics and spatial music now came under the umbrella of signal processing. Just as Pythagorean numeric ratios were enlisted to rationalize the design of religious spaces in the Middle Ages, so, too, computers have now been enlisted to rationalize the design of musical spaces and, indeed, of music itself.

Analysis of other research institutions shows a similar pattern. Thus the mandate of NASA's Advanced Displays and Spatial Perceptions Laboratory, another nationally sponsored institution, is sufficiently broad to allow for research on auditory spatial perception in a wide variety of environments, from the cockpits of airplanes to living spaces. But it, too, must maintain a public image of being socially useful. And although the National Institutes of Health use an expansive definition of hearing research that actively supports research into all forms of auditory spatial awareness, there is an implicit expectation that such research will eventually have some medical utility. In each of these examples, governmental administrators decide whether the mandate of a state-sponsored research institution is to be broad or narrow; that decision, in combination with the interests of researchers, determines what aspect of aural architecture can then be pursued.

As the interests of administrators and researchers shift, so does the focus of the research funded. For almost ten years, the Helsinki University of Technology heavily invested in a large research system for producing virtual audiovisual performances in real time. That activity has now been suspended. During the 1990s, Stanford University's Center for Computer Research in Music and Acoustics actively pursued electronic reverberation as an application of a specific signal-processing technique. That is no longer the case.

In contrast to the formal structure of a governmental money channel, a commercial money channel begins with thousands of anonymous consumers. When individuals attend concerts or buy prerecorded music, when the owner of a recording studio decides which reverberator to buy, or when the producer of a music event selects the performance space, money is injected into a commercial money channel. The difference between governmental and commercial money channels is chiefly in the outer character of the process: commercial decision makers are part of an amorphous matrix of unnamed individuals; governmental decision makers are part of an explicit system of named participants, specified by written laws and public policies.

The length of the commercial money channel becomes apparent when we consider the financing of a spatial synthesizer in two situations: as part of the playback element in a consumer sound system and as the means for adding spatial illusions to music during audio mixing. In the case of consumer audio systems, the money channel connects the consumer to a local electronics store, to the sales distributor, to the manufacturer, to the development group, and finally to the signal-processing engineer. Consumers of equipment are providing salary and equipment for this engineer. In the case of the recorded music, the money channel connects the consumer to the retail store selling music, to the producer of the music, to the professional recording studio, to the equipment manufacture, and eventually to the engineer. There are actually many other decision makers in these long money channels, and they all influence the criteria for sorting among choices and trade-offs.

When designing a reverberator product, as an enlightened engineer, you need to understand the criteria that exist along your channel. How should you view your design choices given the complexity of the process? Should you repackage an old algorithm, should you research a novel concept, should you implement a new but untested idea, or should you merge your self-interest with your colleagues and become a follower of accepted traditions? Answers to these questions are not obvious, yet they have an impact on the monetary resources that eventually flow through the money channel. Expanding resources allows for future research and development with expanding career opportunities; shrinking resources leads to professional stagnation and unemployment. All those involved in a money channel have a stake in the outcome, but their respective risks and rewards are not the same. Researchers, scientists, and developers who ignore sorting criteria do so to their own detriment. Devoting your professional energies to an aspect of aural architecture that is politically unsupported leads to unemployment.

Two contrasting examples illustrate how market forces influence the design of spatial synthesizers. In the 1990s, Lexicon Corporation developed a very powerful surround spatial processor, costing about $10,000, for high-end recording studios; at the same time, Steinberg Corporation developed a plug-in software module, costing only $300, for small-scale project studios, a broad market comfortable with lower quality. Each company was sensitive to the needs, expectations, and resources of its consumers; each identified a well-defined market before designing its products. In another example, the Wenger Corporation created an electroacoustic practice room that simulated a wide range of spatial acoustics, from those of Baroque chambers to those of Gothic cathedrals, for sophisticated music schools. Students with the goal of becoming professional musicians are likely to appreciate the value of experiencing a variety of performance spaces.

If politics is the process of resolving conflicting needs, wishes, and goals among parties, then all forms of priority sorting along a money channel, be it commercial or governmental, are political. Within aural architecture, competing uses for limited funds include building new concert halls, improving hearing prostheses, developing entertainment electronics, inventing safer user interfaces for airplanes with audio display, applying architectural acoustics to public buildings, training the blind to navigate with acoustic cues, and supporting perceptual research in educational organizations. With finite financial resources to be allocated to auditory spatial awareness, society makes its collective decisions based on social values and political power, without necessarily focusing on the intellectual content of its choices.

Robert Frodeman, Carl Mitcham, and Oliver Sacks (2001) argue that politics constitutes a serious force in determining the relevance of knowledge workers, including aural architects: "The political limits of the increased information production are found in the public's increasingly insistent demand that publicly funded research and education clearly show their connection to community needs." Governmental administrators represent the needs of the larger society, and individual consumers represent the needs of smaller subcultures. As the result of the dynamic tension between implementers and their sponsors, specific embodiments of aural architecture enter the society. The specifics depend on the details of the culture and the time period, but our examples argue for the general premise that artistic, intellectual, and scientific activities can only be understood by examining the interplay of political power, money, and knowledge. Our discussion is explicitly modern, but the premise applies as well to twelfth-century Christians designing cathedrals and to ancient Greeks designing open-air amphitheaters.

Conservatism as Cultural Traditions and Career Management

Even with its love for progress, innovation, and experimentation, our culture has been relatively slow to evolve new styles of aural architecture and to adapt new attitudes toward auditory spatial awareness. Because of linkage between architecture and social adaptation to its aural properties, the combination becomes a stable cultural tradition. Current aural architecture is the historical product of conflicts between the forces of innovation and those of tradition. The outcome of such conflicts is neither obvious nor predictable.

By considering how professionals relate to personal risk and reward, we also observe a microcosm of the same tension and conflict between the new and the old. Individuals make decisions to optimize their personal situation. When breaking with conventions and traditions, success and failure are not symmetrical. For example, the mathematician Georg Cantor suffered the destructive consequences of challenging the authoritarian views of his senior colleagues (Aczel, 2000). For every path to success, thousands lead to failure, skewing the ratio of risk to reward toward conservatism. Because society rewards innovators only when successful, a common strategy among less audacious individuals is to choose paths of minimal risk. History focuses on the few successful innovators, while ignoring the influence of the many who, avoiding risk, did not attempt to innovate, indeed resisted innovation. To understand why we have the acoustic spaces we do, we must not ignore that powerful influence.

At the level of society, conservatism manifests itself as cultural traditions; at the individual level, as career management. Building a cathedral, designing a concert hall, and changing an international standard for sound reproduction are expensive activities with large risks and rewards. Earlier, we noted that aural architecture is a product of the tension among nonauditory social forces. In this discussion, we see the same pattern but at the individual level.

A few simple examples from academic research and commercial development will illustrate the relationship between social conservatism and individual career management. There are two basic organizational structures used to evaluate individuals: peer review by respected senior colleagues, and structured hierarchies composed of managers. The former is typical of publicly supported science; the latter, of commercial activities. The outcome is the same but the details differ. In both cases, the organizational structure dispenses rewards and punishments to individuals, who are managing their career like financial investors.

In the case of peer review, colleagues on committees evaluate each other's work. Decisions with regard to publishing articles, receiving grants, and promotions are made with advice from such committees. They are composed of senior professionals and administrators with their own public and private agendas. Any significant professional deviation from the collective wisdom of the peer group is likely to result in rejection. The review process powerfully controls a discipline. Extremely useful in maintaining defined goals and sorting high-quality work from mediocrity, it is also extremely conservative in preserving what Thomas Kuhn (1996) called "normal science." In their worst embodiment, disciplines produce a closed, rigid society (Popper, 1962). By their power to reject manuscripts submitted for publication, grant proposals, and career promotions, peer-review committees enforce a consistent set of values. For an individual, preserving peer respect serves not only practical but also emotional needs for status, acceptance, and appreciation.

In the half century following Helmut Haas's publication (1951) of how sonic reflections increase loudness, thousands of researchers have explored the "precedence effect." For the most part, they examined variations on a theme, expanding our understanding of the effect under various controlled conditions. Research focused on making a model, a low-risk strategy, as opposed to challenging basic assumptions, a high-risk one. For a doctoral candidate working on such research, the appropriate behavior is usually to research a question that leads to a doctorate within a reasonable time, and to support the agenda of senior scientists. Student researchers, who are a major source of scientific labor, only need to demonstrate intellectual competence to their professors; they do not have to prove their utility to the larger society.

Conservatism also results from the need to reduce the stresses of conflicting expectations. Robert K. Merton (1976) articulated a few common examples of inconsistent and ambivalent expectations in the scientific value system. The reward for a major discovery, which may also be taking place simultaneously at other institutions, is sufficiently high that being first has professional value, yet you must not be premature, with preliminary, inconclusive, or inconsistent conclusions. Details are important for reliability and reproducibility, but your concern over minor points must not be excessive or pedantic. You should be open to new ideas, but you should not follow fads. The continuous and subtle stresses that develop from these conflicting demands can lead to the adaptive strategy of a conservative path that minimizes uncertainty.

In contrast to peer-review committees in academic institutions, commercial organizations often use a hierarchy of executives who manage the behavior of their subordinates. Because the success of managers is determined by the behavior of others, they must control others' activities if they wants to optimize their own rewards. The hierarchy is layered, each level managed by a higher level. For the purpose of illustration, let us focus on the lowest level, where individual engineers work as employees of a corporation.

Such engineers, however dedicated and creative, are often treated as serfs by their employers. Ownership of intellectual property in the form of trade secrets or patents drives organizations to severely restrict information distribution in order to protect their competitive advantage. Internally, the need-to-know rule limits communications. As an employee, an innovator is not free to publish without permission, and as a precondition for employment, you are invariably required to assign your intellectual rights to the corporation. In fact, in many countries, the legal definition of an employee is someone whose time, task assignment, and productive output are owned and controlled by an employer. Although employees often identify with their company, the interests of the two parties do not always coincide.

Almost all of the sophisticated reverberation algorithms remain secret, even those primitive algorithms from companies that have long since disappeared. They cannot legally be published. As an editorial advisor to a major engineering society, I observed a truly odd example. An engineer had published an idea in its journal that his previous employer later claimed was proprietary. Rather than become involved with lawyers, the journal expunged the author's article from its archives, ignoring the fact that there were already more than 10,000 copies in circulation. The author eventually prevailed, at great personal expense, but not until he had been professionally punished for his behavior. His career suffered. The company used its legal power to send a message to all professional engineers.

Managers of organizations behave rationally in terms of their self-interest, which may not be in the interest of the investors, customers, employees, colleagues, or the larger community. Creative ideas are sometimes hidden or locked away. The economic consequences of a new development, with the potentially high costs of manufacturing, marketing, and distribution, make the entry threshold for commercialization very high. Even when the rewards for taking risks are potentially large, companies will often choose the status quo, the slow road to death, over the disruptive new technology, whose exploitation may be a life-and-death gamble (Christiansen, 1997). In times of economic stress, short-term survival becomes more important than anything else.

More typically, companies and individuals optimize products judged by the marketplace to be of high quality for their profitability. Because high quality is expensive, and appreciated only by a small percentage of users, however, quality is often redefined as anything that is "good enough." Quality for a few major recording studios is not the same as quality for a thousand project studios. Quality unperceived by consumers has no value to them, even if there is a solid intellectual foundation for it. Thus quality is only one component of the decision-making process, and not always the most important.

Even though every professional is taught that quality, innovation, and creativity are valued, in reality, most advances arise from conflicts that produce winners and losers. Consider the history of electronic reverberation. Digital signal-processing algorithms in the early 1980s made mechanical reverberators (composed of flat plates and helical springs) obsolete. In the 1990s, algorithms optimized for custom integrated circuits made those designed from discrete components obsolete. In today's marketplace, sales of inexpensive commodity software modules of mediocre quality are undermining the economic viability of higher-quality designs. In each case, cost, technology, and quality were in constant flux. Having developed the skills to optimize their algorithms for unique hardware, designers face the difficult choice between preserving their investment in the status quo or acquiring new skills with an untested and potentially risky approach.

There are other choices. Innovators, acting as investors, can start their own companies, only to find it equally difficult to attract needed capital: venture capitalists, though comfortable with risk, invest only in ventures likely to yield a ten- to hundred-fold return on their investment within a few years. Few innovations have the potential to produce that kind of return; most are incremental improvements on existing methods. More than a few venture capitalists have privately commented, however, that their safest choice is often to match the herd behavior of other investors.

Regardless of the organizational structure for research and development, there is an intrinsic conflict of interest between individuals and society. Professionals retain their status only if their expertise is viewed as having perceived value, and that value depends on their skills matching the current technology and market preferences. Major shifts in either undermine that value. It is not wise to specialize in a specific field unless your specialized skills will remain useful and in demand for a long time, either in that field or, if possible, transferred to another.

Subjectivity, Personality, and Cognitive Judgments

Having established that artists, engineers, and scientists have private agendas, which may be rational from each individual's perspective, let us now turn to the role of subjectivity. As the dictionary defines it, subjectivity assumes that an object's attributes have meaning only from the perspective of an observer's thoughts, personality, idiosyncrasies, feelings, beliefs, and opinions. In contrast, objectivity assumes that an object's attributes exist apart from the observer: the facts speak for themselves.

Individuals involved in aural architecture often express a strong belief in their objectivity while actually making decisions based on subjectivity. In fact, as this chapter will argue, every decision, every design choice, and every action has a potentially large and unrecognized component of subjectivity. Consider a few examples. If you are an audio engineer using mathematics as the basis of your reverberator design, you choose to optimize subjectively defined attributes such as quality and efficiency, as well as your career and ego gratification. If you are an acoustic engineer designing a concert hall, you believe that the audience shares your subjective sense of what are important acoustic attributes. Or, if you are a manager of a research group reviewing students' proposed experiments on spatial awareness, you choose to guide the students in ways you subjectively deem likely to minimize the risk of their failing to graduate in a timely manner. We have already explored numerous manifestations of subjective decision making in the discussion on theaters, cathedrals, concert halls, artificial reverberators, and broadcast studios.

The widely shared view that science is as an objective endeavor is simply not true. For a scientist, there is nothing objective in the choice of research questions, the selection of an experimental paradigm, the degree to which assumptions should be tested, and the relevance of theoretical conclusions to real problems. Consider that a researcher studying auditory spatial awareness must make a series of decisions that may include, for example, whether to investigate blind individuals navigating space, whether to apply that knowledge to spatial acoustics, whether to evaluate real spaces rather than simulations of simple spaces, or whether to reconcile results from perceptual studies with neurological studies. Such decisions are all subjective and have a direct bearing on our knowledge of auditory spatial awareness and its application to aural architecture.

If you are the leader of a scientific, engineering, or artistic organization, you spend more time dealing with subjective decision making than with intellectual concepts. Your typical workday is similar to that of a manager in a corporate division, who also must steer the organization with tacit knowledge that cannot be defended as being purely objective. Although there are remarkably few studies that examine the cognitive process of governing scientific and artistic activities, business researchers often analyze the role of subjectivity and personality in making managerial decisions. Because their analysis suggests universal principles that apply to all professions and disciplines, their conclusions are applicable to aural architecture. As a rule, complexity produces subjectivity, and the most complex situations are also the most subjective. Aural architecture is very complex. In considering the decision-making process, Devi Jankowicz (2001) observes that "given the increasingly complex and uncertain environment in which contemporary organizations operate, there is a need for managers to embrace complexity and to learn how to manage uncertainty." Yet few disciplines discuss the relationship between complexity, uncertainty, and subjectivity, even though it can dominate the innovative process.

Complexity theory is part of management education and should be part of all professional training. Outside of the business context, few leaders are even aware that complexity issues exist. Consider, for example, that, as a manager in a research environment, you must recommend projects to graduate students, advise junior scientists on their choice of research directions, allocate resources to competing activities, decide which equipment to purchase, and choose where to spend your time—the most valuable resource of them all. The architect of a space, like a manager, is making choices using tacit knowledge. And the resulting decisions have the potential to produce a large impact on the lives of individuals, the productivity of the group, and the nature of the resulting space. But none of the choices has a clear rational basis. They are value-based decisions that also have a personality component. For example, one leader might value group harmony in resolving conflicting requirements among the supporting staff, and another might autocratically impose a single-minded vision on the group.

Unlike many investigators who study personality, George Alexander Kelly (1955), who developed the concept of "construction alternativeness" as a formal theory, proposed a method to examine subjectivity in real-life situations. His fundamental postulate is that "a person's [cognitive] processes are psychologically channelized by the way that he anticipates events" (Kelly, 1955). Each individual constructs a system based on expectations about the world, and to avoid cognitive dissonance, behavioral and perceptual choices are filtered though that system. We see and hear what we want to see and hear, which makes everything subjective.

Using a formal method designed to make the unconscious basis of subjective decisions visible, Jankowicz (2001) showed how to determine the relative importance of factors such as risk, expense, duration, equipment, complexity, aesthetic quality, return on investment, and audience size in given subjective decisions. In the context of aural architecture, as a designer, you might view religious aesthetics as more important than duration of the project, as was the case for cathedrals. Or you might view higher profits as more important than spatial fidelity, as is often the case for consumer reverberators. Every manifestation of aural architecture has an implicit ranking of attributes, which reflects the subjective evaluation of competing choices.

Jankowicz (2001) applied his method to commercial lending and showed its utility in making tacit knowledge explicit and public. Bank officers are always deciding whose loan applications should be accepted or rejected. A better understanding of the role of subjectivity could increase the institution's profitability. Contrary to Jankowicz's expectation, however, the bankers who participated in his study were extremely uncomfortable having their subjectivity made public; they preferred lower profitability to public exposure of their criteria.

Without training in the cognitive sciences, the contrast between objectivity and subjectivity becomes a dichotomy between "out there" and "in here." Objects and events exist in the external world but cognitive constructs are internal, resulting from the individual's personality and experiential history. Professionals in the arts and sciences are like bankers, feeling vulnerable when their subjectivity is made public. Choices are therefore presented in rationalized, objective, and scientific language without exposing the *internal* cognitive process.

The behavior of disciplines involved in auditory spatial awareness and aural architecture can be traced to the dominant personalities of those who lead the disciplines. In the troika governing the matrix of decisions about aural architecture, the third member, after the larger culture and disciplinary subcultures, is individual subjectivity. Because how you respond to subjectivity depends on your personality, it is useful to have a catalog of personality types. In their study of organizational psychology as applied to motivation, rewards, and authority, Daniel Katz and Robert Louis Kahn (1978) identified three leadership styles: rational-legal, traditional, and charismatic. Using an enlarged set of dimensions, Don Richard Riso (1996) expanded and generalized personality to nine distinct types: helper, motivator, individualist, investigator, loyalist, enthusiast, leader, peacemaker, and reformer. Each type responds to a complex situation in a way that reflects the personality of the leader, which in turn biases the behavior of the group.

Although each of us has degrees of all personality attributes, most of us have a bias toward a few dominant attributes, which then play a stronger role in our decision making and cognitive judgments. We could well imagine that a helper would consider nurturing graduate students to be more important than would a reformer, who would focus on changing paradigms. Enthusiasts use charisma rather than logic to guide them and their group. Peacemakers look for compromise, potentially without regard for intellectual legitimacy of the opposing positions. In this sense, the process of selecting leaders, with their values and personalities, determines the aural architecture of our spaces.

A historical example illustrates how intuition and subjectivity among creative and intelligent scientists led to seminal innovations that proved to be the foundation for artificial reverberation, synthetic spatial simulations, acoustic analysis, and electronically assisted concert halls. For the decades from about 1950 to about 1970, scientists and engineers at the Bell Telephone Laboratories were encouraged to explore any idea that might be an interesting application of computers to audio. Even though the parent organization, American Telegraph and Telephone (AT&T), focused primarily on telephone systems, within the relatively isolated research laboratories, subjective decisions were encouraged in selecting ideas to explore. In this environment, Manfred Schroeder, as one of the many forceful, brilliant, and prolific scientists of the Bell laboratories, made a wide range of contributions, including designing the world's first reverberation algorithm. Yet it was only one of his many recreational demonstrations of signal processing. At this unique moment in history, activities that were unrelated to telephone systems were still consistent with career advancement, economic support, institutional politics, and research productivity. We might assume that the culture at the Bell laboratories reflected the personality and values of the senior executives at that time, who embraced the belief that intellectual freedom and interdisciplinary activities would be useful. But in fact, AT&T had a cultural tradition conductive to the creation of new aural architecture.

Births, Deaths, and Marriages of Disciplines

Rather than having their own discipline, aural architects are scattered among other disciplines, each with its own distinctive bias. Recently, the number of disciplines with some interest in aural architecture and auditory spatial awareness has been growing exponentially. Most of these disciplines were born within the last half century, and some are only a decade old. A hundred years ago, acousticians did not design concert halls; seventy years ago, audio engineers did not think about reverberation; and twenty years ago, musicians did not compose for surround-sound listening. As disciplines appeared and disappeared, the concept of aural architecture evolved, and will continue to evolve.

As an intermediate social unit between individuals and society, disciplines have their own rules and dynamics. Having examined aural architecture in cultures such as ancient Greece and Christian Rome, and discussed the contributions of such individuals as Schroeder and Sabine, let us now consider how modern disciplines provide yet another perspective on the subject.

Like living organisms, disciplines have a natural life cycle that includes birth and death. During their lifetime, they give birth to new disciplines, which then mature and form new families of disciplines. They may be like rebellious relatives, harmonious cousins, or nurturing parents. They may be productive or dysfunctional, contributory or parasitic, fertile or barren, and supportive or predatory. In the simplest cases, the growth and decline of a discipline result from the movement of individuals into or out of a field when they sense shifting career opportunities.

A few examples illustrate the life cycle of some disciplines involved with aural architecture. By the early twentieth century, the physics of sound gave birth to applied acoustics, which included concert hall acoustics. After thriving for a few decades, that discipline reached a stable plateau and is now in modest decline. Simultaneously, the inventions of the radio and phonograph gave birth to audio engineering in the mid-twentieth century, which also initially grew rapidly. It contained the small discipline of artificial reverberation based on mechanically vibrating elements. That discipline disappeared, but before doing so, it married digital signal processing, which gave birth to a new breed of digital reverberators. After thriving for a few decades that activity entered a rapid decline, with perhaps no more than a few dozen experts still practicing. But before old age, the discipline of artificial reverberators married the discipline of concert hall acoustics to produce a new discipline: software spatial synthesizers. By the late twentieth century, a new branch of the spatial disciplines family appeared in the context of multimedia computer games with interactive audio. This will likely be a new home for aural architects, and especially for innovators who try to match aural to visual spaces. The discipline of professional audio recording is now in decline as audio amateurs using home studios thrive, but these amateurs are supporting a new breed of primitive consumer reverberators. These examples represent a few branches of a large family tree.

For a discipline to be viable in the long term, society must provide it economic support, physical facilities, and schools, including teachers, books, and graduation criteria. Depending on the amount of economic sustenance provided, a discipline will increase or decrease the intellectual area it tries to cover. Given the tendency for a group to defend its existing intellectual territory, and to expand it by encroaching on neighboring disciplines, sponsorship determines the size of the territory that can be protected and cultivated. For example, engineering absorbed electrical and material sciences from physics because it had more governmental and industrial support available to transform knowledge into applied technology. Cinema is now absorbing audio because consumers want multimedia, not just audio by itself. The analogy to biological evolution is apparent, as is the analogy to ambitions of nation-states in previous centuries. A discipline is a dynamic organism that adapts to its environment.

To have continuity and long-term stability, a discipline must attract new members, as well as raise its children disciplines to productive adulthood. The dynamic is actually quite simple. Individuals enter a discipline after graduating from school, or emigrate from another discipline, provided, of course, they have the necessary skills and the discipline has career opportunities. As a prerequisite for admission into a discipline, new members are expected to accept the discipline's assumptions, rules, and culture, and to accept advice and guidance from senior professionals at face value. A discipline's cultural values are thus perpetuated through the mutually beneficial relationship between manager and junior engineer or professor and student. The senior member teaches the junior member how to be accepted into the discipline's family, to learn the unwritten rules of behavior, to make a useful contribution, to acquire skills that are useful to survive and thrive, to recognize and accept the discipline's belief system, to accept what it defines as good and as bad. The junior member must be obedient, work hard at supporting the discipline's doctrine and at implementing its goals. At its worst, the transmission of skills and values is little more than indoctrination and tyranny; at its best, it results in nurturing and growth. Mostly, it is somewhere in between.

Unlike ordinary subcultures, disciplines such as aural architecture transcend national boundaries. As an example of a transnational discipline, Sharon Traweek (1988) documented the case of high-energy physicists: "The traffic in students and postdocs strongly resembles the exchange of women between groups though marriage, which serves as a force in, and source of, kinship networks. Roger Keesing has argued that: in [tribal] societies marriage is characteristically a *contract between corporate groups*" (emphasis added). Because the physicist elders normally decide on their positions, students and postdoctoral fellows cannot rebel if they want to continue their careers. Within the global tribe of high-energy physicists, there is a clear ranking of institutions and, by implication, of the individuals working in them. "Knowledge trickles down and students percolate up. The finely structured hierarchy ranks every element of community" (Traweek, 1988). Success in such an environment requires knowledge of the unwritten rules.

A similar dynamic takes place in those disciplines associated with aural architecture. Countries set different national priorities with regard to their support for the arts, science, and empirical research. Some allocate more resources for research about concert halls; others prefer medical prostheses, and still others, electronic music. At the moment, Scandinavia is a major location for professional reverberators, New Zealand actively supports assisted reverberation for acoustic spaces, Finland maintains a productive research facility concerned with virtual spaces, and France has a long-standing tradition for electroacoustic music research. The United States is one of the leaders in multimedia audio. Within these fields, individuals move to host countries if they want to be at the leading edge; otherwise, they meet at international conferences.

Like individuals and countries, disciplines have personalities. Thus disciplines that embrace secrecy, often displaying antisocial personalities, avoid working with colleagues from other disciplines. In his study of the history of acoustics, Frederick Vinton Hunt (1978) noted that Pythagoras (570–497 B.C.), the founder of the science of sound, music, and vibrating strings, created an "an eclectic group that considered it highly improper to reveal the inner secrets of their philosophy to outsiders, or even to the probationers who attended the lectures." Secrecy was the prototype of security. Among aural architects and audio engineers, designers of reverberators are often the most secretive, even avoiding the protection of patents because key ideas must be disclosed. At the opposite end of the spectrum, academic researchers in hearing are often among the most open and accepting. Those who participate in a multiplicity of disciplines are sometimes caught in a crossfire of conflicting values, wanting open access to the knowledge of colleagues from other disciplines while preserving the secrets of their own. Knowledge becomes intellectual property, to be shared, borrowed, or guarded with no-trespassing signs and impenetrable fences.

Within the many disciplines in aural architecture, we find a range of personality attributes, including greed, self-sacrifice, ambition, laziness, openness, and secrecy. As with its individual members, society tolerates disciplines with bizarre personalities if they make productive contributions. In some cases, a discipline uses charm, charisma, and public relations in order to hide dishonesty and intellectual fraud. The risk of discovery further elevates their tribal view that they are superior to all others. Outsiders all too often lack the means to distinguish between an honest attempt to solve a difficult problem and a deceitful cover-up of a barren activity with a low probability of ever being useful. The surround-sound format of the 1970s, using a two-channel matrix to encode four discrete channels, was a perfect example of a dead-end activity that was apparent to only a few at its birth, however apparent to all after its ignominious death. On the other hand, it is seductively easy, and an expression of intellectual xenophobia, to assume that an alien discipline is vacuous and unproductive. For anyone interested in aural architecture, the challenge is in evaluating validity and relevance among the variety of contributing disciplines.

Disciplines and paradigms are sometimes dead ends that offer little to the participants and even less to sponsors and society. Using a philosophy-of-science approach, Thomas Kuhn (1996) asserts that paradigms are rigidly conserved even when no longer productive in answering important questions. Because paradigm shifts are disruptive, with elements of chaos and temporary loss of productivity, there may be an unwillingness to evolve in a more useful direction. Before its lingering death, a dead-end activity still generates a large amount of attention, activity, and appeal. As Hubert L. Dreyfus (1993) explains: "A degenerate research program, as defined by Imre Lakatos, is a scientific [or engineering] enterprise that starts out with great promise, offering a new approach that leads to impressive results in a limited domain.... As long as it succeeds, the research program expands and attracts followers. If, however, researchers start encountering unexpected but important phenomena that consistently resist the new techniques, the program will stagnate, and researchers will abandon it as soon as a progressive alternative becomes available." There is a critical implication here. Before the discovery of a meaningful alternative, but after the program has stagnated, the activity still creates the illusion that it is a thriving field making important contributions. One such thriving dead end in aural architecture is precise simulation of contrived spaces, such as rectilinear boxes, which have neither scientific nor artistic value. Dead-end disciplines, even while alive, seldom produce children disciplines. In contrast, productive disciplines such as scientific acoustics and audio signal processing have produced not only many children, but also dozens of grandchildren.

Having explored aspects of disciplines, we can better understand society's attitude toward aural architecture. We observe that our culture is now more interested in designing virtual than actual acoustic spaces. Yet the two activities are related. Those who now design spatial experiences in multimedia games are the great-grandchildren of acoustic architects who designed concert halls for classical music.

Intellectual Frameworks: Assumptions and Paradigms

Acoustic science, perceptual psychology, audio engineering, and anthropology are all examples of disciplines interested in aural architecture. Each has a unique intellectual framework, which is often incompatible with others. If a discipline is to contribute, its relevant attributes must be extracted and translated into our common intellectual framework for aural architecture and auditory spatial awareness. Earlier chapters first gathered and translated, then fused insights from a wide range of disciplines. This chapter considers the nature of that fusion.

By selecting, extracting, translating, and fusing insights from the disciplines that make up aural architecture, we create interdisciplinary bridges that overcome differences in these disciplines' philosophies, theories, paradigms, methods, and epistemologies. *Philosophies* provide the logical rules for inferences, whether inductive or deductive, and these rules are used to test whether an assertion is tautological, contradictory, or unprovable. *Theories* are compact generalizations of observable data that can predict other results. *Paradigms* are more specific constructs that constrain the domain of inquiry. For example, a paradigm that assumes that acoustic cues determine the aural experience of an acoustic space cannot explain those aspects of a response that originate from free will, experiential history, or emotional mood. Within a given paradigm, there are numerous *methods* for running experiments, such as forced choice, numeric rating, rankings, or multiple choice. Depending on a discipline's intellectual framework, its philosophy, paradigms, theories, and methods are more or less unified, and more or less dependent on each other.

Epistemologies deal with validity of knowledge: under what condition is an assertion contradictory, tautological, consistent, or truthful? The following discussions, as an introduction to the basics of epistemology, explain some of the philosophical choices we need to consider. Rather than evaluating competing intellectual ideas in a search for a single truth, we will use the concept of *epistemic relativism*, which Andrew Webster (1991) defines as "knowledge...rooted in a particular time and culture." Every discipline has its implicit epistemology, and epistemic relativism implies that knowledge depends on culture.

Evolution and Implications of Intellectual Fragmentation

The intellectual foundation of auditory spatial awareness and aural architecture is fragmented among dozens of disciplines. This makes it difficult for individuals in unrelated disciplines to take advantage of each other's knowledge. How should an audio engineer talk with an acoustic scientist, and how should a composer of electronic music talk with the designer of a spatial synthesizer? The answers are not obvious. Transcending the intellectual boundaries of disciplines involves humility and cross-cultural tolerance. This discussion explores how to translate ideas across the boundaries of dozens of disciplines in aural architecture.

The consequences of fragmenting knowledge into isolated islands of specialized expertise become apparent when we consider how fragmentation arose. Long ago, knowledge was not fragmented into recognizable and separable units; intellectual disciplines did not exist. At the beginning of modern Western intellectual history, philosophy represented the grand Greek tradition for understanding everything through observation, contemplation, and discourse, without distinguishing between mental activities and empirical observation. Philosophically arrived at, knowledge revealed the laws of the cosmos. Much later, by the mid-sixteenth century, natural philosophy became the intellectual home for all nontheological knowledge. Then, in the eighteenth century, natural philosophy itself split into fragments, which we now call "disciplines." One family of fragments separated activities of the mind, exemplified by philosophy, from observations about nature, exemplified by empirical science. Another family of fragments separated activities according to subject matter, such as physics, biology, psychology, mathematics, and chemistry.

As knowledge grew in size and complexity, it exceeded the capacity of a single individual to master. Distributing knowledge into disciplines allowed society to retain and propagate its intellectual assets, as if each discipline were responsible for one article (or group of related articles) in one volume of an encyclopedia. Specialization would prove to be an efficient means of advancing knowledge. But specialization also proved to have major liabilities: complementary and contradictory points of view could not be integrated into a composite view, and, more important, the unity of knowledge was split into empirical data, on the one hand, and pure thought, on the other.

By the end of the twentieth century, the number of discernible disciplines exceeded 10,000 (Crane and Small 1992). Their narrowing scope allowed each to evolve a circumscribed domain of legitimate questions, a formal methodology based on an articulated philosophy, and extensive experience with research paradigms that yielded productive results. When resources are focused on a narrow field of interest, rapid progress is possible. But at a cost: narrowing focus creates a knowledge base of great depth yet little breadth. Furthermore, internal divisions within a given discipline create subdisciplines and microsubdisciplines whose boundaries are visible only to insiders. As some sardonic pundits have critically commented using the old cliché: specialists come to know more and more about less and less, until they know everything about nothing.

Modern psychology, for example, has fragmented into more than thirty distinct subdisciplines.[3] As if this fragmentation were not enough, psychology has combined with nonpsychological disciplines to form recognizably distinct hybrid disciplines: psychoacoustics, psychophysics, psychopharmacology, neuropsychology, physiological psychology, psychopathology, psycholinguistics, and psychobiology. Experts in any one of these disciplines will recognize yet further subdisciplines within their particular discipline. For example, psychoacoustics has subdisciplines that focus on localization, detection, dichotics, masking, speech, music, pitch, and so on. And each of these has its *sub*subdisciplines and *micro*subsubdisciplines. There is no intrinsic limit to fragmentation and specialization.

More important than the fragmentation of subject matter, however, is the concomitant decoupling of philosophical rigor from empirical data within all disciplines. Just as pure thought cannot contribute to our understanding of physical phenomena; without observable data, so, too, empirical data have no intrinsic meaning without a philosophical framework for their interpretation. Thus cognitive and perceptual psychology, which involve an internal experience of an external world, are meaningless exercises without an explicit overarching philosophy. Because understanding auditory spatial awareness is an example of this dilemma, building bridges between fragmented but complementary disciplines requires that we understand the philosophy of knowledge in each.

The unresolved conflict within perceptual psychology—between natural science and philosophy—can be traced to its birth. The initial attitude of psychologists revealed a driving desire to distance themselves from philosophical questions. Robert I. Watson (1973) commented, "the rejected child of drab philosophy and lowborn physiology, [psychology] has persuaded itself that actually it was the child of highborn physics." The biases of William James (1890), Wilhelm Wundt (1904), and others, who were instrumental in initiating psychology as an intellectual discipline, are still visible today. As noted by Kathleen V. Wilkes (1984), "physics not only influenced the empirical and theoretical levels of psychology, by setting the 'idealized' standards of morality, method, experimentation, and theory construction; it also provided an ontology and a metaphysics, which, without always being made explicit, supplied a meta-theoretical framework that conditioned theory construction."

As the father of modern psychoacoustics, Hermann von Helmholtz established many of the current data-driven or stimulus-response paradigms in use today (Vogel, 1993). By combining his medical training with his passion for physics and mathematics, von Helmholtz conceived of hearing as comprising three stages: the physical nature of the signal, the sensory detection by the nervous system, and the final transformation into a perception. Perception embodied the internal construction of the external object. Neither he nor other early psychologists, however, could reconcile the external physical properties of stimuli and the internal histories of individual perceivers. Attributing the discrepancies between observations and theory to "unconscious interference," von Helmholtz simply defined *interferences* as aspects of perception (internal experience) that were not in the stimuli. He recognized, for example, that the perception of illusions arose chiefly from the individual's history, not from the stimuli. He also recognized that, because none of the intermediate stages of perception could come into consciousness, there was no access to perception other than from introspective reports.

Before psychology resolved its philosophical dilemma in favor of behaviorism, there were numerous intellectuals who, functioning as psychologists, felt comfortable using paradigms that allowed introspection as a legitimate form of inquiry. Violet Paget (1932), who was a contemporary of William James, but who functioned outside of any formal institution, relied on self-described subjective experiences—introspection by questionnaire—with an emphasis on emotions, imagery, history, and moods. She was challenged by the question "Why should the perception of form be accompanied by pleasure or displeasure, and what determines pleasure in one case and displeasure in another?" (Paget, 1932). For her, this was a scientific question of intellectual grandeur even if there was no mechanism (then or now) to answer it. She openly stated that her goal was to gain understanding from the population of listeners, to let them be her teachers. Rather than seeking constancy between music and listeners' responses, she viewed each listener as a unique individual.

In the early twentieth century, perceptual psychology was both an intellectual battleground over the choice of paradigm and philosophy and a political battleground
over control of the fledgling discipline. The winners, who controlled the direction of
future research, were the behaviorists. As John B. Watson (1913) observed, the concept
of interferences became extinct when psychology was transformed into empirical behaviorism, which simply did not recognize the concepts that Paget had embraced.
By the late twentieth century, however, as they investigated the neurological basis for
perception and consciousness, cognitive scientists realized that there was no independent means for describing internal experiences (Jack and Shallice, 2001). They found
verbal communication to be a convenient and useful means for individual subjects
to describe their private experiences (Ericsson and Simon, 1993). Although a qualita-
tive and subjective technique, introspection proved to be an indispensable source of
information.

In contrast to subjective forms of inquiry, William Thomson (Lord) Kelvin (1894)
articulated the objective, quantitative ideal, "When you can measure what you are
speaking about, and express it in numbers, you know something about it; but when
you cannot measure it, when you cannot express it as numbers, your knowledge is of
a meager and unsatisfactory kind; it may be the beginning of knowledge, but you have
scarcely in your thoughts advanced to the stage of science." But in psychology, especially perceptual psychology, numbers do not in themselves provide meaningful
knowledge, although they may complete knowledge once the philosophical foundation has been established, provided the underlying phenomena are understood and
amenable to numeric evaluation. Thus what would it mean for a listener to assign
a number to reverberation envelopment, or to the perceived size of a musical
instrument?

In his iconoclastic reexamination of the classical psychophysical experiments where
individual subjects numerically rated the intensity of sensory sensations such as loudness, brightness, or weight, Donald Laming (1997) showed that the results were influenced more strongly by the choice of paradigm than by any intrinsic ability of humans
to perform the task. "The evidence so far to hand," he commented "does not support
any intermediate continuum at the psychological level of description which might reasonably be called 'sensation'," and he concluded that "we have cultural expressions for
entities which do not admit [to] objective measurement, not at least in the present
state of scientific knowledge" (Laming, 1997).

Although limited to one specific case, Laming's study serves as a general warning to
those researchers who believe that they can describe internal experiences by observing
external behavior. The assumption that internal sensations are measurable appears so
obvious that psychologists rarely test it. Until Laming reexamined this assumption, it
was believed that basic psychophysical data approached the level of objectivity found
in the hard sciences. Measurability is determined by the phenomenon itself, not by

a researcher's desire to emulate physical sciences. Some aspects of auditory spatial awareness are measurable, others not. As quoted in the epigraph to chapter 6, a sign hanging in Einstein's office said, "Not everything that can be counted counts, and not everything that counts can be counted." Perhaps, then, the experience of enveloping reverberation is not measurable, even though many researchers measure it, and even though it has a strong influence on the spatial experience of listeners.

The basic problem when studying aural experience reduces to a simple but unresolvable dilemma: how to objectively (numerically) measure a listener's aural experience, while incorporating the fact that the choice of perceptual strategies (interferences) is controlled by the listener. Cognitive strategies determine the way we mentally process sound stimuli to create a perception. Consider your mental state as a listener attending a lecture in an auditorium. By choosing a particular strategy, you can select what you want to hear. At any given moment, you can attend to information or emotions. Additionally, as a listener, you can focus on street noise, spatial ambience, a hissing ventilator, a coughing child, or the location of a wall. A skilled phonetician, by using a strategy that emphasizes slight shifts in the phonetic sounds, can focus on the speaker's city of origin. A bored listener can use sounds to stimulate daydreams; an interested listener can acquire new information with a new strategy. Some difficult strategies are unstable, rapidly atrophying without continuous practice. Strategies are not consistent across cultures, or even within cultures. In every case, the listener's choice of strategy determines perception, and an external observer has no means for detecting the current strategy.

The dilemma of cognitive strategies is perceptual psychology's analogue to the Heisenberg uncertainty principle. With an electron, when location is measured, velocity is unknown, and vice versa; knowing one makes the other unknowable. Similarly, with perception, the reliability and applicability of results stand in similar opposition. Using a sanitized context with a simple stimulus in controlled laboratory experiments, reliability is maximized, but at the expense of applicability. When measuring a single aspect of perception, the experimental design intentionally removes or ignores confounding aspects of perception. On the other hand, using verbal communication in the messy world of real life, with its multiplicity of stimuli, emotions, personalities, and cognitive strategies, applicability is maximized, but at the expense of reliability. We call this dilemma the "perceptual uncertainty principle." Either a discipline asks broad questions (real life) without having paradigms that provide numeric answers, or it asks narrow questions (laboratory science) with paradigms that provide numeric answers but with limited applicability. Every intellectual discipline, especially those involved with auditory spatial awareness, finds its place on the continuum between these two extremes; the nature of a discipline's contribution, its questions and answers, depends on its place. For example, those who teach echolocation for spatial navigation emphasize utility and applicability, whereas those who make models of aural localization emphasize accuracy and repeatability. Being at opposing ends of the continuum, such disciplines find communication difficult. Generally speaking, those who create sensory experiences (artists) and those who study the basis for such experiences (scientists) do not readily understand each other.

Ideally, we would like to fuse the insights derived from all locations on the continuum into a single concept, rather than selecting those from a single location. By the early twentieth century, the value of fragmented disciplines was being critically reevaluated in the light of their limitations in dealing with complex problems. As a symbol of a countervailing trend against fragmentation, the term *interdisciplinary* began to appear in the literature during the 1930s (Stills, 1986). Interdisciplinarity emphasizes the aggregated whole rather than its disaggregated components. Like other paradigm shifts, interdisciplinarity was to hold great intellectual promise. But it also faced difficult challenges.

Fusing Intellectual Fragments

While I was conducting a research project at a major teaching hospital, my collaborators demonstrated the difference between an intellectual theory and a compelling picture. When a patient who is clearly ill enters a hospital, the attending physician conducts a patient interview and physical examination and orders a series of laboratory tests that include blood analysis, x-rays, and ultrasound. Although each test produces useful information from procedures and equipment that have been developed over decades of scientific research and engineering development, an experienced physician also knows that diagnostic data contain irrelevant and contradictory information. When wisely fused together, however, noisy diagnostic data from multiple sources produce a coherent picture of the patient's physical condition that has a high probability of being accurate.

In much the same way, by fusing scholarly fragments from different disciplines, this book hopes to produce a coherent, accurate picture of aural architecture. In both cases, fusing is an art that requires an appreciation for both the reliable and unreliable aspects of each fragment. The following discussions explore the intellectual principles that led to such concepts as social, navigational, aesthetic, and musical spaces. Research studies about acoustic spaces (formal science) and the phenomenological experience of space (folk science) both contribute to our collection of fragments. For formal science, robust proofs, rigorous theories, and structured experiments, though useful in specific situations, are not the exclusive means for explaining human experience. Anyone can perform an experiment and generate a theory about a given phenomenon, but the likelihood of its being meaningful depends on how well the phenomenon is actually understood. Sophisticated researchers appreciate that insight often begins from the anecdotal evidence of folk science. When intuition is tested and adjusted by careful experiments, formal theories appear as the final stage in understanding.

Folk science often provides wisdom and intuition through intimate real-life experience with a phenomenon. Consider, for example, the testimonial by marine biologist Robert Earle Johannes (1981), who went to study what Pacific Islanders knew about fish behavior. Johannes stated that his fieldwork with native fishermen advanced the state of knowledge of marine science further in sixteen months than conventional research techniques had in the previous fifteen years. Much like the native fishermen who learned about fish behavior from years of direct experience, there are sensory experts who gained their knowledge of auditory spatial awareness through years of working with acoustic spaces. The value of extensive fieldwork is even greater when complemented by laboratory research using rigorous scientific methods.

Absent a rigorous overarching intellectual framework, however, laboratory investigations of perceptual experience produce a welter of ill-defined conceptual fragments that include consciousness, perception, sensation, awareness, feeling, response, understanding, cognition, thought, alertness, attentiveness, illusion, and image. Each is shorthand for a collection of phenomena with observable manifestations and common properties.

Formal science distances itself from folk science by using intellectual frameworks that are derived from the philosophy of science, which itself has a rigorous framework. Nevertheless, creating formal frameworks still produces endless debates among some of the brightest minds of the century.[4] These discussions, in one way or another, struggle with the concepts of truth, belief, and knowledge. A review of the literature reveals a lack of consensus on even the "obvious" concept of "knowing." We find that assertions are represented as "proof," "theory," "data," "evidence," "consistency," "belief," "justification," "support," "hypothesis," "model," "truth," or "validating simulation."

In contrast to the traditional definition of a theory, which originated in the physical sciences, Andre Kukla (2001) takes the position that "the operative criterion for theory evaluation is not *truth per se* but the *probability of truth*." Imre Lakatos (1970) defines scientific research programs as having a hard-core heuristic, which provides an axiomatic theory, and a set of auxiliary hypotheses, which explain new facts. "A theory without excess corroboration has no excess explanatory power" (Lakatos, 1970). Which is to say, a theory must tell you something about observable data that you did not *already* know, for example, predicting the outcome in an experiment that is more than a degenerate extension of the original experiment. Karl R. Popper (1992) extends this definition by including the requirement that a theory be refutable by empirical data: "testability is falsifiability." Before conducting an experiment to test a theory, the experimenter must be able to contemplate a *plausible* outcome that is *inconsistent* with expectations. Yet, for practicing scientists, especially those without training in epistemology, the concept of a theory is at best ambiguous. Theories often serve only as the means for compacting voluminous data into a summary without having predictive value. Kukla (2001) stated the obvious, but oft forgotten, wisdom: "data do not yield up theories of themselves, nor do theories emerge simply because more data have been added to the lot."

All intellectual disciplines, including those belonging to folk science, ultimately converge on a single issue. Timothy Williamson (2000) comes to the surprising conclusion that *belief* underlies all formulations about knowledge, not unlike axioms in mathematics. To accept belief as the fundamental basis of understanding is inevitable, however unappealing. Each discipline chooses a belief system for its activities on the basis of utility in achieving specific goals. For example, that introspection is a valid source of data is a belief that cannot be proven. Similarly, that numeric data are truly indicative of a phenomenon is also a belief. Earlier cultures believed in visions and voices as a source of truth.

In contrast to science, the legal system explicitly articulates the rules for believing in a truth. For the legal system, there are three formal, working definitions of truth. Depending on the situation, something is proven to be true: (1) if reasonable persons would agree that it is true; (2) if the preponderance of evidence forces a reasonable person to conclude that it is true; or (3) if it is true beyond any reasonable doubt. There is no single truth, and reality is not a relevant consideration. In fact, a given reality can be proven "true" according to one definition and *not* proven "true" according to another.

This immediately leads to the question of who chooses the belief system. Who asks the question is more important than the nature of the answer. For an acoustician, reverberation is a composite of sonic reflections and resonances; for a conductor, reverberation is a means for softening, blending, and enriching musical notes; for a blind person, reverberation is a form of noise that masks the acoustic cues that indicate obstacles; for a perceptual psychologist, reverberation is a measure of the enclosed volume. Each of these individuals "knows" an aspect of reverberation. Physical properties of nature are only indirectly related to the experience of nature. "Truth" becomes flexible when we consider the phenomenon of sound reflections in an enclosed space from the various perspectives of mathematical, empirical, perceptual, and musical acoustics.

Mathematical acoustics, which explores the physical properties of an enclosed space, consists of selecting a formal set of axioms and assumptions, applying the rules of mathematical inferences, and finally, deriving a conclusion. Given a belief system that values a compact, closed-form solution in a single equation, the definition of a problem is massaged until such a solution is possible. As a result, the classical representation of an enclosed space assumes a rectangular enclosure having parallel walls and no sound absorption, a differential wave equation where the velocity of sound is constant, and a gaseous medium that is linear, homogeneous, and time invariant. With all these assumptions, it is easy to prove that the sonic reflection density grows with the square of time, and that the resonance density grows with the square of frequency. Mathematicians can just as easily start with a different set of assumptions, for example, by assuming a randomly shaped space. In this case, a different set of mathematical rules is applied and a different set of conclusions is reached. Even though, in each case, the conclusions are self-consistent, valid, and traceable to axioms, because real spaces deviate from the required assumptions, the conclusions do not necessarily apply to any particular real acoustic space.

Empirical acousticians take real measurements in real acoustic spaces in order to explore whether and how the mathematical conclusions apply. When they work with real life, they encounter problems that prevent them from clearly translating empirical results into mathematics. For example, at high resonance densities, there is no known method for measuring resonances, and at high sonic reflection densities, there is no known method for decomposing fused into individual sonic reflections. Moreover, air is nonlinear, nonhomogeneous, and time varying. Real-life interior surfaces are textured with nooks and crannies. The resulting measurements must be interpreted by taking into account all the vagaries of real life. Empirical acousticians generate a large body of measured data under varying conditions, and then struggle with extracting essential properties. Their conclusions are highly dependent on the spatial details.

In attempting to understand the experience of sonic reflections, perceptual psychologists, like mathematical acousticians, greatly simplify the problem. For example, in exploring the precedence effect, Ruth Y. Litovsky and colleagues (1999) effectively modeled an acoustic space as a single infinite wall having a location and orientation. A listener hears only two sonic events, the direct sound and the first sonic reflection from the wall, each arriving from a particular direction. This research, which is extensive, shows that, within a certain source–wall distance, the sonic reflection fuses with the direct sound to increase perceived loudness, but without destroying aural localization of the sound source. Neurobiologists have also attempted to make models of the auditory cortex that are consistent with these observed results. Note that their model studies are exclusively restricted to manifestations of the phenomenon studied that produce numeric results, for example, the location of the sound source as a function of angle orientation and linear distance. Once measured as numbers, results can be manipulated using statistics, equations, and physical models. As with mathematics, the definition of the problem is massaged to fit the paradigm. A specific paradigm does not reveal other manifestations of the phenomenon, which might be more relevant to other applications.

If we take musical acoustics to refer to the aural experience of music in a concert hall, those who design or use concert halls are mostly interested in the aesthetic and emotional contributions of spatial attributes to the musical arts. Strong aural phenomena are an important part of these arts, whereas weak ones are uninteresting. Understanding the details is less interesting than understanding what is useful and relevant. In this sense, interpreting musical space is a kind of folk science. There may be important insights from formal scientific disciplines, but they remain secondary. Ultimately, this group of listeners is only interested in the way that spatial acoustics influence their experience of the musical arts.

Several observations can be made from this simplified overview of auditory spatial awareness in a multiplicity of disciplines. Each discipline defines the problem to suit its intellectual framework; each chooses questions that are directly relevant to its goals; each has a different definition of a conclusion; and each studies one aspect of a complex phenomenon. Moreover, at our current state of knowledge, interdisciplinary translation is, at best, difficult. But when we fuse the insights from this multiplicity of disciplines, we achieve a reasonable picture of the underlying phenomena.

Even before we arrive at a comprehensive interdisciplinary fusion of contributions from the disciplines interested in aural architecture, we need a way to use knowledge already acquired. There are several possible approaches. All disciplines with an interest in human experience divide their research questions into three classes: "what," "how," and "why." "What" questions catalog observations that describe events as data ad hoc, without extensive interpretation, for example, changes in aural localization against changes in sound stimuli. "How" questions partition the process into a dependent series of components that often result in a predictive model, for example, a mathematical model of auditory perceptual fusion. "Why" questions broaden the scope of the inquiry by drawing implications, for example, the evolutionary implications of a neurological property. To different degrees, every discipline addresses all three classes of questions, but most disciplines strongly favor one of the three. Thus, generally speaking, psychophysics is more concerned with *what* the auditory system is able to do, whereas auditory neuroscience is more concerned with *how* it works. In contrast, cultural and biological evolution are both preoccupied with *why* a particular solution came into existence for a given environmental problem.

The three classes of questions differ in epistemology, or how they define the validity of knowledge. Answers to "what" questions provide hard data but are of little use for inference and application. Answers to "how" questions have greater predictive utility, leading to theories and models, but are often subject to radical revision when the experimental context is slightly changed. Answers to "why" questions, though almost always intellectual speculations, provide a rich foundation for generating hypotheses that can be tested by the other two classes of questions. Each class of questions is dependent on the other two. In general, twentieth-century disciplines were strongly biased toward one, or at most two, of the three classes of question. In our examination of aural architecture, "why" questions are always present, either explicitly or implicitly, as part of our discussions of "what" and "how." To ignore one of the three classes of questions artificially limits the scope of the inquiry.

For our interdisciplinary study of auditory spatial awareness, we have deliberately chosen a weak intellectual framework to allow the inclusion of insights and wisdom from disciplines having a variety of frameworks. We can evaluate all contributions by sorting them into five dimensions:

1. reliability, consistency, and repeatability of results;
2. predictive power and utility of conclusions;
3. strength and intensity of phenomenon studied;
4. breadth of applicability of results to other situations; and
5. numeric quantifiability.

Reliability, consistency, and repeatability reflect the ability to achieve identical results when replicating the experiment. If the results are only weakly repeatable, there are probably a number of uncontrolled variables that have changed, or the research method has been inconsistently applied.

Predictive power and utility reflect the degree to which conclusions can be generalized to explain other situations that are extensions of the reference experiment. Can the conclusions predict data from different experiments that have not yet been implemented? Conclusions with weak predictive power are not very useful.

A strong, intense auditory phenomenon has a large impact on listeners, whereas a weak phenomenon is just barely detectable and only when the context is controlled to neutralize other strong influences. For example, enveloping reverberation is a strong auditory phenomenon, whereas the slight dip in frequency response produced by the array of seats in the audience is a weak one.

Breadth of applicability reflects the utility of the results in other experiments or its relevance to other situations. Results that depend on many other requirements, such as those regarding nonlinear reverberation decay, are narrow, whereas those regarding acoustic arenas are broad.

Numeric quantifiability reflects the ability to represent data with numeric parameters. Even if they embody fundamental insights into the phenomenon studied, anecdotal descriptions and informal observations from uncontrolled life experiences cannot be numerically quantified. Nevertheless, unquantifiable evidence such as phenomenological introspection may represent the most important means of understanding a given phenomenon.

True understanding of a phenomenon occurs when knowledge is reliable, predictive, strong, quantifiable, and applicable to real situations. Many paradigms in the hard sciences come close to the ideal. Unfortunately, with a problem as difficult as auditory spatial awareness, each discipline makes its compromises by emphasizing some dimensions at the expense of others. Because nobody has yet found a way to avoid these compromises, every approach is a trade-off among these five dimensions.

Are we better off if our data are reliable but have limited predictive power? Is a strong auditory phenomenon that can only be described qualitatively as important as a weak one that can be numerically quantified? Is a reliable and predictive theory important even if it cannot be applied to larger questions? Experiments that are highly repeatable and quantifiable generally have less breadth and applicability because their context is most often narrowly controlled and not extensible. On the other hand, experiments with greater breadth and applicability, even though they are closer to real-life experiences, may not be quantifiable or easily repeatable. The trade-off in the intellectual framework of a discipline involves addressing each of the five dimensions by applying the discipline's value system. Each discipline has a different set of values. To quote Niels Bohr, "It is wrong to think that the task of physics is to find out how nature is. Physics concerns only what we can *say* about nature" (Peterson, 1985).

This brings us to a simple conclusion. The human condition is not describable by science alone; rather, science is a derivative of the human condition that happens to be useful for describing some aspects of that condition. The arts and other disciplines that rely on intuition, introspection, and pragmatic constructs, though not formal science, are still highly useful for describing the human condition. Just as by joining together its many islands, we can see the larger archipelago, so, too, by fusing the many dots of knowledge, we can see the larger picture.

Expert Perceivers, Folk Science, and Formal Science

Auditory spatial awareness is curiously different from other kinds of auditory perception because the ability to hear subtle attributes of space varies so greatly among individuals. Those who use this ability as part of their profession or lifestyle, investing thousands of hours of practice, often acquire a high level of proficiency, whereas those without interest or aptitude perform poorly on even the most basic aural spatial tasks. When emotional or monetary rewards for becoming a sensory expert are high, the motivation to invest time and energy increases. Spatial awareness is comparable to recognizing the distinctive sound of a violin, dialect of a speaker, or song of a bird. Some can do it well; most cannot perform at even the most basic level. In a culture that places scant emphasis on the auditory sense, a random collection of individuals taken from the general population will not be representative of what the auditory cortex is capable of achieving.

We argued earlier that there are parallels between experienced native fishermen, who understand the subtlety of fish behavior, and experienced auditory experts, who hear the nuances of spatial attributes. Finding individuals with *innately* heightened auditory spatial awareness in the general population, however, is analogous to performing anthropological fieldwork—to observing real life rather than making measurements in a controlled laboratory environment. Such fieldwork requires a researcher to wander through our culture, like a detective looking for clues. Innate spatial experts are scattered among us, and we need to detect them. Because such heightened auditory spatial awareness is unrelated to formal education and scientific training (explicit knowledge), innate spatial experts are not necessarily found in university classrooms or research laboratories.

Two classes of auditory subcultures have an elevated interest in acoustic spaces: those who study, design, or teach others about them and those who use them. The former examine, create, or explain the auditory context, and the latter participate in it. A few examples illustrate these two classes. Orientation and mobility teachers of the visually disabled create pedagogical tools and training exercises for teaching their students to use acoustic cues to enhance their lifestyle. And these students, if they practice, will become spatial experts, often with a higher performance level than their normally sighted teachers. Similarly, audio engineers use their sensitivity to acoustic spaces when they create spatial illusions in their recording studio, and listeners, if they care, can become aware of those illusions when they spend hours listening in their home theater.

For one reason or another, innate spatial experts rarely publish articles or attend scientific conferences. They may not have a comprehensive language to explain how they sense space. In some cases, they are actually unaware of their ability. Moreover, auditory abilities are not uniform even among such experts. Someone who can recognize a traffic sign by its acoustic signature may hear reverberation chiefly as noise; someone who can hear the difference between early and late reverberation may not hear an opening in a wall. As things stand, absent a formal structure or philosophy, finding and studying such experts amounts to an exercise in folk science: the evidence produced is messy, easily misleading, and hard to defend. Such an exercise can easily degenerate into opinionated nonsense, which only hinders genuine understanding of the matter at hand.

Finding innate spatial experts is essential, however, if we wish to ascertain the limits of auditory spatial awareness, and these limits are central to ascertaining what aural architecture could become if a culture were interested in it. The average auditory spatial awareness of the general population reflects only the average interest in aural architecture. If, however, a researcher poses a question that assumes that the spatial awareness of subjects reflects the auditory abilities of the species, rather than of a particular culture or subculture, scientific formalism becomes problematic. Moreover, most auditory researchers are unaware of auditory subcultures, and are without any means of incorporating them into their research protocols. And when researchers choose tasks and stimuli that are remote from the lifestyle of any auditory subculture, for example, clicks in an anechoic chamber, they are inadvertently suppressing the influence of subculture. And they are also making the results less relevant to people living a normal life because there is no auditory subculture that listens to clicks in dead spaces.

This brings the discussion to an intellectual divide among researchers. When making an assertion about perception, what population is being used to support that assertion?

At one extreme, it could encompass all human beings; at the other, it could apply to a few individuals, carefully selected for their specialized skill. An analogy illustrates the divide. The unusual geography of San Francisco produces widely varying weather patterns in each section of the city. The weather report refers to the microclimates of given neighborhoods, where one is foggy, damp, and breezy, and another is hot, dry, and sunny. We can compute the *average* temperature, which is useful for predicting total energy consumption for the city. But for individual residents, average temperature is meaningless because they select clothing appropriate to the actual temperature of a particular microclimate. The aural architecture of a particular space is a microclimate of auditory spatial awareness.

Because statistical operations, of which averaging is the simplest and most frequent, are central to the discussion of how scientific studies on auditory spatial awareness apply to the larger population, we need to explore this topic in depth. Averaging is a representative example of the broad topic of constraining parameters and manipulating data.

Most psychoacoustic experiments assume that average performance is useful for scientific investigations seeking to answer a specific class of questions. Indeed, establishing auditory performance norms has great utility for detecting medically significant pathologies, mostly by observing degraded performance in patients with specific lesions. There is little diagnostic utility, however, in detecting extraordinary auditory abilities that arise from a choice of lifestyle. Similarly, making a model of a neurological process is implicitly a sweeping generalization about the human species. In both cases, we observe a consistent bias that suppresses the phenomena of unusual auditory ability, either by discarding exceptions that deviate from the norm, or by averaging the data with the hope that exceptions will not influence the conclusions. Specific research goals, which reflect current social and political values, determine the research paradigm, which then constrains the relevance of the conclusions.

Averaging experimental data to improve repeatability is such a common practice that few researchers consider what must be done to make it legitimate. When the underlying process can be quantified as a single numeric parameter or a smooth curve, and when the variability can be reasonably viewed as additive noise, averaging data is both legitimate and desirable. Consider the detection theory model, which was borrowed from radar signal processing. W. P. Tanner, Jr., and J. A. Swets (1954) modeled auditory detection of sound as the application of a decision threshold to a statistical process. Averaging extracts the probability density function. The model is being misused, however, if the assumption of variability as additive noise is not justified. Although the broad category of statistical data mishandling has been well studied (Wang, 1993), in practice, most researchers arbitrarily or unknowingly ignore whether key assumptions are justified. As one of the most common statistical operations, averaging epitomizes the intellectual divide among researchers.

Although many psychological experimenters average as a way to suppress uncontrolled or unobservable parameters, averaging assumes an a priori knowledge about the dominant parameters; it is an experimenter-defined filter that can inadvertently emphasize a minor parameter and suppress a major one. In contrast, multivariate factoring isolates many variables, but is labor intensive and requires a dramatically larger population of subjects to achieve meaningful results. Thus, in the name of efficiency, an experimenter may simply average a single parameter across individuals, or may average multiple data samples from a single individual. The justification is often not provided, and the identity of suppressed parameters remains unknown. If the experimenter selected the wrong parameter as being the relevant one, or if the underlying variation is more important than the mean value, then the averaging operation destroys critical information and allows an incorrect inference to be drawn. For example, the average ability to navigate by echolocation is extremely low, but that observation leads to the incorrect conclusion that human beings do not, and cannot, use that ability. Averaging removes noise, but variations in echolocation ability are not noise.

Averaging is not the only means for suppressing or controlling unwanted parameters. By holding them constant, their influence can be removed from the experimental design. For example, listening tests can be held in a dark environment to remove any influence on hearing from vision; or the light level can be left as an uncontrolled variable and then suppressed by averaging across a variety of light levels. Similarly, the experimenter's view of the dominant parameter is also found in the choice of the sound signal; it limits which aspect of a phenomenon can be observed. To get tractable data, experimenters deliberately keep stimuli simple, with no more than one or two degrees of freedom. Jens Blauert (1997), who explored the catalog of localization issues, made it clear that his research goal was to define the conditions that produce a specific auditory experience. Real life is messier because the number of potentially relevant parameters is vast. For laboratory science, applicability to real life is not necessarily the goal, and the more candid researchers admit as much. Engineers and other non-scientists who read scientific journals often overlook such offhand disclaimers.

After years of research, a discipline is likely to have created a large collection of experimental results, but researchers may not have acquired insight into how they relate to one another or how they combine into a single view of a phenomenon. Most often, this reflects the impossibility of considering all combinations of parameters that reinforce or compete with each other in nonlinear ways, but it leaves open the question of whether and how results can be applied to an uncontrolled environment. Blauert's collected results provided a comprehensive and elegant description of observed spatial behavior under specific laboratory conditions, a compact model, but not necessarily a predictive theory of localization in an unconstrained environment. The model fails when the acoustics are rich and complex, when there are multiple sound sources, when vision influences hearing, and when listeners only partially attend to aural localization, that is, in real life.

Even when experimenters control unwanted *physical* parameters in complex auditory perceptual experiments, other factors can influence the results. These include subject-related factors such as motivation, attention management, learning proclivity, previous experience, genetic predisposition, exposure fatigue, and choice of cognitive strategies. Although Robert A. Wilson and Frank C. Keil (1999) showed that these "extraneous" factors play a significant role in specific research situations, their findings are seldom considered when designing auditory perceptual experiments. Some experiments collapse extraneous factors with one sweeping assertion: "normal" hearing subjects were selected based on their sensitivity thresholds to low-level sine waves. Having met this criterion, subjects are considered to be homogeneous and interchangeable. Such a threshold test simply divides the population into two general groups; it fails to isolate what may be major factors. Other tests would allow experimenters to divide subjects based on motivation, lifestyle, genetics, intelligence, or learning ability, among many other factors. Without engaging in such labor-intensive testing, we cannot know whether and how a particular factor contributes to the results.

In choosing subjects for auditory perceptual experiments, who should represent the species? Many experiments conducted in academic settings use undergraduate students, who are either paid for participating or required to participate as part of their education. These students are readily available. But are they typical? Their motivation is likely to vary dramatically, especially if extensive learning and training are required. A student's internal distractions involving problems with a girl or boy friend or a term paper may show up from time to time; boredom may easily set in when the procedure is long and repetitive; prior to a testing session, a student may have spent hours listening to extremely loud popular music; or, having stayed up all night in a marathon study session, a student may barely be able to keep from falling asleep. For some undergraduate students, being a subject in an auditory experiment is only slightly more interesting than working on an assembly line.

Compare this situation with that of graduate student subjects who have worked together for two years to prove a theory about auditory perception that will lead to their first journal article, who are fully knowledgeable about what they are hearing, and who participate in each other's experiments. Such subjects are highly motivated to invest effort and time in an activity that has a potential for professional advancement, and perhaps, for professional recognition as well. It is extremely unlikely the two groups of subjects would produce similar results in the same experiment.

Our comparison of the undergraduate and graduate subjects clearly illustrates the distinction between dedication and indifference, but not between higher and lower levels of auditory ability, which depend on genetics and lifestyle. Consider, for example, a localization experiment performed at a music school with subjects who just happened to be student conductors, a group known to have enhanced ability to detect sound locations at the periphery of an acoustic space (Münte et al., 2001). Or consider recreational hunters, who are likely to have had more experience localizing sound sources in dense woods, compared with people living in the acoustics of a city. There are many such auditory subcultures that are not typical of the general population. For some auditory perceptual tasks, the influence of an auditory subculture is relatively modest, but for others it is large. I do not know of any studies that have deliberately tried to show the effect of varying the auditory subculture.

This leads to another aspect of the intellectual divide among researchers: inverting the role of average auditory performance and individual variations. Variations measure the degree to which a given auditory perceptual ability is plastic, flexible, dependent on personality, and amenable to training—qualities central to our understanding of that ability. Even simple psychophysical tasks, such as pitch discrimination, can be considered as both a static property of our species and an auditory ability that is learnable. Almost all investigations ignore the latter. As described earlier, Laurent Demany and Catherine Semal (2002) showed that subjects could improve their sensitivity to small changes in pitch by a factor of 3. Although plasticity and variability are aspects of an ability that are harder to investigate, they are more representative of real life. To quote Werner Karl Heisenberg (1958), "what we observe is not nature itself, but nature exposed to our method of questioning." Researchers, like their subjects, are embedded in a culture that biases experience.

For complex abilities of auditory spatial awareness, the dominant parameter is the amount of time that a motivated individual subject spends in focused listening, measured in hundreds and thousands of hours. Genetic proclivity is likely to be an important secondary parameter. The distribution across the population has extremely wide variance, and the mean value can tell us virtually nothing about the innate capacity of the auditory cortex to perceive complex spaces. Consider an athletic analogy. If we wanted to understand athletic ability in a given culture, we would include the culture's attitude toward exercise, the suitability of bone structures for fast movement, and the performance of individuals training for Olympic competition. We would consult coaches, trainers, and athletic experts, who are greatly concerned with variations in athletic ability, seeking to select only the best athletes. We would not measure the average performance of a sedentary population.

We now arrive at an inescapable conclusion: the results of formal scientific research must be considered within their context, most often the laboratory. They are less relevant outside that context and may indeed have little to tell us in contexts that are radically different. In contrast, folk science, though not particularly good at sweeping generalization because it lacks scientific rigor, is better than formal science at exploring the experiential aspects of normal life because it takes place within a rich context. For its part, formal science often provides a post hoc intellectual framework for understanding phenomena that had already been discovered by folk science. For example, consider acoustic scientists who study seventeenth-century violins created by craftsmen in the hope of discovering their science. I doubt that any formal scientific or engineering activity would have led to a recipe for inventing such instruments. The intellectual framework of formal science did not design these violins—folk science did. But formal science is ideal for articulating the properties of those violins, and how to reproduce them.

The worst intellectual charades are perpetrated not by scientists, who are mostly intelligent and humble, but by those who misuse scientific conclusions. In practice, the misapplication of formal science often manifests the same casualness as folk science. There are elements of folk science embedded in the interpretation of formal science. Even when a theoretical foundation is well understood, as it is with some acoustic phenomena, empirical researchers take liberties with assumptions—whether out of ignorance, for convenience, or in the interests of pragmatic efficiency. Sometimes, because there may not be a practical means to test or control assumptions, researchers have simply no choice but to violate or ignore them. However abstruse, formal, and sophisticated formal science may be, its *application* is nevertheless an intuitive art with uncertain generalization and weak formalism—like folk science. Applying results from the laboratory to real life is the art of science.

Confusing scientific models with real life leads to an unconscious belief that abstractions are reality. Heisenberg framed the warning: "Concepts initially formed by abstractions from particular situations or experiential complexes acquire a life of their own." This is called "reifying"—making of abstract concepts something "real." Elegant models that describe extensive laboratory data become a work of art that instills pride in the creators. However, that elegance comes with a price—severely limiting the applicability of model results outside the confines of the laboratory.

In my view, far too little energy has been applied to the problem of scope, assumptions, and applicability of research in sciences involved with human experience. Thus neither the intellectual framework of cognitive science nor its rules for applying research results have advanced to levels comparable to those of the physical sciences. Human beings, who study themselves as part of nature, are far more complex than the physical world. Cognitive science is still an immature science and, like aural architecture, intrinsically interdisciplinary. Intellectual compromises are required. It is therefore appropriate to encourage both rigorous experiments and experiential fieldwork, both formal and folk science. The problems it faces are too difficult for parochialism, which has no utility when the enemy is complexity. The answer is for intellectual diversity to be integrated within an interdisciplinary view. All forms of evidence must be included, always appreciating the differences in their respective intellectual frameworks. None of the disciplines is without its limitations in illuminating truth, and each contributes to our understanding of nature. As a final example, consider that some artificial reverberators have been designed using folk science alone, and some have been designed using mathematics alone, but the best designs artfully combine both frameworks.

To conclude: we must keep in mind that disciplines are human subcultures subject to cultural forces. Both those who study auditory spatial awareness and the conclusions they reach are a responsive consequence of those forces.

8 Auditory Spatial Awareness as Evolutionary Artifact

Among scientific theories, the theory of evolution has a special status, not only because some of its aspects are difficult to test directly and remain open to several interpretations, but also because it provides an account of the history and present state of the living world.
—François Jacob, 1982

Evolution is a useful lens through which we can examine aural architecture, offering the potential of fusing contributions from diverse disciplines into a single picture. Theories about evolution have been successfully applied to broad questions such as the adaptive function of sex and the influence of geography on genetics, and to narrow questions such as the origins of lactose intolerance and sickle-cell anemia. In contrast, traditional disciplines, with their formal paradigms, cannot readily address some kinds of questions. Evolution is fascinating just because it has the potential to offer explanations about phenomena that would otherwise appear to have no explanation.

We begin with the simplified premise that the aural experience of space contributed, at least indirectly, to the reproductive success of our species. From a narrow perspective, our brain evolved specialized auditory substrates that could incorporate spatial attributes into awareness. But from a broader perspective, auditory spatial awareness also contributes to our ability to thrive in socially complex groups. Although we have already analyzed aural space from the perspective of art, science, cultures, and sub-cultures, we have only alluded to the dominant role that social cohesion plays in all aspects of aural architecture. All known cultures reinforce social cohesion by social, musical, or religious rituals, which take place in spaces, often spaces dedicated for particular ritual functions. Similarly, those who design or select spaces, as well as those who listen to those spaces, are also responding to their social context. We can therefore examine the evolution of auditory spatial awareness and the resulting aural architecture from the perspective of social cohesion.

Evolution is, however, a post hoc theory—using the same evidence twice—for both constructing and validating a hypothesis. According to Karl R. Popper (1959), that makes evolution a prescientific theory. Consider the discovery of a previously unknown island with a unique terrain, inhabited by a tribe whose members have flat feet. An evolutionary argument reverses the observation so that it becomes "flat fleet were a genetic adaptation that allowed the tribe to survive in their unique terrain." It cannot be disproved. Stephen A. Gould and Richard C. Lewontin (1979) noted that it is relatively easy to create intellectual constructs when there is no independent means for testing their assertions. Evolution provides answers that are, at best, plausible and useful and, at worst, intellectual fictions. Scholars have in fact rather low confidence in the validity of specific evolutionary conclusions offered as the *most likely* explanations for what is observed.

Nevertheless, it is tempting to conclude our discussion on auditory spatial awareness by exploring the larger story of why we came to be what we are. Curiosity about our origins appears in virtually every culture, and evolution is the most recent explanation, often replacing myth and religion. In part, the motivation to understand our origins is driven by an apparent lack of rational and predictable behavior on the part of individuals, groups, and cultures. Behaviors that appear to be illogical or irrational become more comprehensible if we assume they are artifacts of adaptation to earlier ecological niches. Similarly, unusual auditory perceptual abilities, such as in-head aural localization when listening with headphones, may be nothing more than artifacts of older adaptations to ancient environments.

Every species of social animals, including human beings, developed its own sensory and social approach to surviving in its niche, which itself dynamically changed with shifts in weather, geography, and the adaptive choices of other competing species. Had environmental stresses and opportunities appeared later or earlier during our evolution, then the history of our adaptation would have been different; our auditory spatial awareness would have evolved other properties, which would have then influenced our aural architecture.

Evolution contributes five themes to our earlier discussions. First, as a subset of hearing, auditory spatial awareness allows us to perceive and locate physical obstacles in space (navigational spatiality), as well as to compensate for the influence of spatial acoustics on communications (social spatiality). Second, social cohesion is a core component of evolutionary theories; all aspects of aural architecture are based on assumptions about the function of social groups in a space. Third, our modern brains are an evolutionary solution to older problems; biological trade-offs over millions of years determined the properties of our auditory and cognitive cortices. Fourth, the wide variations among individuals in auditory spatial awareness and in the ability to enhance that skill with practice are explained by the diversity in the physical environments of our ancestors, on the one hand, and in our social environments as developing children, on the other. And fifth, the human brain did not necessarily evolve with the ability to understand itself or its properties; there is nothing in our understanding of evolution to suggest we should be conscious of how and why we use auditory spatial awareness in spaces.

Although we can actually observe how modern humans experience their acoustic spaces, we can also consider how our prehistoric ancestors might have experienced theirs. Evolution provided a neurological solution to their problem of surviving, and that solution is still with us in modern society. Researchers attempting to understand how early humans responded to stimuli and situations, extrapolate from modern humans and modern society. But humans were never designed for modern society, which is only a dot on the evolutionary timescale. Consider a mechanical analogy. An engine originally designed to run on alcohol is now fueled with gasoline because of its greater availability. If we are to understand how the engine is meant to perform, however, we need to explore its original design context, rather than focusing only on its current one. So it is with human beings. In some respects, our response in a modern environment is nothing other than an artifact of how our prehistoric ancestors responded in their environments. Our aural architecture has its roots in the prehistoric past.

Since Darwin first introduced the idea of evolution 150 years ago, it has come to permeate both popular and scholarly literature. Unfortunately, the popular literature often trivializes complex issues with specious conclusions, and the scholarly literature is riddled with arcane arguments of interest only to academic researchers. This chapter examines the basic principles of evolution as applied to auditory spatial awareness. When incorporating evolution into our interdisciplinary perspective, we find plausible explanations for some issues, and unexplainable mysteries for others.

Evolutionary By-Products Define Modern Humans

Within the human animal, there are both special-purpose, hard-wired biological structures, such as the external ear and its low-level neurological processing of sounds, and general-purpose, soft-wired creative, cognitive, and perceptual structures such as learning to appreciate aural architecture. These solutions represent the two extremes of the continuum of possible responses to environmental stresses and opportunities. Hard-wired structures reflect how the gene pool of the species shifts toward a new phenotype (general properties of the species) over hundreds of generations. Soft-wired structures reflect how individuals learn new behaviors, perceptions, and cognitive strategies (unique properties of individuals) over hours, months, or years. All species display some degree of both.

Paradoxically, brain substrates designed to learn are actually a hard-wired solution that optimizes the trade-off between the efficiency of a hard-wired solution and the flexibility of a soft-wired solution. A substrate designed to learn from experience is like a general-purpose computer that does nothing until programmed, whereas a substrate designed for only one function is like the specialized computer in a cell phone that works immediately. Since both types of solutions exist within the same animal, and since all biological and neurological structures have an energy and space cost, over thousands of generations, evolution carefully balanced their respective contributions of long- and short-term adaptation. For human beings, evolution favored the learning solution using self-modifying brain substrates, which is why culture plays a significant role in determining human nature.

Auditory spatial awareness is a perfect example of both types of adaptation. Unlike some species of bats, we do not have a hard-wired specialization to aurally visualize the world entirely by hearing sonic reflections from synchronized vocalization. But we have the hard-wired ability to fuse early sonic reflections with the direct sound, presumably because all acoustic spaces have a sound-reflecting floor. We have the soft-wired ability to learn to hear space as demonstrated by blind bicycle riders, and we adapt to spaces that vary from jungles to enclosed rooms.

The learning function has been optimized for acquiring abilities that contribute to survival—most notably, motor dexterity and pattern perception—all of which relate to the external world. From an evolutionary perspective, learning to recognize patterns was a fundamental survival skill. Interpreting animal tracks, forest sounds, weather patterns, star formations, soil texture, edible grains, and so on had immediate practical value; patterns are unique to each ecological niche. Traditionally, experienced elders taught their young by example. Learning was based on repetition. Eventually, the young males would become hunters, foragers, navigators, and ultimately, decision makers. Richard C. Lee (1979) describes a group of expert hunters who could identify an animal's sex, age, health, and eating patterns simply by examining its tracks. Because newly acquired abilities were often a matter of life and death, individuals who displayed unique learning intelligence would become leaders, inventors, innovators, and, most important, successful parents to the next generation.

Learning to interpret physical clues left by an animal is similar to learning to aurally visualize a space by listening to auditory cues. The method for learning both tasks, repeatedly studying numerous examples, is similar. Had you grown up in an aural "tribe," you would have become an expert at recognizing acoustic cues, and interpreting their relationship to those spatial "animals" that created them. As an adolescent eager to learn new skills from aural "elders," you would have been taken through thousands of spaces in the "forest" of soundscape niches. Many years of such training would have refined your auditory spatial awareness to a high art form. Because each ecological niche offers unique patterns, your ability to learn to recognize those important patterns would have contributed to your survival and to your tribe's survival.

Neurological Specialization Hides Self-Awareness

What does "learning to hear space" actually mean? Among other things, it involves pursuing a lifestyle that provides multiple opportunities for attending to those aural cues provided by spatial acoustics. Attending to auditory spatial awareness over time changes individuals both internally (privately) and externally (publicly). Privately, individuals may think, perceive, or feel differently because they sense spatial details aurally. Publicly, individuals may display an improved ability to navigate a space in the dark. Illustrating this distinction, language is both public and private; perception is mostly private; motor skills are public; emotions are an odd mixture of public (body language, voice tone, behavior), and private (mood, arousal, and attentiveness). From the perspective of evolutionary survival, only behavioral manifestations of learning would have been relevant because they alter individuals' relationship to the external environment. Private learning is irrelevant unless, however indirectly, it eventually produces some external consequences, which can then be observed.

Auditory spatial awareness is a perfect example of the problem of linking internal experience to observable ability. Behavior can be observed in carefully designed laboratory experiments or in everyday life, as when we watch a blind individual riding a bicycle. When spatial awareness produces an emotional reaction, however, such as a change in mood in a certain kind of acoustic space, we must find a way to make such private experiences observable. Failure to find a technique does not mean that there is no private awareness.

This leaves us with two intertwined issues. To what degree are we actually aware of acoustic spaces? And to what degree can that awareness be either observed or communicated? There are three possibilities: an absence of private awareness, unobservable awareness, and observable awareness. It is difficult, if not impossible, to distinguish the first two cases; even in the third case, external behavior may be only weakly or indirectly linked to internal awareness. Thus we need to explore the larger question of how internal states can be observed. In this respect, auditory spatial awareness cannot be studied without also assuming the existence of brain substrates that adequately translate internal states into behavior, consciousness, or observable manifestations. Recent research has begun to shed light on the intrinsic philosophic problems embedded in concepts such as awareness, perception, and consciousness, all of which are directly relevant to studying the human experience of aural architecture.

Let us now examine what is known about brain neurology to shed light on the problem of interpreting the experience of aural architecture. The eminent neurobiologist Michael S. Gazzaniga (2000) has stated: "the human brain is a bizarre device, set in place through natural selection for one main purpose—to make decisions that enhance reproductive success." One manifestation of neurological optimization is the evolution of specialized neurological substrates, each of which is optimized for specific functions. A brain that is organized into two distinct halves is an obvious example of an optimization because each half can contribute different functions, which increases efficiency of a limited quantity of neurological resources within a fixed head volume. The corpus callosum, which connects the two brain hemispheres, allows each half to share information from separate functions in the other half. Because each hemisphere acquired special abilities, only relatively important information is communicated between them. The left hemisphere does not transmit information that is not needed by the right hemisphere, and vice versa. Neurological communication is biologically expensive, and unneeded communication reduces the efficiency of specialization. Communication and specialization, working in opposite directions, evolved a delicate balance.

Extensive study of patients with split brains over a period of 40 years has made the roles of the two hemispheres more apparent. Orrin Devinsky (2000) observed that lateral specialization, where each half of the brain takes on special functions, exists in many species. But in humans, that specialization has been further extended in order to support language in the left hemisphere, while leaving the right half available for traditional functions. As a generalization, the left hemisphere deals chiefly with language, reason, planning, and logical thought, whereas the right hemisphere deals chiefly with moods, emotions, visceral body states, and the affective meaning of external perceptions. The right hemisphere, which provides processing for attention, visuospatial, body schema, and emotional functions, supports self-awareness, body image, and their relationship to the social environment. In addition, Devinsky implicates the right hemisphere as the dominant locus for the perception and the expression of emotion, including comprehension, gestures, facial expression, intonation, as well as contextual inferences from nonverbal speech, music, pain, and the affective meaning of cartoons. Dahlia W. Zaidel (2000) argued that the asymmetry in processing between the two halves of the brain also appears in the concepts of meaning systems: the right hemisphere is home to novel and noncultural metaphors, whereas the left hemisphere is home to stereotypical and cultural metaphors.

Hemispherical specialization implies that various substrates are only partially aware of what other substrates are experiencing. In fact, what we think of as the unity of consciousness is not unified at all—it just appears that way. By implication, auditory spatial awareness is not a conscious and unified experience of an external environment. Awareness is actually the result of activities in many substrates in both hemispheres, and their ability to communicate with each other is, at best, imperfect. Gazzaniga (2000) commented that the left hemisphere is "driven to generate explanations and hypotheses regardless of circumstances. The left hemisphere of a split-brain person does not hesitate to offer explanations for behaviors that are generated by the right hemispheres. In neurologically intact individuals, the interpreter does not hesitate to generate spurious explanations for sympathetic nervous system arousal. In these ways, the left hemisphere interpreter may generate a feeling in all of us that we are integrated and unified." If nothing else, this analysis explains the difficulty in verbalizing our affective states, and our reaction to affect laden stimuli, like music. Emotions, mood, and visceral sensations, which represent our body states, influence how an individual responds to the environment, but those states may not necessarily be describable.

"When the foregoing research is taken together," Gazzaniga (2000) concludes, "rather simple suggestions are appropriate. First, focus on what is meant by 'conscious experience.' The concept refers to the awareness human beings have of their capacities as a species—awareness not of the capacities themselves but of our experience of exercising them and our feelings about them. The brain is not a general-purpose computing device, it is a collection of circuits devoted to these capacities." The illusory unity of consciousness and the holistic nature of language mask the existence, limitations, relationships, and contributions of those separate neural circuits.

Patients who, for one reason or another, have had their corpus callosum cut, so that their two brain hemispheres operate independently (Baynes and Gazzaniga, 2000), exhibit truly bizarre behavior under controlled conditions, illustrating the segmentation of experience. For example, stimuli presented to the left visual field, which maps to the right hemisphere, can produce strong emotional responses but without the corresponding ability to describe the visual scene. The emotional experience is real, and often intense, but without the awareness of the stimulus that gave rise to the emotion, much like being unconscious of perception. To a far lesser extent, and also without realizing it, normal individuals also have segmented hemispheres, although their normal brains still provide better communications, however imperfect, between the hemispheres than do split brains.

The phenomenon called "blindsight"—seeing without seeing—also illustrates the partitioning of awareness. In her review of this topic, Petra Stoerig (1996) provided numerous examples of lesions in the primary visual cortex that prevented sight but where the individual could still experience the visual object. For example, when a monkey was made (totally) blind by having a bilateral occipital lobectomy, he still learned to navigate and discriminate visual objects (Humphrey, 1974). Removing the top picture level does not necessarily remove connections that allow other neural structures to remain aware of the external world. Awareness exists with and without consciousness. Think of our brain as a complex machine that happens to be equipped with indicators, which act much like windows of consciousness. Some states of the machine are accurately reported by these indicators, but others remain hidden.

Nicholas Humphrey (2000) goes further with the argument that evolution progressively shifted sensory awareness of external stimuli from publicly observable reactions to private experiences. Primitive organisms always respond publicly, whereas complex organisms most often respond privately. The degree to which such private experiences can be represented in words may reflect the degree of neurological connection between the language center and the particular perceptual substrate. When there is a modest degree of connection, some aspects of perception can be verbally communicated, often better communicated with practice, although only up to a point. In contrast, when there is no connection, no amount of effort will allow an experience to be communicated. On the other hand, if the perception changes your body state, such as making your stomach muscles tighten, you can sense that change and report it, although there is a considerable difference between communicating a sensory experience itself and reporting its reactive consequence to your body state. However rich its vocabulary, English is still at a loss for words when communicating internal experiences, especially those which produce affect. Moreover, we may not be conscious of an internal state, even though it is being publicly broadcast, such as the involuntary blush that signals our feeling embarrassed.

In recognition of the importance of emotions in the aural architecture of musical spaces, Glenn KnicKrehm (2004) searched for those performance spaces that had the reputation for producing "heart pounding, raised neck hairs, goose bumps and tears of joy." Having visited some 600 musical spaces constructed between the eleventh and the nineteenth centuries, and having identified the ones with this reputation, acoustic scientists then only measured their acoustic properties (Bassuet, 2004). Other than simply asking listeners, there is no formal way to sort performance spaces into those of high and low affect. Many, perhaps most, emotional aspects of auditory spatial awareness are not readily observable. This view is entirely consistent with neurological research from an evolutionary perspective. Interpreting the impact, function, and meaning of aural architecture is therefore correspondingly difficult, limited, and anecdotal. Because evolution did not provide us with a reliable mechanism to observe and communicate affect, using scientific experiments to understand the aural experience of spatiality is fraught with risks and uncertainty.

Learning as an Adaptation to the Environment

Some individuals are able to interpret spatial details by listening, and they are more likely to develop this ability if they grow up in an environment where such learning is stimulated and motivated. Rich aural environments encourage the acquisition and development of auditory spatial awareness, whereas aurally barren environments do not. Earlier, we observed significant differences in the aural architecture of cultures, and those differences are passed down from generation to generation through children. Even though explicit research on the connection between neurological development and cultural values is relatively sparse, many principles are now understood. Brain development and culture cannot be separated.

The archetype of learning is the baby. Jean-Pierre Bourgeois (1999) suggested that the dynamics of synaptic growth in the newborn, driven by genetic triggers and environmental stimulation, create an individual with unique abilities, propensities, intelligence, and perceptual sensitivities. Aspects of synaptic growth that are common to all individuals represent what we call "human nature"; aspects that differ substantially lead to the uniqueness of each individual, which includes, not just the obvious personality differences, but also subtle perceptual predispositions, sensory preferences, learning modes, cognitive strategies, and emotional temperament.

Several months before birth, the neuron inventory of the human fetus is fully developed but without stable connections; the wiring is still primitive (Bourgeois, Goldman-Rakic, and Rakic, 2000). Although the quantity and density of synaptic connections grow rapidly in a "biological exuberance" beginning shortly before birth and continuing during the next year, these newly formed connections are like an overgrown garden that needs to be shaped and pruned. Nature has chosen to grow massive connections and then to refine their properties by removing excess connections rather than adding new ones. This pruning process stabilizes after puberty, and continues into the third decade of life.

We can understand many of the principles of neurological change in the developing infant from studies of other species. Marcus Jacobson (1969) has identified how environmental exposure changes neural wiring. Individual neurons develop highly specific synaptic connections, but in the early developmental stages, these connections are modifiable. The window of flexibility depends on the type of neuron. Michael M. Merzenich and Christoph E. Schreiner (1990) found idiosyncratic differences in the auditory cortex among adults of the same species. Every individual shows some degree of auditory neurological adaptation, providing distinctive differences in aural spatial ability.

Especially for human beings, the division of labor among brain substrates for each sensory system is far less categorical than implied by the names that scientists give to specific substrates. For complex substrates, the names are only an indication of an embryonic understanding of both neurological functions and developmental dynamics. Anne G. De Volder and colleagues (2001) found that auditory and tactile imagination activated visual substrates. Similar cross-modal reorganization has been found in deaf individuals. Norihiro Sadato and colleagues (2002) observed a similar reorganization of the visual cortex for tactile discrimination tasks, but only for individuals who lost their sight before age 16. Disruptive magnetic stimulation of the visual cortex degraded Braille reading in early-onset, but not in late-onset, blindness (Weeks et al., 2000). These observations indicate that the susceptible period for this form of functional cross-modal plasticity[1] does not extend beyond age 14. Observations of early neurological plasticity in a wide variety of surgically blinded animals are fully consistent with such observations in humans (Kahn and Krubitzer, 2002). Brain substrates are highly flexible in terms of the functions they serve.

Neurological growth and environmental learning both involve changes to the brain. Jean-Pierre Bourgeois, Patricia S. Goldman-Rakic, and Pasko Rakic (2000) sort the pattern and timing of changes in the brain into three categories of learning that distinguish the contribution of environmental exposure to brain growth. *Experience-independent* (hard-wired) learning is innate, requiring limited environmental exposure; *experience-dependent* (soft-wired) learning is achieved only with significant environmental exposure; and *experience-expectant* (soft- and hard-wired) learning is mainly achieved during an innate window of opportunity, when the environment is the teacher.

In this third category, the hard-wired predisposition to extract information when exposed to the environment is largely restricted to a critical stage of brain development, after which learning is more difficult, if not impossible. Language acquisition is perhaps the best example of experience-expectant learning. From about age 2 to puberty, the brain acquires language by extracting semantic and syntactic rules from the exposure to *any* spoken language. When the brain is in this special learning mode, even a minimal exposure to speech will result in acquisition of language. Similarly, sensory integration, where the visual, auditory, and tactile senses of space are made to align, is also based on exposure during a critical window, which begins at birth (Stein, Wallace, and Stanford, 2000).

Among the likely many experience-expectant kinds of learning, auditory learning remains essentially a mystery except for the few cases that have been studied. A. H. Takeuchi and S. H. Hulse (1993) suggested that absolute pitch recognition cannot be learned after age 6, and D. A. Pearson (1991) describes a window between ages 9 and 11 for learning an auditory attention-switching task. A child's ability to suppress the influence of early reflections (the precedence effect) approaches the adult level only after age 5, when the child has been exposed to rich acoustic environments (Litovsky, 1997; Burnham et al., 1993). The ability of the blind to use echolocation to aurally visualize acoustic space or of music enthusiasts to become sensitive to the subtleties of a concert hall also has a learning window that starts to close at a young age.

Most studies of hearing assume, probably incorrectly, that perceptual abilities are predominantly experience independent (innate), such as the ability of newborn infants to crudely localize sounds without significant exposure to the sound world (Aslin and Hunt, 1999), or experience dependent, such as the ability to evaluate spatial simulators or to identify objects with sonar, which requires training.

Even when the experience-expectant learning window closes, some amounts of neurological plasticity must be preserved to accommodate changes in the environment, both internal and external. For example, the ability to aurally localize using binaural cues remains sufficiently plastic to keep visual and auditory space images aligned despite size changes produced by head growth or the use of eyeglasses (Shinn-Cunningham, Durlach, and Held, 1998). Equipped with visually shifting prisms for an extended duration, an owl modified its cognitive map of space to compensate for the perceived spatial change. But when the prisms were later removed, there was neurological evidence that the original map had been preserved, although temporarily deactivated (Zheng and Knudsen, 1999). The owl had created dual cognitive maps. Similarly, Paul M. Hofman and A. John van Opstal (1998) described how lateral location ability was dramatically disrupted in human subjects when the shape of their outer ear was modified, but after some practice, performance steadily improved. Richard Anderson and colleagues (1997) described the neural structures by which auditory location cues in a head-centered frame of reference are transformed into eye-centered coordinates after compensation for eye fixation. Ultimately, subjects achieve a single external representation, with all sensory sources reconciled for consistency. The need for consistency implies some degree of plasticity.

Because the human brain is so adaptive, assertions about human nature or innate perceptual abilities are philosophically problematic. Such assertions fail to recognize the importance of culture: the microculture of the infant, the miniculture of the adolescent, and the macroculture of the adult. As a general conclusion, Dean V. Buonomano and Michael M. Merzenich (1998) explain: "the cortex can preferentially [re]allocate cortical areas to represent selected peripheral inputs. The increased cortical neuronal population and plasticity-induced changes is the coherent response... thought to be critical for certain forms of perceptual learning." Learning is an adaptive response to the environment; our brain is a manifestation of culture.

With regard to the impact of culture on auditory perception, Georg von Békésy (1960) reported an experiment in which a male Rom subject showed normal pitch discrimination but extremely poor loudness discrimination. Because his musical tradition considered pitch rather than loudness as being dominant, he could hear loudness but discarded it as having no significance. Thus not attending to an auditory attribute is, in effect, equivalent to not experiencing it, like irrelevant background noise. Just as a bushman, having lived in the forest for his entire life, would find it difficult to recognize and interpret the acoustics of enclosed spaces, so would an academic researcher, having lived and worked almost entirely in enclosed spaces, find it difficult to navigate the acoustics of a forest.

The problem in studying auditory spatial awareness is that the dominant aspect of learning does not take place under controlled conditions of a school or laboratory. Most learning is woven into life, be it listening to a mother's lullaby during the first nights of life in a nursery or attending weekly concerts and religious services in a church. In the school of life, it is usually not obvious what is being learned. By age 20, an individual has spent over 100,000 hours in a wide range of acoustic environments, which vary greatly across individuals and cultures. Olivier Deprès, Victor Candas, and André Dufour (2005) suggest that the improved auditory ability to localize found among those with myopia arises from the need to use auditory information during ordinary living. Rather than studying the biological properties of our species, scientists who explore auditory spatial awareness are actually observing culture. And depending on that culture, some individuals have more or less auditory spatial awareness than others.

Individuation by Learning and Genetic Predisposition

When members of a culture learn new skills or new applications for old ones, culture takes a new direction. The evolution of aural architecture thus depends on the degree to which individuals learn to appreciate auditory spatial awareness. With enough interest among enough people, a culture is likely to invest resources in aural architecture, which then provides an environment for the next generation to acquire similar spatial abilities. This process builds on itself, either encouraging or discouraging an appreciation for aural space.

What, then, does it mean to "learn to hear space"? Is such learning similar or dissimilar to acquiring other perceptual skills? Is learning simply a catchall concept that depends strongly on what is being learned? The answers to these questions are fragmentary and inconsistent. Although we know much about language acquisition, we know little about acquiring auditory spatial awareness. Nevertheless, some suggestive patterns emerge from a diverse body of research literature that is somewhat peripheral to our topic.

Consider the ability to identify or produce a pitch without a reference pitch. The consensus is that this ability represents, not special hearing, but rather the association of a currently heard pitch with a remembered one. This conclusion is consistent with the lack of correlation between the ability to discriminate pitches and the ability to identify them; indeed, those who can readily discriminate pitches may not be able to identify them. There are other examples. Training audio engineers to hear subtle differences in timbre also requires the acquisition of vocabulary, as well as a reliable auditory memory (Letowski, 1985). Auditory spatial awareness is very much like auditory awareness of timbre and pitch. Dramatic variations in a seemingly homogeneous population are explained by differences in auditory memory and labeling strategies, as well as by enhanced auditory acuity for spatial attributes. Listening is more than hearing; it is more than sensing, detecting, and discriminating sounds. Listening is the act of making sense out of an aural experience by incorporating all that has been remembered from previous experiences.

Even if auditory perceptual learning is not yet well understood, modern researchers can often detect changes in brain activity that result from the acquisition of a particular skill. For example, compared with nonmusicians, musicians show stronger neurological responses to timing errors as small as 20 milliseconds (Rüsseler et al., 2001), to slightly impure chords (Koelsch, Schroger, and Tervaniemi, 1999), and to perturbations in melodic patterns (Tervaniemi et al., 2001). Using magnetic resonance imaging as a measure of neurological activity in the auditory cortex, Christo Pantev and colleagues (1998) examined the difference between musicians and nonmusicians. When subjects were listening to piano notes, the region of neurological activity was 25 percent larger for musicians than for nonmusicians. Even though all the musicians in the study had been actively involved with music for the previous five years, practicing on average 25 hours per week, the size of the neurologically active region correlated with the age of initial musical training. Musicians who began practicing before age 12 had the largest active region. Similarly, using magnetic resonance imaging, Gottfried Schlaug and colleagues (1995) found that musicians with absolute pitch showed a marked increase in brain asymmetry in those regions associated with the auditory cortex. Moreover, violin and trumpet players showed different neurological responses when listening to the same notes on their own and on others' instruments (Pantev et al., 2001b).

As an extreme example of learning an even narrower class of sounds, Laurent Demany and Catherine Semal (2002) trained subjects to distinguish a 3,000 Hz tone from those of slightly lower or higher frequencies. In the course of ten training sessions, the subjects' average performance improved by at least a factor of 2. More interesting, the newly acquired skill did *not* apply to discrimination at a markedly lower or higher frequency (1,200 Hz or 6,500 Hz). This result is consistent with a comparable study in monkeys that revealed an enlargement of the region of the auditory cortex responsible for the particular frequency used in training (Recanzone et al., 1993). Auditory perceptual training at specific frequencies had an observable manifestation in the auditory cortex. Neurons that matched the frequencies used in training showed greater tuning ability and increased latency. Demany and Semal also demonstrated that enhanced discrimination of pure tones did not transfer to complex signals with rich timbre, even at the same pitch. Pitch discrimination was timbre specific, namely, discriminating the pitch of sine tones only loosely correlates with discriminating the pitch of instrument tones. This result exposes the fallacy that sounds having the same pitch are neurologically equivalent. Substrates adapt to highly specific attributes of sound, and corresponding auditory learning is much more specific than initially expected.

Although the long-term neurological adaptation displayed by musicians arises from thousands of hours of practice, short-term neurological adaptations are found in the general population. Pantev and colleagues (2001a) tested the neurological response of subjects listening to music through a filter that suppressed the region between 700 Hz and 1,300 Hz. In three sessions lasting three hours each, they observed a short-term reduction in activity for neurons exposed to 1,000 Hz tones. When a specific neural region was no longer active, neurological changes were already taking place.

These are only a few examples that demonstrate neurological adaptation to the spatial aspect of sound. Consider an organ, whose pipes produce harmonically related notes at separate locations. Because pitch depends on location, organists do not fuse multiple pitches in the same way other musicians do (Brennan and Stevens, 2002). Similarly, conductors strongly experience the spatial components of music because the musicians they conduct are dispersed across the stage. Using neurologically evoked potentials, Thomas F. Münte and colleagues (2001) compared the spatial acuity of conductors, musicians, and nonmusicians at peripheral locations. Even though musicians and conductors have comparable exposure to and training in performed music, conductors are significantly better at peripheral localization. It is not only exposure to a situation that drives learning but also the motivation to benefit from that learning.

Like auditory training, repeated motor activities also change the brain. The somato-sensory representation in the motor cortex for fingers was significantly changed in vio-linists compared with subjects in the control group (Elbert et al., 1995). When carried to an extreme, extensive practice can produce a hand disability called "focal dystonia" (Pujol et al., 2000). With this condition, musicians experience loss of control and deg-radation of skilled hand movement. Because intense exercise of a given finger expands its corresponding neurological region, and because the regions for the fingers are ad-jacent, they compete with each other for the same resources. There are no free neu-rological resources to be allocated. As they encroach on each other, adjacent regions produce the equivalent of a short circuit (Elbert et al., 1998). This pathology is gener-ally irreversible, and musical careers have been ruined by focal dystonia.

Just as specific perceptual and motor skills may be unrelated so, too, emotional responses to sound may be unrelated to its perception. For example, Isabelle Peretz, Lise Gagnon, and Bernard Bouchard (1998) found evidence that recognizing the struc-tural components of music was separate from evaluating the affective component along the continuum of happy–sad. In other words, perceiving music and experiencing mu-sic are neurologically separate, albeit related, processes. To the degree that this is true, sensitivity to musical attributes is unrelated to appreciation of the emotional and cul-tural meaning of music. There are anecdotal examples of listeners who have a height-ened awareness of musical subtleties, but show little affect response, and conversely, there are listeners who experience an intense emotional response to music, but with little conscious awareness of its subtleties. We might also expect that listeners can re-spond emotionally to aural architecture without being consciously aware of acoustic attributes.

Unquestionably, extensive auditory training and exposure to sonic or acoustic environments alter the brain, with corresponding improvement in observable auditory abilities. But can every brain be trained on every auditory task? We all know that a child lacking either fine-motor coordination or sensitivity to sound will never become a brilliant musician. We also observe that some individuals who are blind from birth acquire the ability to navigate space by listening, whereas others do not. Even when we ignore the influence of motivation, we observe that some individuals are simply born with enhanced ability to both learn and perform certain tasks. By studying the degree of correlation in ability between genetically related individuals, Robert Plomin and L. A. Thompson (1993) attempted to form a probabilistic measure of abilities that have a strong or weak component of inheritance, but, unfortunately, these did not in-clude auditory spatial awareness.

Howard E. Gardner (1999) has identified the following specific types of intelli-gence: linguistic, mathematical, musical, kinesthetic, spatial, interpersonal, intraper-sonal, spiritual, moral, existential, and naturalist. Rosamund Shuter-Dyson and Clive Gabriel (1981) have shown that musical intelligence is further divided into melody, harmony, pitch memory, and rhyme intelligence, among several other kinds. Correlations between these abilities are only modestly positive. Some individuals are good at recognizing a melody but only average at sensing time. We assume that skilled musicians are probably those who are good at all the required abilities that are part of performing music, a statistically unlikely outcome. I have no doubt that auditory intelligence is yet another specific kind of intelligence, distinct from visual, olfactory, and tactile. I also have no doubt that within auditory intelligence there are an equally large number of separable abilities, of which auditory spatial awareness is but one. And within that ability, there are still further separable abilities that include the ability to aurally localize, as well as the abilities to discriminate, respectively, sonic reflection densities, spectral colorations, spatial gradients, and so on. Auditory spatial awareness is actually only a label for a group of independent perceptual abilities involved in hearing space.

Few studies demonstrate auditory giftedness because most researchers ignore the large differences in auditory ability when studying specific tasks. Some evidence, however, is available concerning different levels of performance. Seymour Shlien and Gilbert A. Soulodre (1996), using only a few subjects, found a tenfold difference in frequency modulation sensitivity and a threefold difference in detecting the duration of a 2-decibel gap. Some listeners had exceptionally high sensitivity to small differences in loudness, whereas others had exceptional memory for tonal and rhythmic sequences. In a related study, Kristin Precoda and Teresa A. Meng (1997) found that listeners repeating tasks, though consistent with themselves, were not consistent with each other. In his study of gifted listeners, Shlien (2000) confirmed that the ability to detect specific audio degradations varies significantly across the population.

Unlike psychologists, who study learning paradigms in controlled laboratory situations, educators must work in the real world with real people. Describing a school environment, David A. Sousa (2000) asserts that learning depends on sense and meaning, which is to say, it depends on the ability both to detect an attribute (sense) and to assign it personal relevance (meaning). It is difficult to learn a skill when the experience is felt to be emotionally irrelevant. In addition, educators recognize that some people are mainly auditory learners; others, mainly visual or kinesthetic learners. Sousa warns of a mismatch between the preferred sensory modality and the nature of the experience; visual and kinesthetic learners do not readily attend to the subtlety of auditory information, be it emotional or verbal.

If we assume that the previous research applies equally to auditory spatial awareness, and if we also assume that the lack of evidence of such awareness essentially reflects our culture's indifference to it, we come to at least three important conclusions. First, scientists will never detect the subtle and varied differences in auditory spatial awareness abilities among individuals unless they design suitable protocols that will uncover those differences. Second, for as long as there is little encouragement or opportunity to acquire, let alone develop, auditory spatial awareness, our society will surely have an impoverished aural architecture. And third, to produce a subculture of aural architects with a high level of such awareness requires exposing children to a rich aural environment, which encourages those with any genetic disposition to become experts and leaders.

Children who spend most of their time watching television and adults who sit at a desk for thousands of hours each year have far less exposure to acoustic variety than those in many earlier cultures who grew up hunting in forests and on mountains, tending farm animals after dark, navigating noisy towns with low illumination, or, in more recent centuries, attending dozens of opera and concert performances. Our children are acquiring their aural attitudes from the spatial and sensory legacy of now several generations of aurally impoverished listeners. It is up to us enrich that legacy.

Awareness as a Composite of Emotions and Perceptions

A sound that is meaningful, by definition, produces an emotional or affective response. For aural architecture to be meaningful, it, too, must produce an affective response in the listeners who avail themselves of it—an emotional response to space. How should we explore such an elusive topic? The *Oxford English Dictionary* defines emotions as "mental feelings," which sheds little light on the matter. Drawing on findings from the fields of evolution, perception, and neurobiology, cognitive scientists have begun to explain why such a ubiquitous concept as emotion remains so elusive. Emotions are everywhere, like water for fish. Whenever we care about what we are perceiving, an affective component must be present. Thus emotions become an amorphous concept for everything that gives meaning and texture to our perceptual experiences. If, however, we want to understand how aural architecture produces spatial experiences that have impact and relevance, we need to examine the affective attributes of acoustic spaces.

When Zoltan Kövecses (1990) examined the semantic content of emotional concepts in common language, he concluded that there was, in effect, a relatively consistent underlying model. Emotions, like a fluid filling the body, are expressed in a body language. Rather than being just a literary tool, metaphors provide clues about the internal representation of emotions. Consider, for example, metaphors for anger: "You made my blood boil" (heated fluid); "Her face was scarlet" (hot container surface); "I blew my stack" (excessive internal pressure); "He bottled up his emotions" (strong container). But where we have a closed container model for anger, we have an open container model for sadness, melancholy, affection, love, and similar emotions. Visceral changes to the body, which can be sensed, are the only window into this subterranean process. Emotions relate to physiological changes: "He felt it in his gut"; "She had a heavy heart"; "It took my breath away." Thus, as Kövecses sees the folk language of emotion, the body is the container holding emotions, and the body surface displays them.

Why would evolution have produced a brain that responds to stimuli without being consciously aware of doing so? Conscious awareness is not always advantageous. The limited computational capacity of an animal's brain is best used if the animal attends only to the most important tasks of the moment, devoting all its biological resources to the most essential aspects of the environment. Selective attention solves the problem of being overloaded with irrelevant data that would confuse or otherwise delay choosing the appropriate response. Because an animal needs to react to danger, not simply to think about the meaning of lurking predators, the auditory thalamus feeds in parallel the amygdala (automatic feelings) and the neocortex (conscious thought) to give both systems an opportunity to evaluate sound on its own terms. Being a contemplative and flexible process, rational thought is too slow to produce a rapid response to high-impact stimuli. In his simplified model of fear, Joseph E. LeDoux (2000) represents awareness of sound as resulting from multiple inputs: from the amygdala, which extracts the emotional affect of the stimulus; from the hippocampus, which remembers the associations of previous experiences; and from the auditory cortex, which processes features of the signal.

The emotional brain, which is sometimes labeled the "limbic system," has no direct representation in brain consciousness, though it is a major, or perhaps *the* major, contributor to perception. As with many brain substrates, we observe only how it influences aspects of our experience, like seeing a shadow but not the object that cast it. The emotional brain controls the degree of arousal, which largely determines how much effort to invest in attending to the outside world, selecting aspects of the stimuli that are worth focusing on. The emotional brain also provides associations to those stored memories of historical experiences that are relevant to the current situation. In summary, the emotional brain determines which aspects of an experience are worth remembering, what meanings to ascribe to the components of the experience, and when to draw upon those experiences in the future.

Even without knowing the neurological details, we have enough insight to place emotions in a social context. We can view self-awareness of emotions as the result of the brain broadcasting to itself information that is directly relevant. For example, as long as we are consciously aware of needing food, there is no utility in our being similarly aware of low blood sugar, which is the hidden neurological response to a visceral state. We observe our stomach making noises, we observe ourselves staring at a steak, and we observe ourselves experiencing an irritable response to a neutral situation. Low blood sugar is reported indirectly without the need for a sensor. Kent C. Berridge and Piotr Winkielman (2002) argued that "liking" is also an unconscious emotion whose only observable manifestation is in the way that it influences immediate or planned behavior. Do we choose something because we like it, or do we like something because we observe ourselves wanting to choose it? "I like concert hall 'A' better than 'B' because I prefer to hear music there." The feeling of liking something is also indirect.

This suggests that *feelings* are conscious awareness of the body's relationship to the environment, whereas *perceptions* are conscious awareness of the environment itself. I see the book (perception); I contemplate taking it (conscious awareness of an action); I interpret the sequence as coveting (feeling the pleasure of possession). Timo Järvilehto (2000, 2001) goes further by treating consciousness as the awareness of the unitary organism-environment. Positive emotions result from increased harmony between the individual and the environment, and negative emotions result from increased disharmony. In this view, emotional activities in the brain are distributing information about the organism-environment relationship to parts of the brain and body that need to know how to plan a response to the harmony, or lack thereof. From this perspective, only certain information is relevant to providing a conscious interpretation of the environment, perhaps to planning a strategy for improving harmony. Only some information is communicated to the language center for articulating the motivation for a planned action, and that information may not fit into natural language; language did not evolve to support introspection. Like the iceberg, 90 percent of our emotional life is hidden from view. Moreover, our emotions evolved to handle the important organism-environment relationship in prehistoric environments, not in modern society.

Broadcasting your emotional state to other individuals in your group has survival value because that knowledge can be used to influence the behavioral choices of others. To you as an individual, other members of your group are part of the environment. For instance, when you are not feeling aggressive, your smile communicates that state to others, who then form their response based on that knowledge. Emotional broadcasting is a language that supports group cohesion. Many aspects of emotional displays are, in fact, involuntary forms of body language without awareness. Self-awareness is irrelevant if a broadcast is automatic, not requiring consciousness or voluntary action. Karen L. Schmidt and Jeffrey F. Cohn (2001) view the face as a biological adaptation that provides a low-cost and spontaneous means of emotional signaling—for publicly and reliably showing a state of fear, joy, disgust, sadness, anger, excitement, deference, grief, or comfort.

An external stimulus connects you as an individual to your environment, and produces a combination of public, private, and hidden reactions. Some aspects of your current relationship to the environment are communicated to consciousness, some aspects are communicated to your group, some aspects are communicated to your emotional brain and to other brain substrates, and some are communicated to specific biological organs. In each case, you are "aware" of the stimulus but your awareness depends on your personal and biological agenda. Your emotional state is the *adaptation process* for interpreting the stimulus in relation to that agenda. Your emotions bias the choice of action, reaction, coping strategy, perception, and attention. In short, your emotions are the meanings assigned to your relationship to the environment that produced the stimulus. Under this broad definition, virtually all aspects of experience are emotional, and every experience has an emotional component, albeit often weak and unconscious.

There is very little agreement on how terms such as *affect, feelings, emotions,* and *moods* should be defined. Joseph P. Forgas (2000) proposed the following: as a broad concept, *affect* refers to both moods and emotions; *moods* refers to the relatively low-intensity, diffuse, and enduring affect states that have no obvious cause and little cognitive content; in contrast, *emotions* refers to short-lived, intense affect states that usually have an obvious and direct cause. An auditory stimulus that contains an affective component can change the affective state of listeners, whether this refers to short-lived and intense emotions, to longer-term and diffuse moods, or to both.

We are finally in a position to integrate the concept of affect into our discussion of auditory spatial awareness. Like all forms of art, aural architecture can change the affective state of listeners, perhaps with only a subliminal shifting of mood, or perhaps with an overwhelming emotion. When aural architects function as artists, they are intending to influence the affective state of the listeners within the spaces they design. Many spaces, even those without a designer, still qualify as aural architecture if their acoustics influence the affective state of listeners. We now understand why musical and religious spaces are the most prominent forms of aural architecture. Such spaces emphasize the affective experience, and listeners may feel that the space has personal significance. Finally, our inability to explain how, why, or even whether listeners experience a change in affective state does not mean that such changes have not occurred. Although there are countless anecdotes of paintings and music bringing viewers or listeners to tears (Elkins, 2001), there are few such anecdotes for visual or aural architecture. Some art forms simply have less affective content than others, and weak affect is difficult to observe or communicate. Moreover, even when the affective components of aural architecture are strong, few recognize the origins of their mood changes and emotional shifts.

Hearing as a Means for Navigating and Communicating

It is impossible to know how or why our species evolved an auditory system sensitized to certain aspects of spatial acoustics. Instead, we can explore how our and various other extant species are still adapting to their current environments. Two themes emerge. Hearing is the sensory means for receiving vocalized communications among conspecifics, and hearing is also the means for sensing the soundscape, which is composed of the physical environment and the sonic events contained therein. These two themes relate to each other in different ways for each species.

Why did mammals evolve an auditory system that detects sound vibration? There are other ways for sensing the environment. Some bony fish are aware of weak electric fields in water (Nelson and Maciver, 1999); some birds are aware of the earth's weak magnetic field (Diebel et al., 2000); some amphibians are aware of humidity gradients; and some animals can create a map of their environment using infrared heat (Bullock and Cowles, 1952) or ultrasonic sound (Mann et al., 1998). With sensory acuities beyond those of human biology or advanced technology, many animals can sense an imminent earthquake (Tributsch, 1982). Hearing is only one of the many ways animals sense and navigate an environment.

In addition to sensing the environment, individuals of a species must communicate with one another, and vocalization combined with hearing is an excellent means of doing so. Regardless of the social structure, every social animal must advertise sexual availability, demonstrate healthy genes, and find a mate. In order to maintain social connection with the genetic pool of appropriate partners, an individual animal must broadcast the appropriate signals. One common signaling choice is vocalization, which has a more controllable and often wider geographic range than visual displays or chemical pheromones. It is not dependent on illumination and visibility. It is relatively immune to being blocked by physical obstacles. And it is biologically efficient to maintain. From this perspective, the auditory system would have adapted to receiving vocal broadcasts. Yet the human voice rarely extends above 6,000 Hz, whereas an undamaged ear can hear frequencies above even 20,000 Hz. Clearly, the extra bandwidth can be used to detect, recognize, and localize other kinds of sounds. Hearing has been optimized for something beyond communications among conspecifics.

W. C. Stebbins (1980) argued that the auditory system evolved in early mammals in order to exploit nocturnal niches that were free of large diurnal predators. In a dark environment, smell and hearing would have been the chief means for detecting other animals. And hearing would have been the only means for detecting distant objects and surfaces. The extinction of the predator population then allowed mammals to radiate into a much larger diurnal environment, where sensory evolution progressed. Only then did vision become a major contributor to survival.

Although the details differ, all mammalian species developed a similar auditory system, composed of the external pinna and ear canal, and the internal tympanum, three-bone ossicular chain, coiled cochlea, and auditory cortex. Because of this similarity among mammals, Richard R. Fay and Arthur N. Popper (2000) asserted that all auditory variations are the evolutionary modifications inherited from a common ancestor. Modern human beings, however unique our adaptations, are just one of many branches from that ancestor. Douglas B. Webster, Richard R. Fay, and Arthur N. Popper (1992) argued that differences among those branches reflect the unique environmental stresses and biological constraints in each ecological niche. The modern human auditory system is therefore a result of the evolutionary path taken by mammals, primates, and early *Homo sapiens*.

Over time, hearing evolved to become useful in a wide variety of situations: receiving vocal signals from conspecifics, monitoring those from prey and predator species, detecting dynamic sonic events in the environment, and aurally visualizing physical objects and geometries from their acoustic perturbations. Each of these uses played an important role in determining how a species competed for a stable ecological niche. Because air is a common resource, which can become overtaxed, cluttered, or degraded by nature and other users, auditory strategies adapted to particular environments. In addition, strategies evolved to balance the advantages of successful communication with conspecifics against the disadvantages of both warning prey and attracting predators. In general, the biological machinery for hearing and vocalization evolved as part of a composite system. How we hear aural architecture is the result of this process. Human beings also evolved while being embedded in a social environment, which itself was embedded in a physical environment.

Adapting to Acoustic Environments

Although our knowledge about how early humans adapted to their acoustic niches is speculative, we do have a large body of information about how extant birds, primates, and other species are still adapting to acoustic environments. Acoustic evidence about early habitats, which were mostly forests, jungles, and savannas, is available in a few regions where civilization has yet to intrude. By combining these fragments of information, we observe some general principles. We know that any given acoustic environment hosted competing species with competing solutions to the task of survival. Each species evolved its specific solution in response to the solutions adapted by other species.

A summary of the basic acoustics of early environments illustrates the evolutionary complexity of adapting the auditory system and vocalization strategies to a local ecology. Evolution provided both genetic and learning adaptation so that small groups could respond to the specifics of local environments. Both solutions are evident in a variety of species.

We begin by examining acoustics of natural spaces. In an open plain with no obstacles, sound intensity decreases in proportion to the distance between the sender and receiver. Upon encountering an object, such as a tree branch, a sound wave is absorbed, reflected, transmitted, or any combination of the three. Each of these processes is itself complex. For example, flat planar surfaces produce coherent echoes, whereas highly irregular surfaces disperse sound waves in many directions. Dense vegetation absorbs high frequencies but the ground surface reflects low frequencies. Wind, which moves branches and leaves, both creates turbulent noise and modulates the intensity and frequency of sound. Ambient noise, which then interacts with objects, is nonuniformly distributed in the environment. Some regions are quieter than others, some have less high-frequency sound, and some have diffused sound fields. Finally, air is far from an ideal transmission medium. Its turbulence disperses high frequencies; its molecules absorb them. Its thermal layers create both dead zones, with no sound, and hot spots, with focused sound. All of these forms of acoustic degradation increase with distance.

Each species in these early environments evolved a vocalization strategy that was consistent with these specific acoustic degradations. And each choice, whether long steady tone, short burst, or chirp with modulated variations in amplitude or frequency, had its trade-offs in terms of the amount of information carried and the distance over which it radiated. Information in the sound was often more fragile than the sound itself. In modern terms, more often than not, you might understand that someone was talking but not what was being said.

In a forest, the signal path from sender to listener includes a multiplicity of small sonic reflections from every surface of every tree, which reflects sound back into the forest, arriving at the listener well after the direct sound. At greater distances, the direct sound weakens, but the sonic reflections remain as forest reverberation. In their study of the acoustics in deciduous forests, Douglas G. Richards and R. Haven Wiley (1980) illustrated how a 25-millisecond tone burst at 1,000 Hz becomes a 150-millisecond diffuse pulse at a distance of 25 meters (80 feet), but an 8,000 Hz pulse does not spread as much because its sonic reflections are absorbed by foliage. Similarly, a steady 2,000 Hz sine wave was received with large random amplitude fluctuation (acoustic degradation) at a distance of 60 meters (200 feet) from the source, and fluctuation intensity was strongly dependent on the wind velocity near the ground.

Weather alters the acoustic geography, enhancing or degrading the transmission of sound. In the middle of the day, when the heat of the sun produces stratification layers, atmospheric turbulence increases, acoustically degrading transmitted sound. As the distance between sound source and listener increases, phase fluctuations gradually become more random, making the received signal sound less and less like the original. On the other hand, a thermal inversion, hot air layer above cold, enhances sound transmission because sound waves are channeled through a sonic conduit, with upward radiating waves redirected by the boundary between the two layers of air back toward the ground (Humphreys, 1940). Indeed, Peter M. Waser and Charles H. Brown (1984) observed thermal inversions in the Kenyan rain forest with a 4-decibel gain at 100 Hz.

Each acoustic environment has a different capacity to transmit sonic information, and this varies according to the time of day, the season, and the weather. More important, animals in the forest compete for the most desirable auditory channels. Like food and land, air is a limited common resource that is contested as well as shared. Animals with strong voices may dominate those with weak ones, but loud broadcasts also invite predators. Forest reverberation may prevent predators from aurally localizing a prey animal, but it also prevents the animal's conspecifics from doing so, a decided disadvantage for a distressed monkey calling for help. Thus selecting an appropriate auditory strategy and acoustic environment involves both evolution and social learning.

In their comprehensive compendium of the ecology and evolution of acoustic communication in birds, Donald E. Kroodsma and Edward H. Miller (1996) illustrated the adaptive process by which numerous species balance information capacity, predatory risk, sonic competition, and the acoustic constraints of the environment. Evolutionary flexibility becomes readily apparent when examining the wide range of adaptive responses among birds. Some species sing special songs at dawn, when atmospheric conditions favor the long-distance transmission of sound. Given that acoustic degradation increases with distance, Marc Naguib (1995) suggested that Carolina wrens listen for reverberation and high-frequency attenuation independently to determine the distance of a singing fellow wren. Those same birds adapt their song to the acoustics of the native environment in order to create a recognizable group identity, much like a social dialect (Gish and Morton, 1981). Birdsongs show greater acoustic degradation in alien than in native habitats. Birds in forest habitats sing at lower frequencies (in the neighborhood of 2,000 Hz) than birds in grasslands because forests have less noise at these frequencies, thereby increasing the transmission distance of their songs (Ryan and Brenowitz, 1985). All aspects of the acoustic environment put evolutionary pressure on a given species to take advantage of local acoustic properties. To optimize vocalization to the nuances of a habitat, individual birds learn the details of their song rather than being born with a predefined song.

Nor are birds unique in adopting auditory strategies. A comparative analysis shows that the blue monkey is 18 decibels more sensitive to low-frequency tones than the semiterrestrial rhesus monkey of about the same size in the vocalization region of 125–200 Hz (Brown and Waser, 1984). By adapting their vocalization and auditory thresholds to this relatively noise-free frequency region, blue monkeys are able to increase their calling range by a factor of 16. The macaque monkey has evolved a vocalization and auditory sensitivity to a class of sounds that serves to identify members of the group, thus producing social cohesion in an acoustic environment where nonmembers are otherwise similar. Specialized use of a particular auditory channel, however, depends on a particular group, not on the generic properties of the species (Dittus, 1988).

Similarly, bats in the Amazon valley shifted their vocalization from the more typical 100,000 Hz region to 8,000 Hz because the high humidity of the tropical rain forests rapidly attenuates ultrasonic signals (Griffin, 1971). Male short-tailed crickets can increase the area of their calling song by a factor of 14 by perching in treetops instead of on the ground (Paul and Walker, 1979). Fish take can take advantage of the high-frequency cutoff of shallow water to avoid detection by predators but still maintain communication with their conspecifics (Forrest, Miller, and Zagar, 1993). As these examples clearly illustrate, animals are more than merely aware of their particular acoustic environment. They use that awareness to evolve more useful communication strategies within a shared competitive auditory channel.

Interpreting the meaning of vocalization reveals several philosophical problems, which also apply to studies of human hearing. Meredith J. West and Andrew P. King (1996) summarized one problem: "labeling a communications system by its predominant sensory modality may be misleading, and bias us towards too narrow a view." In an alternative construction, William M. Mace (1977) suggested that the role of hearing is found by "asking not what's inside your head but what your head is inside of." Local habitat ultimately defines the nature of the animal. James J. Gibson (1966) shifted the focus to an instrumental view: "Animals [including human beings] do not perceive or communicate for the sake of perceiving or producing a display but for the sake of managing the social environment." A goal-directed view of hearing and vocalization is more informative and predictive than a mechanical one. Perception and vocalizing are a means to an end: surviving in social groups.

The range over which conspecifics can hear each other's vocalizations, which in chapter 2 was called the "acoustic horizon," determines the geographic area of a cohesive social unit. Like the French village of the nineteenth century, where the acoustic horizon of the church bell determined membership in the village, the size of the social group is determined by the choice of listening and vocalization strategies, in combination with acoustic geography and population density. In the evolution of many species, these four factors aligned: group size, vocalization strategy, population density, and acoustic geography. For our species, there was more flexibility in that alignment, but the same four factors still play a role. Acoustic geography influences social geography (culture).

When the nature of acoustic degradation is consistent, predictable, or familiar, the auditory cortex can extract sufficient information from it to enhance survival. For example, because acoustic degradation is proportional to distance, modeling the type and amount is a means of determining the distance to the source. Marc Naguib and R. Haven Wiley (2001) concluded that perceiving distance implies both a neurological model of the acoustic degradation and a neurological model of the sound source before being degraded, which can then be compared to each other. Similarly, binaural processing using two ears, when combined with a model of acoustic degradation, is another means of reducing its influence. Such neurological solutions probably evolved to overcome the limitations of early soundscapes.

Assuming that our auditory cortex evolved the means to model the acoustics of forests, jungles, and savannas, we are now using old evolutionary solutions to hear new spaces. There is no better example of this than how we aurally experience reverberation. What does it mean to hear a concert hall with ears designed for a forest?

Characterized by a high density of low-level sonic reflections from surfaces typically found in the environment, tree trunks, branches, and leaves, forest reverberation is generally limited in duration to about 200 milliseconds, compared with many seconds for large enclosed spaces. When we earlier analyzed the perception of concert hall reverberation, we observed that early sonic reflections fuse with the direct sound, whereas later ones form sustained, enveloping reverberation. The boundary between these two aspects of reverberation is the same order of magnitude as the duration of forest reverberation, about 100 milliseconds. We hear the early reverberation of an enclosed space with brain substrates that evolved for forest reverberation. The adaptation of fusing early sonic reflections into a perceived single source would have been useful to our early ancestors. Moreover, what we now call "apparent source width," which is a property of early reverberation, appears to be nothing more than an auditory awareness of the degradation of sound by forest acoustics.

Except for the occasional cavern, there was never any historical counterpart of an enclosed space that was capable of spreading sound energy over a long duration, what we call "sustained reverberation." Moreover, early humans did not make caverns their natural habitat. Similarly, except for an occasional cliff or steep embankment, there was no mechanism for creating a sonic reflection with sufficient delay that could be heard as a discrete echo, a discrete event. Forests, jungles, and savannas do not produce echoes. From a survival perspective, it would have been critical for an early human to distinguish between a single event with multiple sonic reflections and multiple sonic events from different sources. It is therefore not surprising that sound arriving well beyond the fusing interval is heard as distinct from the direct sound, either as a coherent echo or as diffuse reverberation. Both are perceived as if originating from different sources, even though we, as modern humans, know that the spatial acoustics create the second event from the first. The linking of reverberation with the direct sound is cognitive rather than perceptual.

Another difference between indoor and outdoor acoustics is the degree to which sound transmission is static, without time-varying changes to the auditory channel. Even in large enclosed spaces, such as a cathedral, air is relatively stable and homogeneous, at least compared with natural environments. The auditory cortex expects to hear spectral and temporal variations produced by turbulent air, shifting thermal layers, and the surfaces of moving objects. Animal vocalizations also contain similar variations. Even without acoustics, sound sources were never spectrally pure, perfectly periodic, or reliably repeatable. The need for such variation, in order to sound natural and pleasant, is exemplified by the musical tradition of vibrato and tremolo, explicit changes to pitch and amplitude. But the same is true of spatial acoustics. Our auditory cortex is designed with the expectation of variability when perceiving space. Reverberators using a static algorithm sound more mechanical, whereas those which include the appropriate random modulations sound more natural.

In addition to segmenting sound into discrete sonic events, the ability to aurally localize the source of a single sound exists in almost every mammal. It is easy to understand why localization is useful. Where are the prey or predators, and which path leads to food or safety? Many parts of the auditory cortex and the resulting auditory perceptions they produce are entirely consistent with the need to aurally localize. The affective component of aural localization, which contributes knowledge about the sender's location, is separate from and independent of the affective component of the direct sound, which contributes knowledge about the sender's message and emotional state. If aural localization of important sounds is that critical to survival, we can assume that inability to aurally localize would make an animal attentive, uncomfortable, and perhaps anxious. Such would then be the case for diffused reverberation, where sonic reflections lacking a strong direct component arrive from all directions.

To localize a sound source, the auditory cortex suppresses the irrelevant information in early sonic reflections while extracting the difference in time and amplitude for the direct sound, which arrives before the reflections. Perceptual and neurological scientists, who have been studying this ability for years, call it the "precedence effect," whereas audio engineers, who use this in the design of public address amplification systems, know it as the "Haas effect." In its simplest form, localization of the direct sound remains stable even when followed by a single discrete sonic reflection in the time window from 2 to 20 milliseconds after the direct sound. Because this effect is robust, stable, and consistent across the population, it invites an evolutionary explanation. Fusing the numerous early sonic reflections in a 100-millisecond time window is consistent with forest reverberation. What is the analogy for a single strong sonic reflection in a 20-millisecond time window? The only flat hard surface that could consistently produce a large specular sonic reflection is the ground, and the delay between the direct sound and a sonic reflection from the ground would be on the order of 5 milliseconds for a primate standing at a modest distance; for primates living in trees, the delay would be significantly longer. A single sonic reflection from the ground is a universal property of all spaces that existed for ancient and modern animals.

We in modern society still experience acoustic spaces in a way consistent with our inherited legacy from our early ancestors in their prehistoric spaces. Scientists can observe aspects of this aural inheritance in their laboratories; composers incorporate aspects in their musical creations; aural architects embed aspects in their spatial creations; and listeners hear aspects when attending a concert or conversing in their living rooms.

There are unexpected modern auditory experiences that have no evolutionary antecedents, for example, static acoustics. The acoustic world of early humans was never static. Perhaps the most dramatic example, however, is in-head aural localization, which occurs when listening to sound with headphones: the spatial contradictions between the sounds heard by the right ear and those heard by the left have no natural spatial counterparts. Unable to determine a location in the external world using its inherited rules, the auditory cortex simply places the source inside the head, thus seeming to preserve the reliability of detecting real sound from real events in real spaces. That placement is itself an evolutionary optimization. Faced with spatial contradictions, the auditory cortex, rather than making a best guess about location, simply removes the sound source from the external world. This is consistent with the difficulty that engineers have in creating signal-processing algorithms that consistently produce externalized sounds.

Spatial Imaging Using Echolocation

With the realization by Donald R. Griffin (1944, 1958) and others in the mid-twentieth century that several species could navigate with their ears, scientists began to realize that this ability was more than an odd curiosity. For many species, nature had indeed evolved an aural means for sensing objects and geometries by the way that they influence sonic attributes.

Sound illuminates a space in the same way that light does; ears as well as eyes can sense illuminated objects. Like a built-in biological flashlight, vocalization is a means to illuminate the environment, and the clicking tongue or tapping cane of a blind person walking down the street is an acoustic flashlight. By attending to how the environment changes the sound, an aural image of an acoustic space can be created, a process called "echolocation." But when an animal uses background sounds from other sources, the process is auditory spatial awareness. The distinction, which is often ignored, depends only on the origin of the sound source.

In some species, evolution elevated the importance of auditory spatial awareness, and in a small percentage of these, additional evolutionary pressure matched their vocalization to their auditory cortex. This linkage between hearing and vocalization for echolocation, such as in bats, dolphins, and a few other kinds of animals, is relatively rare; originally, the auditory system developed for communications between conspecifics, and for decoding sonic events in the soundscape. Michael J. Novacek (1985) and M. Brock Fenton (1985) suggested that ultrasonic echolocation in bats evolved from its initial use for communications, only later becoming a specialized means for sensing the physical environment and replacing vision. But George D. Pollack (1992) makes the additional point that the auditory foundation for echolocation is nothing more than an enhancement of the generic capacity for auditory spatial awareness, but optimized jointly with the evolution of specialized vocalizations. The neurological foundation for auditory spatial awareness, perhaps in vestigial form, is not unusual among a wide variety of species. The auditory spatial awareness of bats and dolphins is only a specialized extension of a common ability. A less specialized version of auditory spatial awareness exists in rats, hamsters, and shrews, as well as in humans.

Golden hamsters have the ability to locate a shallow platform by the way it changes the acoustic ambience, a skill consistent with their nocturnal activities and their underground habitat (Etienne et al., 1982). The shrew, also with an underground habitat, has a weakly developed visual system that is limited mostly to sensing light intensity. Although it relies mostly on touch to gain information about its surroundings, echolocation is its only means to acquire remote information before approaching an object. E. R. Buchler (1976) found that the wandering shrew (*Sorex vagrans*) uses echolocation chiefly as means of exploring unfamiliar environments: it increases its rate of ultrasonic transmissions to as high as 20 per minute when placed into a new maze, but then decreases them to about 1 per minute after exploring the maze for six minutes.

Other species of shrews, namely, the masked shrew (*Sorex cinereus*), the American shrew (*Sorex palustris*), and the short-tailed shrew (*Blarina brevicauda*) also show some evidence of echolocation ability when exposed to strange surroundings. They emit short pulses in the frequency range of 30,000–60,000 Hz (Churchfield, 1990). These shrews can discriminate between an open and closed tube at a distance of 20 centimeters (8 inches) using ultrasonic vocalization (Forsman and Malmquist, 1988). Thomas E. Tomasi (1979) observed that individual shrews showed differential echolocation ability on the different tests. Some were good at distance detection, others at discriminating small openings, and still others were able to detect objects around corners.

Griffin (1986), who studied the larger topic of acoustic orientation of animals, describes a class of nocturnal birds that use echolocation in much the same way as bats. The oilbirds fly quite confidently deep within the fully darkened caverns of Guácharos in Venezuela by emitting 1-millisecond sound pulses at about 7,000 Hz. When they leave the caverns for an illuminated environment, these pulses cease. With blocked ears, the birds flew directly into the cavern walls. In Griffin's view, any nocturnal bird, indeed, any bird inhabiting a dark environment, is an evolutionary candidate for echolocation.

James Gould and Clifford Morgan (1941) showed that the rat could easily detect high-frequency auditory signals; John W. Anderson (1954) demonstrated that rats could vocalize at high frequencies. Donald A. Riley and Mark Rosenzweig (1957) showed that rats could detect, entirely from acoustic cues, an alley blocked by a vertical barrier. Although a rat is capable of consistently producing and hearing high-frequency vocalization, and although it can hear a barrier, such skills are rarely observed in the laboratory. Rather, rats are far more often observed to use numerous other mechanical noises to illuminate the space: sniffing, sneezing, loud clicking of teeth, scratching the floor. When traditional experiments involving rats in a maze were reexamined, earlier results were questioned and challenged because experimenters had not considered the rat's ability to detect spatial properties using the auditory channel.

Although difficult to study, the dolphin is one of the best examples of how evolution fused auditory and visual imaging as a means for navigating the complexity of an underwater environment. Adam A. Pack and colleagues (2002) observed that dolphins could match objects perceived visually with those perceived by echolocation. Moreover, rather than just detecting specific attributes of an object with sight or hearing, the dolphin experiences the external world holistically; vision and echolocation make equivalent contributions to the representation of objects. Barry E. Stein and M. Alex Meredith (1993) described multisensory neurons that responded to both visual and auditory stimuli as a possible explanation for a fused image. As a large mammal with highly developed cognitive skills, the dolphin is an example of a species that uses sight and hearing interchangeably. In contrast, most mammals favor one of these senses over the other for navigating the environment. From this perspective, the dolphin is unique.

Studying echolocation and auditory spatial awareness is fraught with methodological difficulties, uncertainties, and ambiguities. Without carefully controlled experiments that mimic the appropriate social and physical niche, an animal may simply choose not to use echolocation. Individual animals have the ability to selectively choose a strategy based on their immediate needs. There is no reason to believe that an animal would display behavior that reveals it unless there was a need to. Because of the increased risk of predatory attack when vocalizing, animals would have evolved selective uses of sound generation when needed, and only when the risk-reward ratio was favorable. Generated sounds may serve either to signal conspecifics or to echolocate. We cannot read an animal's mind, and we cannot determine if, or when, auditory spatial imaging is taking place. We can only interpret observed behaviors that are sometimes consistent with a navigational strategy using hearing.

We now know that the ability of a species to hear space ranges from nonexistent to highly refined, depending on the evolutionary path taken by its ancestors as they adapted to the stresses and opportunities in a unique sequence of environmental niches. A species acquired its ability only when certain brain substrates were allocated for decoding acoustic cues from objects and spatial geometries. Because these substrates may be artifacts or vestiges of other evolutionary optimizations that only partially served the function of hearing space, auditory spatial awareness may be a primitive supplement to refine visual awareness. Only in rare cases have auditory substrates been optimized for echolocation.

Even without understanding the details of evolutionary optimization in brain substrates, we know that there are several adaptation mechanisms. Although multiple abilities compete with one another for sparse neurological resources in a fixed brain volume, different abilities may share a resource, performing double duty without a corresponding extra cost. In fact, the ability to isolate a single voice in an environment containing spatially distributed sound sources is a close cousin of auditory spatial awareness. In both cases, the auditory cortex builds an auditory model of the environment. That model is simply used in different ways. In the first case, the model suppresses the influence of acoustics on the perception of the target voice; in the second, the model provides information about the space itself. The neurological ability to suppress acoustic degradation can just as easily be used to decode spatial acoustics. With humans, decoding speech and aurally visualizing acoustic spaces share brain substrates because both depend on similar types of acoustic cues.

Along with shared substrates, there are also vestigial ones that were used in the remote past, but have less relevance now. At some time, a group of individuals may have evolved an ability to use auditory spatial awareness during extended periods of darkness in protective caves, or for tracking unseen game through complex forests. Dense forests are acoustically different from open grasslands, which are yet different from complex mountain ranges, tundra, and coastal fishing regions. When a group of individuals remained in a particular environment for dozens of generations, some adaptation is likely to have taken place.

Our ability to hear spatial attributes or to learn to hear them may thus depend on the degree to which that ability helped our ancestors survive and propagate themselves. This may in part explain why only some of us demonstrate an ability to learn echolocation. Auditory spatial awareness may depend on the lifestyle of our particular ancestors. The same evolutionary pressure that led individual species to optimize their auditory cortex for different functions also operates on small groups of individuals living in their particular soundscape niche.

Evidence, unfortunately unrelated to spatial hearing, shows that isolated human populations acquire a degree of biological specialization within a few dozen generations. Consider some examples. Because the ability to digest lactose foods as an adult is based on a genetically controlled enzyme, this adaptation has been traced to populations that had a history of living with domesticated farm animals (Johnson, Cole, and Ahern, 1981). Light-skinned individuals can trace their ancestors to the northern climates where light skin pigmentation favors the ability to absorb vitamin D from limited sunshine (Loomis, 1967). In contrast, near the equator, there is greater need for melanized skin protection against ultraviolet rays, which destroy folic acid (a critical B complex vitamin) and injure sweat glands, disrupting thermoregulation (Jablonski and Chaplin, 2000). A human gene has been identified that correlates with improved athletic endurance in high-altitude mountaineers (Montgomery et al., 1998). Because of the differences in increased heat loss with increase in body surface area, most populations from the tropics have longer and slimmer body shapes than do populations from the Arctic (Jones, 1992). The ability to function at high altitudes at low oxygen levels has been described in terms of both individual and genetic adaptation (Hochachka, Gunga, and Kirsch, 1998). We may speculate that those with an enhanced auditory spatial awareness had ancestors living in an environment where that ability had survival value.

Aural architects, musical composers, and scientific researchers are therefore taking advantage of vestigial abilities as they discover how to apply prehistoric solutions to modern life. As a species, we were not designed for our current environment of enclosed spaces and complex soundscapes. Robin Fox (1997) summarized the inconsistency between humans and our institutions: "In some sense [spaces] are human because they are human inventions. But it is one of the paradoxes of an animal endowed with intelligence, foresight, and language, that it can become its own animal trainer: it can invent conditions for itself that it cannot handle because it was not evolved to handle them." The aural architecture of our modern spaces trains those of us who occupy or inhabit them.

Interdependence of Biology, Nature, and Culture

When we trace the common themes in aural architecture back to their origins, we find them inevitably intertwined with social, cultural, and biological evolution. How then can evolution explain aural architecture? Like many creative and intellectual endeavors, aural architecture is an extension of earlier evolutionary solutions, which allowed our species to survive through thousands of generations. Although the specific spatial designs and our experience of them are unique to each social situation, common themes transcend specifics. Individuals formed social groups because they improved their chances of propagating. These groups then constructed their aural architecture as a manifestation of the social properties. Using social cohesion as a framework, let us then connect auditory spatial awareness of architecture to the survival value of our evolutionary trajectory. Despite being speculative (and appearing to be a digression), the following arguments and explanations provide answers that are, at the very least, consistent with early discussions. We are applying the concepts advanced by Peter J. Richerson and Robert Boyd (2005) to aural architecture: genes and culture shared an interdependent evolutionary trajectory.

The subspecies *Homo sapiens sapiens*, which first appeared in Europe and Asia around 50,000 years ago as the modern human, descended from the archaic human *Homo sapiens*, which first appeared in the fossil record 250,000 years ago. There is fossil evidence that this evolutionary antecedent descended from *Homo erectus*, who appeared at least a million years ago. There are some 50,000 generations between early and modern humans. Although the evolutionary path has been long, complicated, and mostly unknown, several milestones can still be seen to influence the aural architecture of our species in the twenty-first century.

Each of the thousands of biological properties that define a species is subject to evolutionary pressure, but once a property changes, the context changes. The auditory system exists within the context of an external physical environment as well as an internal biological one. A small change in one internal biological system then changes the context for all other such systems. Carl Gans (1992) explains: "The structures involved in vertebrate audition reflect parallel shifts in various biological roles, such as ventilation, ingestion, and the perception and production of sounds. Understanding of the shifts requires a parallel consideration of the physical principles and functional morphology of the systems, as well as the ecology and behavior of the organism."

Consider an illustration. When the visual system of a species evolved a small high-acuity region, its auditory system came under adaptive pressure to evolve a wide-field localization ability to provide steering information to the visual system—ears telling eyes where to look. In contrast, if its visual system had evolved uniform acuity over a large field of view, its eyes could have detected important objects without requiring steering information; there would have been less environmental pressure for its auditory system to develop accurate localization ability. But now consider what would have happened if predators entered its environment. The species might have taken refuge in dark caves, thereby avoiding the predators, but that change would also have shifted the balance from vision to hearing. Vision would have been useless in a dark cave. The visual system might then have atrophied, putting yet more pressure on the auditory system to provide a comprehensive aural image of the environment using auditory spatial awareness. If, a thousand generations later, the predators disappeared, the species might have moved back to open spaces and continued to evolve, but from a very different evolutionary starting point. The auditory and visual systems of the species leaving the caves would be very different from those of the species that first entered them. If the species had developed echolocation while in the caves, it might become a nocturnal predator, or its echolocation ability might atrophy, leaving only a vestigial residue. Thus the temporal sequence of adaptive responses heavily influences the evolutionary trajectory taken. Optimization is local, not global; a species evolves as a sequence of minor design responses to a continuously changing environment.

When geography limits the mobility of individuals, thereby preventing breeding among distinct groups, the genetic pool of each group evolves along its particular trajectory. Individuals within an isolated group become more homogeneous, even as each group diverges from other isolated groups. With respect to auditory spatial awareness, one group might be the progeny of thousands of generations of adaptation to forest acoustics, whereas another group might have had ancestors that adapted to the acoustics of an open expanse of tundra, or to the strong echoes of craggy mountains. When individuals respond to stimuli in a spatial awareness experiment or to the aural architecture of a space, their experience is necessarily influenced by the social and environmental history of their ancestors.

Social Intelligence of Enlarged Brain Creates Culture

Most manifestations of aural architecture provide communal spaces. Audiences who listen to music in concert halls are participating in the shared experience of a group. Large spaces are expensive to build, thus requiring groups of investors. Scientists who study auditory spatial awareness are members of professional groups. Throughout the earlier discussions, we repeatedly referred to culture, which is the largest group that shares values. To expand our understanding of aural architecture, we must explore not only how and why our species creates social groups and cultures, but also how our social and biological evolution relate to one another, and how all of this influences our aural architecture. Answering these questions will give us some insight into the function of aural architecture in our modern context.

We begin with bipedalism, the ability to move on two feet, which is considered one of the major biological shifts that initiated the evolutionary transition to human beings. Bipedalism, combined with other adaptations, played a crucial role in determining the requirements of social groups, especially with regard to energy balance: nutritional intake and energy expended. Kevin D. Hunt (1994) suggested that chimpanzee bipedalism and australopithecine anatomy both originate from the same adaptive pressure, to collect low-hanging fruit. David R. Carrier (1984) argued that the morphology and physiology of human bipedal locomotion became specialized for long-distance running to hunt animal prey by relentless pursuit. The shift to better-quality food sustained larger social groups, the antecedent for culture.

Nina G. Jablonski and George Chaplin (1993) postulated that bipedal displays and mock fights would have served as a noninjurious and socially ritualized method of resolving intragroup conflicts, thereby reducing the mortality rate. One of the critical issues in groups of mammals is the mechanism by which conflicts over limited resources, territory, and sexual partners are resolved. The survival value of individuals that could efficiently hunt as a collective unit, and the survival value of individuals that could ritualistically fight without injuring each other, were important precursors to primate and eventually human societies. Individuals that evolved a predisposition toward the social intelligence necessary for working within a group had a better chance of reproducing.

Following bipedalism, the human brain became larger and more complex, elevating the importance of mental abilities. As an alternative to speed, agility, and strength, thinking shifted the balance from physical to mental processes. Smarter individuals had a better chance of outthinking and outwitting prey, predators, and sexual rivals. Moreover, when combined with the coordinated activities of groups, elevated intelligence became somewhat like a large distributed brain in a dispersed organism. Bipedalism expanded range and mobility, and enlarged brains made those activities more efficient. With multiple brains, ears, eyes, arms, and legs joined by social cohesion, hunting parties became a potent force. The elevated intelligence of individuals contributed to the elevated intelligence of the collective group.

When nonhuman primates first learned that participating in small groups provided better survival value than working alone, they also elevated the value of group cohesion. Without cohesion, internal conflicts over allocating tasks, resources, and sexual partners would destroy the group, forcing individual members to focus on their own survival. Yet even within a group, individuals are still competitors. Social intelligence balances individual needs against the benefits of deferring to the group. Understanding that balance, as well as enforcing group cohesion, required social intelligence. Nonhuman primates evolved along an evolutionary branch that elevated the importance of social intelligence in forming complex social groups, and human beings went still further in forming complex societies.

Once having followed the evolutionary branch of functioning within large complex groups, each individual human could contribute particular skills, abilities, and intelligences, such as hunting, cooking, navigation, farming, tool building, and so on. Intellectual diversity had more value to a cohesive group than to an individual. Today, we still observe diverse forms of mental ability: some of us are better at auditory pattern recognition, others at mathematical logic, still others at conflict resolution, and so on. From an evolutionary perspective, uniformity in intelligence would have been a weaker choice than diversity, and that is still true.

From an evolutionary perspective, advanced intelligence is not straightforward. Large brains have a biological cost that must be balanced against their contribution to survival. Although it accounts for only 2 percent of total body weight, an adult human brain consumes 20 percent of total energy intake (Aiello and Wheeler, 1995). Relative to body weight, which is the relevant metric, the human brain is three times larger than any other species (Passingham, 1982). Leslie C. Aiello and Peter Wheeler (1995) argued that, compared with other primates of comparable weight, human beings increased the energy available to their brains by decreasing the energy available for digestion, rather than by increasing their total energy needs. Over the past 4 million years, the hominid brain has increased in volume from 400 to 1,400 cubic centimeters, with a corresponding decrease in energy budget of the digestive system.

The evolutionary consequence of a large brain is even more apparent in the large head of a human infant. A newborn consumes upward of 70 percent of its caloric intake to maintain its brain metabolism. By age 10, a human child will have consumed more than a million calories provided by others. Whether measured in calories or money, human children have always been expensive. Not only are the adults in a social group supporting the nutritional cost of their large brains, but they are also supporting the costs of their children's developing brains. Social cohesion supports these costs.

Statistically analyzing a primate database, Tracey H. Joffe (1997) showed that larger brains correlate with an extension of the time period spent as a juvenile, during which the intricacies of complex social life must be learned. Unlike the infants of other mammals, the human infant is subjected to strong social forces and environmental interactions while its brain is completing its growth. Edward F. Adolph (1970) showed that the developmental order of growth stages in a fetus is the same for the twelve species of mammals that he considered, except that a significant part of human growth takes place after birth, whereas for other mammals, growth is essentially complete at birth. Using the ratio of brain to body weight as a measure of growth, A. Barry Holt and colleagues (1975) came to the same conclusion: among mammals, the growth and development of human infants are unmatched for their slowness. At birth, our brains are still growing at fetal rates; some neurological and cognitive abilities are not fully developed until well into the third decade of life. The basic development of a human child is being completed at the same time that the child is acquiring extensive experience in a particular environment, and that experience influences how development will transform the child into an adult. Because the brains of developing children are still plastic while being molded by culture, culture evolves synchronously with biological evolution.

Stephen A. Gould (1977) summarized the essence of our species:

Human evolution has emphasized one feature of this common primate heritage—delayed development, particularly as expressed in late maturation and extended childhood. This retardation has reacted synergistically with other hallmarks of hominization—with intelligence (by enlarging the brain through prolongation of fetal growth tendencies and by providing a longer period of childhood) and with socialization (by cementing family units through increased parental care of slowly developing offspring). It is hard to imagine how the distinctive suite of human characteristics could have emerged outside the context of delayed development. This is what Morris Cohen (1947), the distinguished philosopher and historian, had in mind when he wrote that prolonged infancy was "more important, perhaps, than any of the other anatomical facts which distinguish *Homo sapiens* from the rest of the animal kingdom."

For Louis Bolk (1929), genetically determined extended childhood is the driving force of society. Even as culture molds the child, the needs of the child mold the culture.

Raising children is natural selection operating at the level of genetically based psychology. Humans with personality attributes that were antisocial or ultraindividualistic, hence not child centered, did not produce as many descendants as those who put their energies into families. Culture is simply an efficient mechanism by which individuals can find sexual mates to produce viable children, to supply families with an adequate supply of nutrition, and to protect them from predators such as wolves, hyenas, and wild cats and from other dangers. Humans and their primate cousins use culture in the same way—because it provides reproductive advantage.

Individuals of all social species evolved groups of particular sizes and properties, and such groups then became the environment within which individual members lived and propagated their genes. Our species is no different. The previous discussion characterized the original properties of our particular type of social unit, which was dominated by the delayed development and extreme dependency of our infants. As a species, our gene pool evolved within this social environment, creating generations of social animals that could live and thrive in these social units. Although modern society includes a few individuals who prize isolation and individuality, and although most of us prize participation in multiple groups of varying sizes and properties, we all carry a genetic proclivity to form social units similar to those of our early ancestors.

Aural architects and those who experience aural architecture are part of this evolutionary trajectory. Regardless of our artistic, scientific, or intellectual talents, we carry with us the survival value of delayed brain development and the social intelligence to function within a cohesive group. Even today, we can see the evolutionary importance of group harmony in how we design spaces, and how we use them. Some aural architects and research scientists may claim to be independent of their cultural biases, but, as a species, we evolved as social animals. Our aural architecture is by and for such social groups.

Culture as an Evolutionary Invention

Appreciating the nature of human culture is easier if we explore some of the universal patterns of animal cultures. Nearly all known animals exist in groups of conspecifics for a multiplicity of benefits: defense against predators, cooperative food acquisition, division of labor, and nurturing and educating the next generation. For all our uniqueness as a species, nearly every attribute of modern human society can be found, albeit in a less complex form, within some animal culture. Animal cultures, like their human counterparts, serve to train their young to survive in specific ecological niches, including adapting to local soundscapes. Because sounds and acoustics vary from region to region, learning is still a better evolutionary strategy than a fixed biological solution.

For example, avian species, even with their bird-sized brains, are genetically endowed with the ability to create a primitive culture that is passed from generation to generation. That culture includes an oral-aural tradition of songs that are adapted to the acoustics of the environment. Species of birds living in one region produce songs that are different from those in neighboring regions, and those differences increase with distance. European blackbirds teach naive conspecifics to use mobbing calls to indicate when a dangerous predator is nearby (Curio et al., 1978). Like birds, colonies of Weddell seals living in fiords only 20 kilometers (12 miles) apart each have unique vocalizations (Morrice, Burton, and Green, 1994), whose differences are learned. Similarly, isolated groups of male elephant seals each use different threat vocalization dialects, which persist from generation to generation (Le Boeuf and Peterson, 1969). With their more complex cultures, velvet monkeys teach their infants to differentiate types of birds by using eagle alarm calls for six predatory raptors (Seyfarth and Cheney, 1986). At least in controlled laboratory conditions, a male chimpanzee who had learned sign language was observed actively teaching it to his son (Fouts, Fouts, and van Cantford, 1989).

The major difference between human beings and other primates is that we followed an evolution branch that led to a rich vocalized language, whereas other primates did not. Except for that difference, Duane D. Quiatt and Vernon Reynolds (1993) confirmed what every visitor to a zoo observes: primates are very similar, both physically and behaviorally, to humans. However, the primates do not have an efficient way of passing on their experiences to the next generation. Some communication methods allow a limited amount of information to be transmitted, but that information is insignificant compared with what a human child learns in only a few years. Because of this inefficiency, chimpanzee cultures do not display any "ratcheting" by which each generation can cumulatively build on what they inherited from previous generations. There is simply too much information lost to reach the critical stage where information accumulates.

The development of language, and its role in communication between generations, is therefore central to discussions about human culture, for it dramatically expands the complexity and depth of information. Public language is the mechanism by which the human mind extends beyond the scope of what an individual brain can muster. "Once people communicate with language," Steven Mithen (2000) observed, "it makes little sense to conceive of mind as being constituted within the body of a single person, as each person draws upon, exploits, and adds to, the ideas and knowledge within other people's minds." Language, especially written language, binds generations. Our evolutionary branch of primates, endowed with genetics to support a collective mind, is the only species that supports cultural evolution. Although the inclusion of cultural evolution as a manifestation of individual genetics is relatively recent, it supports the observation that human culture originates from human evolution, not independently of it (Barkow, Cosmedes, and Tooby, 1992). Certainly language is the major link between biological and cultural evolution.

For all its power to propagate cultural knowledge, because of its weak ability to represent auditory spatial awareness, language is not particularly useful for communicating aural architectural traditions. In this respect, the cultures of humans are actually similar to those of birds and monkeys; each kind of animal adapts its particular culture to local acoustics and social needs. In itself, aural architecture is more a secondary than a primary component of human culture, where space serves the derivative function of supporting social cohesion.

For accumulated knowledge to be passed on to other members of a community and, more important, to the next generation, there needs to be a stable social structure that preserves the community. Members need a sense of communal obligation, assigned tasks, and an appreciation for the mutual gain of staying together. Myths, religion, rituals, song, dance, music, traditions, rites of passage, and other such activities, serve to bind individuals together in larger units. In communities lacking such stability, communal knowledge decays and future generations are more vulnerable to environmental challenges. Thus artistic religious expression has high value rather than just being art for its own sake. The word *religion* derives from the Latin *ligare*, which means "to bind"; religious institutions bind individuals together. Bernard Grant Campbell (1998) summarized the implications: "In fact, all rituals may be described as religious, for not only do they bind individuals to the core of social knowledge, but, by performing them, individuals are bound to each other in a common activity often requiring much skill and effort." Such rituals require special spaces with properties matched to them; hence, aural architecture has its roots in binding rituals.

Like those of other species, each human culture evolved along its own social path, with specific values, rituals, and organization. Cultures that grew and thrived expanded their scope, successfully competing for resources. Cultural niches expanded and contracted as they encountered one another, and those better adapted to the sociophysical environment absorbed, overpowered, invaded, or destroyed weaker cultures. Consequently, over the millennia, we evolved from a multitude of isolated and small groups of hunter-gatherers into a single, massive global community linked by efficient trade and communications.

In parallel with this social expansion, the size of our aural architecture expanded to accommodate larger audiences, which progressed from a few dozen (ritual caves), to hundreds (early Greek temples), to thousands (open-air amphitheaters, cathedrals, and concert halls), to millions (recorded and broadcast virtual spaces). The size of spaces supporting performed music and religion as forms of social cohesion evolved with the size of cultures. And because the cohesive power of music and religion arises in part from its emotional content, the influence of spatial acoustics on emotions becomes a critical component in aural architecture.

Properties of Social Cohesion in Small Groups

Besides examining our larger culture as a means of understanding aural architecture, we also need to explore culture on a small scale. Like other primates, human beings were not designed to function in social units comprising millions of people. As societies and their cultures grew in size, following the genetic imperative to form traditional cultures, people created smaller social units, which we call "subcultures," comprising perhaps no more than a few hundred individuals and resembling primate and early hominid societies. When we examine the details of aural architecture, we clearly observe that acoustic spaces are the creation, not of the larger culture, but of subcultures, sometimes on behalf of the larger culture, and sometimes independently of it. Although the larger culture may support science, those actually studying auditory spatial awareness are a small group of researchers who work together in an auditory subculture as an extended family or tribe. Those who share an appreciation for acoustic nuances, such as audio mixing engineers creating virtual spaces or the blind navigating a space using echolocation, also form auditory subcultures.

Having considered the evolutionary process in terms of brain substrates, individuals, and the larger culture, let us now consider subcultures, the original form of human society, and the layer intermediate between the larger culture and the individual. A subculture exists in an environment containing other subcultures, like multiple tribes sharing a forest. Subcultures also adapt to their environmental stresses and opportunities. We can learn much about the origins of aural architecture by examining the behavior and properties of modern subcultures. We can observe how subcultures operate within the larger culture, and how individuals behave within their particular subculture, which is more homogeneous than the larger culture. Subcultures are small enough to explain individual behavior, yet large enough to respond to the larger culture. Moreover, as the natural social unit, the subculture provides consistency over our history as human beings: the general properties and behavior of any given subculture, originating from genetics, provide a stable framework that does not depend on details.

The analogy between traditional older societies and modern subcultures is not perfect because a modern individual usually belongs to many subcultures, or occasionally to none at all. Nevertheless, there is much to be learned by exploring those aspects of the analogy that have explicative value. The following discussion amplifies the description of auditory and professional subcultures in chapter 7. Such subcultures have a secondary set of properties that often overpowers their primary goal of building and analyzing acoustic spaces. To survive, a subculture also needs its version of social cohesion. For individual members of a subculture, the necessary social skills to survive within the subculture are as important as the architectural skills to build spaces.

The first human social groups, which existed thousands of years ago, are not available to study, but our close cousins on neighboring branches of the primate tree still exist. Humans, bonobos, and chimpanzees share a close common ape ancestor dating back 7 million years, unlike the Old World monkeys composed of gorillas, baboons, and macaques, which split off from our evolutionary line of descent much more than 30 million years ago (Sibley and Ahlquist, 1984). The human species shares 20 million more years of common history with the chimpanzees than either species does with the Old World monkeys. For this reason, chimpanzees provide a reference for understanding our common ancestors.

The world's largest captive chimpanzee colony, at the Arnhem Zoo in the Netherlands, has been studied for over a decade by a team of primatologists led by Frans de Waal (1998). The depth and longevity of this study allowed researchers to assign each individual chimpanzee a unique name, personality, history, family, and relationship to every other chimpanzee in the colony. This integrated study illustrated the complex working society that displayed elements of cooperation, alliance, confrontation, deception, and reconciliation not unlike those found in other primate societies, including our modern human subcultures. The variety and complexity of these social dynamics required a particular kind of animal intelligence to achieve group cohesion and social stability. More important, understanding the social tools used to maintain the intimate bonds between individuals having to resolve physical aggression, dominance conflicts, and competition over sexual partners provided a model of how the benefits of group membership balanced the costs of group living.

Because primate societies are a major research focus, numerous examples illustrate how social skills are critical, especially during periods of instability or stress. For example, when an alpha male attacked a female, others came to her defense, and shortly thereafter, the conflict was resolved with a reconciliation kiss. Bonobos use sexual release for pacification, especially at feeding time, when the potential for conflict increases (de Waal, 1989). There are numerous instances in the 1989 de Waal study where a third chimpanzee intervened to bring peace between two fighting opponents, and then withdrew when peace was achieved. Because chimpanzees are working within relationships that have a past, present, and a future, conflict resolution repairs the damage already done and avoids more serious future damage that could result if harmony were not restored. An unresolved conflict can cost a friend and companion, with the resulting loss in mutual support benefits, which are substantial.

In a later study, de Waal (2000) described primates as having a natural heritage of conflict resolution. Filippo Aureli (1997) proposed that postconflict anxiety reduction, rather than the alternative pragmatic choices of tolerance and avoidance, motivated reconciliation, which is more reliable than just a temporary truce. Only harmony reduces the uncertainty about what will happen when opponents meet again; fear of future confrontation also has a cost. Even though we cannot penetrate the primate mind by observing behavior, professional primatologists are convinced that other primates also have an internal emotional life (Dittrich, 1992). Our knowledge of primate societies suggests an evolutionary pressure to acquire both the mental processing skills for reading subtle behavioral cues and the emotional communications skills to influence the outcome of a conflict.

In chimpanzees, the most common form of conflict reconciliation involves physical contact, such as kissing, grooming, touching, and sexual release. Mutual grooming, cleaning fur of plant debris picked up during normal travel, serves more a social function, based on age, sex, rank, and kinship, than a hygienic one, often consuming 20 percent of the animals' time. Grooming is most intense when solidifying an unstable relationship. Friends show greater behavioral tolerance, support one another in encounters with others, protect one another's status against assertive threats, and ensure better access to reproductive partners. At the biological level, grooming as a form of soothing correlates with the release of endorphins, natural opiates (Dunbar, 1996). Monkeys who have been groomed show higher levels of these hormones than those who have not actively engaged in grooming behavior, which induces relaxation and a mild form of euphoria, reducing social tensions. In contrast, social deprivation during the critical learning period destroys the ability of the individual to function in a group (Russon, 1997); high levels of harassment in marmosets prevent young females from undergoing puberty. Social harmony has survival value in all primates, including human beings.

We can clearly observe not only overt sharing and conflict, but also the subtler forms of dishonest signaling by withholding or actively falsifying information among other primates (Hauser, 1997b). Depending on its hunger level, which of its fellows is watching, and the quality of its food, a rhesus monkey may choose not to vocalize a food call to the group in order to have more for itself. Moreover, even though those with positive social skills are likely to profit from the expertise of others by teaching and sharing knowledge, they could just as easily behave like beggars and scroungers to acquire the fruits of another's expertise without any effort (Russon, 1997). Since deception is a choice, individuals try to protect themselves from being victimized by looking for signs of duplicity, seeking to confirm reliability, and identifying the status of the caller, and by the trust implicit in the relationship to the caller. In general, primate deviousness parallels behavior found in children at the earliest stages of socialization (Whiten, 1997).

Within this complex world of multiple social interchanges, genetically based social intelligence determines the number of interactions. Each relationship requires a detailed model of interactive history, individual preferences, and an understanding of individual psychologies. Those with the keenest social skills develop a wider network of potential collaborators. Modern political leaders, like their primate counterparts, are often those with a highly developed ability to communicate emotionally, convincingly, and manipulatively with a large number of individuals, without necessarily having an enhanced standard of morality or honesty. On the other hand, in smaller egalitarian groups, peer pressure is extremely effective at limiting the power of any leader; leaders are followed by choice, not by enforceable power. Christopher Boehm (1993) argues strongly that counterdomination behavior allows subordinates to neutralize the nominal power of the alpha individuals. In fact, Mark Bekoff (2001) believes that morality was the direct result of experiencing the advantages of trust, fairness, and cooperation in small groups.

Intelligence to exploit the physical world of inanimate objects, as we know from experience with some antisocial experts in specialized professions, is altogether different from the social skills required to make friends, to be accepted and supported by the group, or to attract a sexual partner. Although modern society often focuses on scholastic intelligence and displays an ambivalent attitude toward social skills, the value of emotional intelligence is now being recognized as the best predictor of life success (Goleman, 1997). The old adage "It is not what you know but who you know" still rings true in modern subcultures.

A large number of researchers now agree that social intelligence was the major component in human evolution that enabled ever larger groups to form, even if there are conflicting theories to explain the details (Whiten and Byrne, 1997). Language skills and emotional intelligence are two obvious tools that allow individuals to exercise their influence in a social situation. Richard Dawkins and John R. Krebs (1978) view vocal signaling as being a highly efficient application of a low-energy force to manipulate the behavior of the listener. By explicitly rejecting the informational interpretation of vocalization, with its assumption of a sender speaking to a listener, they substitute the concept of the actor's impact on the reactor, not unlike electronic amplification, where a small energy input can produce a disproportionately large response. An observable behavioral interaction may as often serve to manipulate as to share information.

Primate groups typically comprise some 30 individuals, with limitation on group size arising from the exponential growth in the number of one-to-one relationships that need to be maintained with a given set of tools. Robin Dunbar (1998) extended the earlier Machiavellian hypothesis of Richard W. Byrne and Andrew Whiten (1988), which viewed mental ability as social intelligence, by showing a strong relationship between the size of the neocortex and the size of the social group. Simply stated, smart individuals can interact with many friends and enemies. Using the data from a multiplicity of primate species, Dunbar showed that larger-brained species function in larger groups. That relationship is strong for both the larger definition of the social group, based on the number of potential coalition partners, and the smaller definition of clique, based on the number of intimate grooming partners. A further analysis (Kudo and Dunbar, 2001) asserted that the size of the neocortex determines the individual's ability not only to store knowledge or learn mechanical skills, but also to manipulate complex social information.

For modern humans, the predicted maximum size for a collaborative group is about 150, using the average size of the human neocortex as the variable from Dunbar's research. Numerous examples, whether prehistoric, historical, or modern, support that number. Neolithic villages from 6500 B.C. in Mesopotamia contained 25 dwellings with an average of 6 people per dwelling. Hutterites in communal farms in South Dakota consist of some 110 individuals, East Tennessee rural mountain communities have roughly 200 residents, and professional armies, modern as well as ancient Roman, employ fighting units of 150 soldiers. Beyond a size of 150 individuals, a formal and stratified hierarchy with authoritarian figures is needed to preserve social stability. Dunbar interpreted the working maximum size of 150 as reflecting the cognitive limit on the number of relationships that could be maintained at a sufficient depth to provide mutual support.

The implications of Dunbar's basic thesis are profound. Defined as a social skill, not as intellectual knowledge, social intelligence comprises elements of alliances, friendships, feuds, seduction, physical fighting, deception, and manipulation. With insufficient social intelligence, relationships become unstable, and participation in the group produces emotional stress and mortal dangers. In many primate societies, death by group members accounts for the highest mortality rate—essentially murder when peaceful solutions have failed to solve a problem.

Dunbar (1993) observed that the need for efficient bonding among humans would be served by language because it provides a way to keep emotional connections. Individuals share gossip to stay informed about the activities of others. Listening to the conversation in university common rooms, he observed that no more than 25 percent of the conversation was devoted to matters of intellectual, political, scientific, or cultural issues. The remainder was devoted to social subjects. Gossip still survives in modern society as a marker of group inclusion. If you are not in the gossip network, you are not part of the group (Barkow, 1992). Moreover, when used responsibly, gossip serves as a social control mechanism to regulate individual behavior (Wilson et al., 2000). It is increasingly apparent that much of human social intelligence involves sensitivity to subtle relationships, and the ability to manipulate those relationships.

The importance of social intelligence and social cohesion aligns with the observation throughout this book that aural architecture and the subcultures of aural architects depend on social cohesion. Small groups of individuals are responsible for creating spaces used by other small groups. Neither an individual nor a larger culture designs and builds a cathedral, concert hall, or spatial simulator. Rather, subcultures, often with power disproportionate to their size, create spaces for the larger culture. The design of such spaces is driven by the social dynamics within subcultures of architects with knowledge about aural architecture. But that knowledge coexists along with other (unspoken) goals. Aural architects are also social animals.

Creating an acoustic space is only a means to an unrelated end; we manipulate spaces for social reasons. Spatial design never exists outside of its social context, which is composed mostly of specific subcultures. Plastic is the more scientific concept for properties that can be molded by environmental pressures.

9 Concluding Comments

Architecture begins where engineering ends.
—Walter Gropius[1]

Architecture is to make us know and remember who we are.
—Geoffrey Jellicoe, 1989

As human beings, we interact with both our social and our physical environment by using all our senses, thereby becoming aware of events, objects, and other people, as well as the spaces within which these are embedded. Each of our senses plays a unique and complementary role in creating our internal experience of the external world.

In contrast with preliterate societies, however, our modern technological society tends to devalue hearing, smell, taste, and touch, preferring sight as the principal means for sensing the environment. Traditionally, we therefore both create and experience architecture visually, rather than with all our senses.

Although we think of hearing primarily as how we sense such active sound sources as speech, sirens, or snapping twigs, it is also how we sense the passive acoustics of our environment. Walls and open doorways change active sounds in a perceptible way, as do enclosed spaces. If we listen carefully, we can sense a wall or an open doorway by the presence or absence of an echo, and the depth of a cave by its resonances. We can also hear how the acoustics of a space, whether bathroom or concert hall, changes the way a voice or a violin sounds.

Even when we are unable to form an aural image of a space, and even when we are unaware that a space changes sound, spatial acoustics, whether of a living room, a concert hall, an office building lobby, or a cathedral, can influence our mental state. Despite its importance, however, auditory spatial awareness remains subtle, often unconscious, and seldom recognized outside of those professional disciplines which focus on aural architecture.

The aural architect often cannot be identified because the design, selection, or creation of an aural space is distributed among a wide variety of individuals—including those actually using the space—who are not aware of their contributions, and because the acoustic attributes of the space are often an accidental by-product of impersonal, socioeconomic forces. In essence, an aural architect is more of an abstraction than a person. A single discipline cannot claim ownership of aural architecture because it is far more fluid and dynamic than visual architecture. Yet, even without a professional owner, aural architecture influences us all.

As modern humans living in familiar spaces with omnipresent electric illumination, most of us can see little use in having auditory spatial awareness. But should the lights fail, the need to navigate or orient in a space by listening will remind us of how useful it is to sense space without vision. Similarly, we sometimes "feel" someone approaching from behind even when that someone is silent. These uses of auditory spatial awareness most likely served our prehistoric forebears well. Living in often ill-lit environments without reliable light sources, and facing ever-present danger from predators, they would have found hearing a valuable complement to seeing space and objects.

The developing brain of an infant, and specifically its auditory cortex, responds to acoustic exposure by adapting to specific soundscapes provided by the family subculture. A rural farm with a rich natural soundscape, especially at night without artificial illumination, provides children with opportunity and motivation to experience a complex aural environment by listening. Similarly, a musical family encourages its children to use hearing as a primary means for social and emotional connections. In contrast, children who grow up in noisy metropolitan apartments with acoustically porous walls may experience auditory overload, and therefore ignore auditory spatial awareness. Extensive use of television, video games, and headphones connected to portable audio devices does not provide opportunities for hearing space.

As individuals, we can enhance our ability to hear space by choosing to exercise that ability in our daily lives. As a culture, we can create social and architectural opportunities to encourage our fellow citizens, especially our children, to acquire spatial awareness. When many of us attend to our environment by listening, our culture is more likely to invest in spaces that have a complex and socially desirable aural architecture. In turn, a rich aural architecture is more likely to stimulate development of that auditory spatial awareness. And so on. The process is self-reinforcing.

Our review of aural architecture in different periods and cultures showed that choices were mostly artifacts of socioeconomic forces unrelated to acoustics. Nonaural economic, religious, political, and social imperatives determined how an acoustic space was selected or designed, and only afterward did the aural architecture become apparent. Even when constructing musical spaces, where the importance of acoustics was clearly recognized, other imperatives often had a strong and competing influence on the final design.

Regardless of its origin, when an acoustic space is repeatedly used for a specific purpose over a long period, the culture begins to associate its aural personality with that purpose. Association leads to tradition; musical spaces have their tradition, as do religious, political, and social spaces. These aural traditions are bound to other traditions in the culture with a stable and enduring interdependency. Acoustic spaces and their social functions evolve together, mutually influencing each other. As Christianity has evolved, so have its churches; as public music has evolved, so have its concert halls; as technology has evolved, so have its virtual spaces. Revolutionary changes in aural architecture are usually the result of corresponding revolutionary changes in other parts of the culture. Aural architecture both expresses and supports culture. With advances in technology, and a corresponding interest in virtual spaces, aural architecture may now begin to lead the culture, as the visual arts have for so long.

Our concept of aural architecture resulted from interpreting and integrating the contributions from dozens of disciplines, each with its limited view of aural space, and each with its unique philosophy, language, assumptions, paradigms, and cultural biases. In order to fuse these disparate views into a coherent picture of aural architecture, we have borrowed interdisciplinary techniques that have previously proved useful for reconciling diverse perspectives. There is no single truth about the nature of auditory spatial awareness and its role in perceiving space. Differences among disciplines that address the subject are mostly the result of having a limited perspective.

The unifying theme in our discussions on aural architecture has been its influence on social cohesion. Over the centuries, aural spaces have been created or selected to provide environments for a variety of groups and individuals. And the aural qualities of these spaces can either impede or support social cohesion over social distances that range from intimate to public.

Because the nature of social cohesion shifts as culture evolves, so, too, does aural architecture. Historically, small towns in warm climates actively encouraged cohesion by embracing aural connections through open windows, public commons, large churches, and outdoor living. Currently, modern advanced cultures embrace independence and privacy, supporting cohesion by means of the telephone and the Internet, the electronic fusion of remote acoustic arenas. The current generation frequently experience aural architecture with the virtual spaces of manufactured music. The difference between then and now is nothing other than an evolution of cultural values.

Besides being a subject of scholarly interest, the principles of aural architecture can easily be applied to ordinary life. When my family and I moved into our house, we decided to remove all the doors from the rooms on the ground floor, thus making a relatively large public acoustic arena. But by introducing extensive plush (sound-absorbing) rugs and furniture into the large volume of this single space, reverberation was dramatically reduced, thus lowering the noise level. This increased the acoustic arena resources of the space: the multiple small arenas made possible by sound absorption gave those within them aural privacy without physical boundaries. The rugs and furniture also served to reduce sharp resonances, thus improving the quality of music reproduction. And by removing small inaudible objects from paths used for walking, we enhanced navigational spatiality, reducing the likelihood of an accident with dim lighting. In the garden, a high wooden fence created a partially private acoustic arena, isolating that space from the noise of local traffic. Yet, because the acoustic arena created by the fence extended vertically into the trees, we still had access to the sounds of nature. In short, our family has had the benefits of aural architecture, without incurring significant cost or effort. We were our own aural architects.

By incorporating, even tangentially, music, poetry, literature, and sculpture into our relationship with the world around us, we gain a richness that enhances the quality of our daily lives. Aural architecture also adds its richness, but unlike other art forms, we cannot escape the influence of aural architecture because we live inside it. Whether intentionally designed or accidentally selected, our aural spaces influence our moods and behavior. Learning to appreciate aural architecture by closely attending to auditory spatial awareness is one way we can control, and thus improve, our personal environment.

Notes

Chapter 1

1. In some restaurants, the conflict between aural and visual architecture may be intentional. Aural unpleasantness induces diners to leave sooner, rather than lingering, which then increases the economic return for the owners.

Chapter 2

1. The psychology literature generally uses the word *unconscious* to mean hidden feelings and thoughts that subtly influence behavior, as, for example, in Freud's model of the ego, id, and superego. In contrast, the cognitive literature uses the word *nonconscious* to mean the absence of any perceptual awareness of an external stimulus, even if the brain responds to the stimulus. For our purposes, however, *unconscious* is understood to include both meanings.

2. The English language provides a rich vocabulary for visual experiences. But because there is no appropriate word for "flooding a space with sound," we will borrow *illuminate* to serve that function, even though it is a visual word. Similarly, when listeners form an impression of spatial details by listening, they will be said to *visualize* the space, a word also borrowed from vision.

3. Absent an established word to that effect, the term *aurally visualize* will be used for auditory imaging of objects and space.

4. Use of the first person singular pronoun in running text throughout volume refers to the first author, Barry Blesser.

5. Ideally, the shape of the acoustic mirrors should be a segment of an ellipse rather than of a parabola.

6. On the subject of silence, which is as complex and interdisciplinary as aural architecture, see especially Jaworski, 1993, 1997; Tannen and Saville-Troike, 1985.

Chapter 3

1. Unlike *visualize*, *auralize* refers to an *external* process, namely, to a spatial simulation that produces real sound; the coinage is less than a decade old. Accordingly, and as explained in chapter 2, note 3, the term *aurally visualize* will be used to refer to the internal process.

2. To appreciate the physical scale of these ancient temples, see Daniel Cilia's archaeological review (2000) of the megalithic temples of Malta.

3. To appreciate, in detail, the explosive creativity in musical form and space during the eighteenth and nineteenth centuries, see Forsyth, 1985.

Chapter 4

1. Theoretical musicology uses the word *space* to represent the dimensions of music itself, such as pitch, timbre, tempo, and so forth. For us, *space* is the experience of a real or virtual environment, and *musical space* is the experience of spatial attributes when music is the sonic illumination.

2. Many instruments can only play an arpeggiated chord, a rapid succession of notes that belong to it, but spatial reverberation converts them into a true cord composed of simultaneous notes.

Chapter 5

1. Ignoring the narrow and formal definitions that are usually applied to particular musical styles, we will use the term *contemporary music* to mean present music as well as avant-garde, postmodern, and experimental music of the twentieth century, especially music that involves manipulating space.

2. For an analysis of the role of audio as part of video, which follows different rules from a space used only for music, see Chion, 1994.

Chapter 6

1. Quoted from a sign hanging in Einstein's office at Princeton.

2. For an extensive review of methods as applied to spatial perception, see Berg and Rumsey, 1999; Zacharov and Koivuniemi, 2001; and Bech, 1999.

3. For more about the musician's perpective, see also Gade, 1989, and Nakayama, 1984. These studies assume that concert halls should be designed with distinctly different acoustic properties for audiences and musicians, each with a different criterion for quality and each occupying a different space.

4. For a comprehensive treatment of statistical acoustics, see Cremer and Muller, 1982.

Chapter 7

1. Counted among the pursuits and disciplines of aural architects are musicology, archaeology, neurobiology, mathematics, anthropology, sonic ecology, psychophysics, medical science, acoustic physics, cultural evolution, spatial modeling, biological evolution, cognitive psychology, perceptual psychology, religious ceremonies, sound mixing, hearing-aid design, theatrical sound, audio engineering, music composition, film-sound editing, multimedia games, musical performance, acoustic architecture, echolocation training, virtual space simulation, and auditory displays.

2. For more on interdisciplinarity, see Klein, 1900, 1996; Kline, 1995; Finkenthal, 2001; and Messer-Davidow, Shumway, and Sylvan, 1993.

3. The subdisciplines of psychology now include perceptual, interpersonal, gestalt, clinical, social, environmental, behavioral, differential, comparative, evolutionary, ecological, organizational, transactional, performance, comparative, architectural, humanistic, ethical, educational, motivational, learning, spatial, industrial, experimental, existential, emotional, criminal, theoretical, aesthetic, philosophical, military, metaphysical, applied, and folk psychology, among others.

4. On the philosophy of science, see Boyd, Gasper, and Trout, 1991; on the philosophy of psychology, see O'Donohue and Kitchener, 1996.

Chapter 8

1. *Plastic* is the more scientific designator for properties that can be molded by environmental pressures.

Chapter 9

1. Quoted in Heyer, 1993.

References

Ackerman, D. (1990). *The Natural History of the Senses*. New York: Random House.

Aczel, A. (2000). *Mystery of the Aleph: Mathematics, the Kabbalah, and the Search for Infinity*. New York: Four Walls Eight Windows.

Adolph, E. (1970). Physiological stages in the development of mammals. *Growth* 34: 113–124.

Aiello, L., and Wheeler, P. (1995). The expensive tissue hypothesis: The brain and the digestive system in human and primate evolution. *Current Anthropology* 36: 199–221.

Alexander, C. (1979). *The Timeless Way of Buildings*. New York: Oxford University Press.

Allen, J., and Berkeley, D. (1979). Image method for efficiently simulating small room acoustics. *Journal of the Acoustical Society of America* 65: 943–950.

Anderson, J. (1954). The production of ultrasounds by laboratory rats and other mammals. *Science* 119: 808–809.

Anderson, R., Snyder, L., Bradley, D., and Xing, J. (1997). Multimodal representation of space in the posterior parietal cortex and its use in planning movements. *Annual Review of Neuroscience* 20: 303–330.

Aristotle (350 B.C.). On the Soul. In J. Barnes (ed.), *The Complete Works of Aristotle*. Princeton, N.J.: Princeton University Press, 1984.

Arnold, D. (1959). The significance of "Cori Spezzati." *Music and Letters* 40: 4–14.

Ashmead, D., Wall, R., and Eaton, S., Ebinger, K., Snook-Hill, A., and Yang, X. (1998). Echolocation reconsidered: Using spectral variations in the ambient sound field to guide locomotion. *Journal of Visual Impairment and Blindness* 9: 615–632.

Aslin, R., and Hunt, R. (1999). Development, plasticity, and learning in the auditory system. In C. Nelson and M. Luciana (eds.), *Handbook of Developmental Cognitive Neuroscience*. Cambridge, Mass.: MIT Press.

Atagi, J., Ando, Y., and Ueda, Y. (2000). On the effects of time-variant sound fields on subjective preferences. *Journal of Sound and Vibration* 232(1): 71–77.

Augustine, Saint (387). De Musica Translated by R. C. Taliafero. In *The Writings of Saint Augustine*. Vol. 2, *Fathers of the Church*. New York: Cima Publishing, 1948.

Aureli, F. (1997). Post-conflict anxiety in nonhuman primates: the mediating role of emotion in conflict resolution. *Aggressive Behavior* 23: 315–328.

Bacon, F. (1626). *Sylva Sylvarum; or, A Naturall Historie and New Atlantis*. Edited by William Rawley. London: William Lee.

Bäder, K., and Blesser, B. (1977). Digitaltechnik im Studio: Ein elektronisches Nachhallgerät. *Fernseh- und Kinotechnik* 31: 443–445.

Bagenal, H. (1951). Musical taste and concert hall design. *Journal of the Royal Musical Associaton* 78: 11–27.

Bagenal, H., and Wood, A. (1931). *Planning for Good Acoustics*. New York: Dutton.

Bahn, P. (1998). *Cambridge Illustrated History of Prehistoric Art*. New York: Cambridge University Press.

Barkow, J. (1992). Beneath new culture is old psychology: Gossip and social stratification. In J. Barkow, L. Cosmides, and J. Tooby (eds.), *The Adapted Mind*. New York: Oxford University Press.

Barkow, J., Cosmedes, L., and Tooby, J. (eds.) (1992). *The Adapted Mind*. New York: Oxford University Press.

Barron, M. (1983). Auditorium acoustic modeling now. *Applied Acoustics* 16: 279–290.

Barron, M. (1984). Impulse testing techniques for auditoria. *Applied Acoustics* 17: 165–181.

Barron, M. (1988). Subjective study of British symphony concert halls. *Acustica* 66: 1–14.

Barrow, J., and Tipler, F. (1986). *The Anthropic Cosmological Principle*. Oxford: Clarendon Press.

Bassuet, A. (2004). Acoustics of early musical spaces from the 11th to the 18th centuries: Rediscovery of the acoustical excellence of medium-sized rooms. Abstract. *Journal of the Acoustical Society of America* 115(5): 2582. Available at http://www.acoustics.org/press/147th/Bassuet.htm (accessed 29 March 2006).

Bauck, J. (2001). A simple loudspeaker array and associated crosstack canceller for improved 3D audio. *Journal of the Audio Engineering Society* 49(1/2): 3–13.

Bauck, J., and Cooper, D. (1996). Generalized transaural stereo and applications. *Journal of the Audio Engineering Society* 44(9): 683–705.

Bauer, B. (1971). Directional ambiguity of quadraphonic matrices. *Journal of the Audio Engineering Society* 19(4): 315–316.

Bauman, R. (1983). *Let Your Words Be Few: Symbolism of Speaking and Silence among 17th Century Quakers*. Cambridge: Cambridge University Press.

Baynes, K., and Gazzaniga, M. (2000). Consciousness, introspection, and the split brain: The two minds/one body problem. In M. Gazzaniga (ed.), *The New Cognitive Neurosciences*. 2nd ed. Cambridge, Mass.: MIT Press.

Bech, S. (1999). Methods for subjective evaluation of spatial characteristics of sound. In *Proceedings of the Sixteenth International Conference of the Audio Engineering Society: Spatial Sound Reproduction*.

Beck, G. (1993). *Sonic Theology: Hinduism and Sacred Sound*. Columbia: University of South Carolina Press.

Begault, D., and Pittman, M. (1996). Three-dimensional audio versus head-down traffic alert and collision avoidance system displays. *International Journal of Aviation Psychology* 6(1): 79–93.

Begault, D., Wenzel, E., and Anderson, M. (2001). Direct comparison of the impact of head tracking, reverberation, and individualized head-related transfer functions on the spatial perception of a virtual speech source. *Journal of the Audio Engineering Society* 49(10): 904–916.

Behrmann, M. (2000). Spatial reference frames and hemispatial neglect. In M. Gazzaniga (ed.), *The New Cognitive Neurosciences*. 2nd ed. Cambridge, Mass.: MIT Press.

Bekoff, M. (2001). Social play behavior: Cooperation, fairness, trust, and the evolution of morality. *Journal of Consciousness Studies* 8(2): 81–90.

Beltrán, A. (ed.) (1998). *The Cave of Altamira*. New York: Abrams.

Beranek, L. (1960). *Noise Reduction*. New York: McGraw-Hill.

Beranek, L. (1962). *Music, Acoustics, and Architecture*. New York: Wiley.

Beranek, L. (1988). Boston Symphony Hall: an acoustician's tour. *Journal of the Audio Engineering Society* 36: 918–930.

Beranek, L. (1992). Concert hall acoustics. *Journal of the Acoustical Society of America* 92(1): 1–39.

Beranek, L. (1996a). *Concert and Opera Halls: How They Sound*. New York: Acoustical Society of America.

Beranek, L. (1996b). Acoustical and musical qualities. *Journal of the Acoustical Society of America* 99: 2647–2652.

Berezan, J. (2000). ReTurning. *Grace Millennium* 1(1): 14–19.

Berg, J., and Rumsey, F. (1999). Identification of perceived spatial attributes of recordings by repertory grid techniques and other methods. Paper presented to the 106th Convention of the Audio Engineering Society, Munich, May 8–11. Preprint 4924.

Berg, J., and Rumsey, F. (2001). Verification and correlation of attributes used for describing the spatial quality of reproduced sound. In *Proceedings of the Nineteenth International Conference of the Audio Engineering Society: Surround Sound: Techniques, Technology, and Perception*.

Berkhout, A. (1988). A holographic approach to acoustic control. *Journal of the Audio Engineering Society* 36(12): 977–995.

Berkhout, A., de Vries, D., and Boone, M. (1980). A new method to acquire impulse response in concert halls. *Journal of the Acoustical Society of America* 68: 179–183.

Berridge, K., and Winkielman, P. (2002). What is an unconscious emotion? The case for unconscious "liking." *Cognition and Emotion* 17: 181–211.

Besmer, F. (1983). *Horses, Musicians, and Gods: The Hausa Cult of Possession-Trance*. South Hadley, Mass.: Bergen and Garvey.

Blauert, J. (1997). *Spatial Hearing: The Psychophysics of Human Sound Localization*. Revised ed. Cambridge, Mass.: MIT Press.

Bledsoe, C. (1980). Originators of orientation and mobility training. In R. Welsh and B. Blasch (eds.), *Foundations of Orientation and Mobility*. New York: American Federation of the Blind.

Bleek, W., and Lloyd, L. (1911). Doings of the springbok. In *Specimens of Bushman Folklore*. London: Allen.

Blesser, B. (2001). An interdisciplinary synthesis of reverberation viewpoints. *Journal of the Audio Engineering Society* 49(10): 867–903.

Blesser, B. (2006). Artificial ambiance processing system. U.S. Patent 7,062,337. Issued June 13, 2006.

Blesser, B., and Bäder, K. (1980). Electric reverberation apparatus. U.S. Patent 4,181,820. Issued January 1, 1980.

Blesser, B., and Lee, F. (1971). An audio delay system using digital technology. *Journal of the Audio Engineering Society* 19(5): 393–397.

Bliven, B. (1976). Annals of architecture: A better sound. *New Yorker*, November 8, pp. 51–135.

Blumlein, A. (1933). Improvement in and relating to sound-transmission, sound-recording and sound-producing systems. British Patent 394,325.

Boehm, C. (1993). Egalitarian behavior and reverse dominance hierarchy. *Current Anthropology* 34: 227–234.

Boehm, C. (1997). Egalitarian behavior and evolution of political intelligence. In A. Whiten and R. Byrne, (eds.), *Machiavellian Intelligence II: Extensions and Evaluations*. New York: Cambridge University Press.

Bolk, L. (1929). Origin of racial characteristics in man. *American Journal of Physiological Anthropology* 13: 1–128.

Boone, M., and Braat-Eggen, P. (1994). Room acoustic parameters in a physical scale model of the new music centre in Eindhoven: Measurement method and result. *Applied Acoustics* 42: 13–28.

Boorstin, D. (1983). The rise of the equal hour. In *The Discoverers*. New York: Random House.

Borg, I., and Groenen, P. (1997). *Modern Multidimensional Scaling*. New York: Springer.

Borish, J. (1984). Extension of the image model to arbitrary polyhedra. *Journal of the Acoustical Society of America* 75: 1827–1836.

Borish, J., and Angell, J. (1983). An efficient algorithm for measuring the impulse response using pseudorandom noise. *Journal of the Audio Engineering Society* 31: 478–487.

Bork, I. (2000). A comparison of room simulation software: The second round robin on room acoustical computer simulation. *Acustica* 86(6): 943–956.

Born, G. (1995). *Rationalizing Culture: IRCAM, Boulez, and the Institutionalization of the Musical Avant-Garde*. Berkeley: University of California Press.

Boulanger, R. (ed.) (2000). *The Csound Book*. Cambridge, Mass.: MIT Press.

Boulez, P. (1971). *Boulez on Music Today*. Cambridge, Mass.: Harvard University Press.

Bourgeois, J. (1999). Synaptogenesis in the neocortex of the newborn: The ultimate frontier for individuation. In C. Nelson and M. Luciana (eds.), *Handbook of Developmental Cognitive Neuroscience*. Cambridge, Mass.: MIT Press.

Bourgeois, J., Goldman-Rakic, P., and Rakic, P. (2000). Formation, elimination, and stabilization of synapses in the primate cerebral cortex. In M. Gazzaniga (ed.), *The New Cognitive Neurosciences*. 2nd ed. Cambridge, Mass.: MIT Press.

Bowman, W. (1998). *Philosophical Perspectives on Music*. New York: Oxford University Press.

Boyd, R., Gasper, P., and Trout, J. (eds.) (1991). *The Philosophy of Science*. Cambridge, Mass.: MIT Press.

Boyle, R. (1662). Air as a medium for the transmission of sound. In *New Experiments, Physico-Mechanical, Touching the Spring of the Air*. Oxford: T. Robinson.

Braat-Eggen, P., van Luxemburg, L., and Booy, L. (1993). A new concert hall for the city of Eindhoven: Design and model tests. *Applied Acoustics* 40: 295–309.

Bradley, J., and Soulodre, G. (1995a). The influence of late-arriving energy on spatial impression. *Journal of the Acoustical Society of America* 97(4): 2263–2271.

Bradley, J., and Soulodre, G. (1995b). Objective measures of listener envelopment. *Journal of the Acoustical Society of America* 97(5): 2590–2597.

Bregman, A. (1990). *Auditory Scene Analysis: The Perceptual Organization of Sound*. Cambridge, Mass.: MIT Press.

Brennan, D., and Stevens, C. (2002). Specialist musical training and the octave illusion: Analytical listening and veridical perception by pipe organists. *Acta Psychologica* 109: 310–314.

Brown, C., and Waser, P. (1984). Hearing and communications in blue monkeys (*Cercopithecus mitis*). *Animal Behavior* 32: 65–75.

Brant, H. (1967). Space as an essential aspect of musical composition. In E. Schwartz and B. Childs (eds.), *Contemporary Composers on Contemporary Music*. New York: Holt, Rinehart and Winston.

Buchler, E. (1976). The use of echolocation by the wandering shrew (*Sorex vagrans*). *Animal Behavior* 24: 858–873.

Bulfinch, T. (1964). *Bulfinch's Mythology*. London: Spring Books.

Bullock, T., and Cowles, R. (1952). Physiology of an infrared receptor: The facial pit of the pit viper. *Science* 115: 541–543.

Buonomano, D., and Merzenich, M. (1998). Cortical plasticity: From synapses to maps. *Annual Review of Neuroscience* 21: 149–186.

Burnett, C. (1991). Sound in the Middle Ages. In C. Burnett, M. Fend, and P. Gouk (eds.), *The Second Sense: Studies in Hearing and Musical Judgment from Antiquity to the Seventeenth Century*. London: University of London.

Burnham, D., Taplin, J., Henderson-Smart, D., Earnshaw-Brown, L., and O'Grady, B. (1993). Maturation of precedence effect thresholds: Full- and pre-term infants. *Infant Behavioral Development* 16: 213–232.

Burns, E., and Viemeister, N. (1981). Played-again SAM: further observations on the pitch of amplitude-modulated noise. *Journal of the Acoustical Society of America* 70: 1655–1660.

Byrne, R., and Whiten, A. (eds.) (1988). *Machiavellian Intelligence: Social Expertise and the Evolution of Intellect in Monkeys, Apes and Humans*. Oxford: Clarendon Press.

Cage, J. (1961). *Silence: Lectures and Writings*. Middletown, Conn.: Wesleyan University Press.

Campbell, B. (1998). *Human Evolution*. 4th ed. New York: de Gruyter.

Campbell, N. (1992). *Principles of Sensory Training*. Newton, Mass.: Carroll Institute for the Blind.

Campos, G., and Howard, D. (2000). On the computational time of three-dimensional digital waveguide mesh acoustic models. In *Proceedings of the 26th Euromicro Conference* 2: 332–339.

Camras, M. (1968). Approach to recreating a sound field. *Journal of the Acoustical Society of America* 43(6): 1425–1431.

Carlson-Smith, C., and Wiener, W. (1996). The auditory skills necessary for echolocation: A new explanation. *Journal of Visual Impairment and Blindness* 90: 21–35.

Carothers, J. (1953). *The African Mind in Health and Disease: A Study in Ethnopsychiatry*. Monograph Series, no. 17. Geneva: World Health Organization.

Carpenter, E. (1955). Eskimo space concepts. *Explorations* 5: 131–145.

Carpenter, E., and McLuhan, M. (1960). Acoustic space. In E. Carpenter and M. McLuhan (eds.), *Explorations in Communications*. Boston: Beacon Press.

Carrier, D. (1984). The energetic paradox of human running and hominid evolution. *Current Anthropology* 25(4): 483–495.

Carroll, L. (1871). *Alice in Wonderland*. Reprint, New York: Norton, 1971.

Cavanaugh, W., and Wilkes, J. (1999). *Architectural Acoustics*. New York: Wiley.

Charles, R., and Ritz, D. (1978). *Brother Ray*. New York: Dial Press.

Chion, M. (1994). *Audio-Vision*. Translated by C. Gorbman. New York: Columbia University Press.

Christensen, C. (1997). *The Innovator's Dilemma*. Boston: Harvard Business School Press.

Chu, W. (1980). Statistical properties of energy spectral response as a function of direct-to-reverberant energy ratio. *Journal of the Acoustical Society of America* 68(4): 1208–1210.

Churchfield, S. (1990). *The Natural History of Shrews*. Ithaca, N.Y.: Cornell University Press.

Churchill, W. (1943). Speech delivered to the House of Commons, October 28, 1943.

Cilia, D. (2000). The Megalithic Temples of Malta. Found online at http://web.infinito.it/utenti/m/malta_mega_temples/index.html (accessed 12 April 2004).

Classen, C. (1993). *Worlds of Sense: Exploring the Senses in History and across Cultures*. New York: Routledge.

Clottes, J., and Lewis-Williams, D. (1996). *The Shamans of Prehistory: Trance and Magic in Painted Caves*. New York: Abrams.

Cocchi, A., and Farina, A. (1990). Reliability of scale-model researches: A concert hall case. *Applied Acoustics* 30: 1–13.

Cohen, M. (1947). *The Meaning of Human History*. La Salle, Ill.: Open Court.

Cohen, S., Silverman, A., Bressler, B., and Shmavonian, B. (1965). Problems in isolation studies. In P. Solomon, P. Kubzanski, P. Leiderman, J. Mendelson, R. Trumbull, and D. Wexler (eds.), *Sensory Deprivation: A Symposium Held at Harvard Medical School*. Cambridge, Mass.: Harvard University Press.

Coleridge, S. (1907). *Biographia Literaria*. Edited by J. Shawcross. Oxford: Clarendon Press.

Cooper, C. (1974). The house as symbol of the self. In J. Lang, C. Burnette, W. Moleski, D. Vachon (eds.), *Designing for Human Behavior*. Stroudsberg, Pa.: Dowden, Hutchinson and Ross.

Cooper, D., and Bauck, J. (1989). Prospects for transaural recording. *Journal of the Audio Engineering Society* 37(1/2): 3–19.

Corbin, A. (1998). *Village Bells: Sound and Meaning in the Nineteenth-Century French Countryside*. Translated by M. Thom. New York: Columbia University Press.

Cott, J. (1973). *Stockhausen: Conversations with the Composer*. New York: Simon and Schuster.

Coulson, D., and Campbell, A. (2001). *African Rock Art*. New York: Abrams.

Cowan, N. (1984). On short and long term auditory stores. *Psychological Bulletin* 96: 341–370.

Cox, T., Davies W., and Lam, Y. (1993). The sensitivity of listeners to early sound field changes in auditoria. *Acustica* 79: 27–41.

Cox, T., and Shield, B. (1999). Audience questionnaire survey of the acoustics of the Royal Festival Hall, London, England. *Acustica*, 85(4): 547–559.

Crane, D., and Small, H. (1992). American sociology since the seventies: The emerging crisis in disciplines. In T. Halliday and M. Janowitz (eds.), *Sociology and its Publics: The Forms and Fates of Disciplinary Organization*. Chicago: University of Chicago Press.

Cremer, L., and Muller, H. (1982). *Principles and Application of Room Acoustics*. London: Applied Science.

Cuff, D. (1989). Through the looking glass: Seven New York architects and their people. In R. Ellis and D. Cuff (eds.), *Architects' People*. New York: Oxford University Press.

Culhane, J. (1983). *Walt Disney's "Fantasia."* New York: Abrams.

Curio, E., Ernst, U., and Vieth, W. (1978). The adaptive significance of avian mobbing: 2. Cultural transmission of enemy recognition in blackbirds: Effectiveness and some constraints. *Zeitschrift für Tierpsychologie* 48: 184–202.

Dahl, L., and Jot, J. (2000). A reverberator based on absorbent all-pass filters. In *Proceedings of the COST G-6 Conference on Digital Audio Effects (DAFX-00)*. Verona.

Dalenbäck, B. (1995). The importance of diffuse reflection in computerized room acoustic prediction and auralization. *Proceedings of the Institute of Acoustics* 17: 27–33.

Dalenbäck, B. (1996). Room acoustic predictions based on a unified treatment of diffuse and specular reflection. *Journal of the Acoustical Society of America* 102: 899–909.

Dattorro, J. (1997a). Effect design: 1. Reverberator and other filters. *Journal of the Audio Engineering Society* 45(9): 660–684.

Dattorro, J. (1997b). Effect design: 2. Delay-line modulation and chorus. *Journal of the Audio Engineering Society* 45(10): 764–788.

Dauvois, M., and Boutillon, X. (1990). Études acoustiques au Réseau Clastres: Salle des peintures et lithophones naturels. *Bulletin de la Société Préhistorique Ariège-Pyrénées* 45: 175–186.

Davenport, W., and Root, W. (1958). *Introduction to the Theory of Random Signals and Noise*. New York: McGraw-Hill.

Davidson, J. (2005). Place matters: Where we hear music influences how we hear it. *New York Newsday*, January 2.

Davies, W., and Cox, T. (2000). Reducing seat dip attenuation. *Journal of the Acoustical Society of America* 108(5): 2211–2218.

Dawkins, R., and Krebs, J. (1978). Animal signals: Information or manipulation? In J. Krebs and N. Davies (eds.), *Behavioural Ecology: An Evolutionary Approach*. Oxford: Blackwell Scientific.

de Bruin, S. (1958). The "electronic poem" performed in the Philips Pavilion at the 1958 Brussels World Fair: The electronic control system. *Philips Technical Review* 20: 45–49.

Declercq, N., Degrieck, J., Briers, R., and Leroy, O. (2004). A theoretical study of special acoustic effects caused by the staircase of the El Castillo pyramid at the Maya ruins of Chichen-Itza in Mexico. *Journal of the Acoustical Society of America* 116(6): 3328–3335.

De Forest, L. (1921). Means for amplifying currents. U.S. Patent 1,375,447. Filed June 24, 1913.

de Gelder, B., Böcker, K., Tuomainen, J., Hensen, M., and Vroomen, J. (1999). The combined perception of emotion from voice and face: Early interactions revealed by human brain response. *Neuroscience Letters* 260: 133–136.

DeLong, T. (1980). *The Mighty Music Box: The Golden Age of Musical Radio*. Los Angeles: Amber Crest Books.

Demany, L., and Semal, C. (2002). Learning to perceive pitch differences. *Journal of the Acoustical Society of America* 111(3): 1377–1388.

Deprès, O., Candas, V., and Dufour, A. (2005). Auditory compensation in myopic humans: Involvement of binaural, monaural, or echo cues. *Brain Research* 1041: 56–65.

de Sousa, R. (1997). *The Rationality of Emotions*. Cambridge, Mass.: MIT Press.

Devereux, P. (2001). *Stone Age Soundtracks*. London: Vega.

Devereux, P., and Jahn, R. (1996). Preliminary investigations and cognitive considerations of the acoustical resonances of selected archaeological sites. *Antiquity* 70: 665–666.

Devinsky, O. (2000). Right cerebral hemisphere dominance for a sense of corporeal and emotional self. *Epilepsy and Behavior* 1: 60–73.

De Volder, A., Toyama, H., Kimura, Y., Kiyosawa, M., Nakano, H., Vanlierde, A., Wanet-Defalque, M., Mishina, M., Oda, K., Ishiwata, K., and Senda, M. (2001). Auditory triggered mental imagery of shapes involves visual association areas in early blind humans. *NeuroImage* 14: 129–139.

de Vries, D., and Hulsebos, E. (2001). Spatial fluctuations in measures of spaciousness. *Journal of the Acoustical Society of America* 110(2): 947–954.

de Vries, D., Hulsebos, E., and Baan, J. (2000). Spatial fluctuation of spaciousness measures in auditoria. Paper presented to the 108th Convention of the Audio Engineering Society, Paris, February 19–22. Preprint 5147.

de Waal, F. (1989). *Peacemaking among Primates*. Cambridge, Mass.: Harvard University Press.

de Waal, F. (1998). *Chimpanzee Politics*. Revised ed. Baltimore: Johns Hopkins Press.

de Waal, F. (2000). Primates: A natural heritage of conflict resolution. *Science* 289: 586–590.

Diderot, D. (1749). Letter on the blind. In *Early Philosophical Works*. Reprint, New York: Lenox Hill, 1972.

Diebel, C., Proksch, R., Green, C., Neilson, P., and Walker, M. (2000). Magnetite defines a vertebrate magnetoreceptor. *Nature* 406: 299–302.

Dittrich, W. (1992). Is the monkeys' world scientifically impenetrable? *Behavioral and Brain Sciences* 15: 152.

Dittus, W. (1988). An analysis of Toque macaque cohesion calls from an ecological perspective. In D. Todt, P. Goedeking, and D. Symmes (eds.), *Primate Vocal Communication*. New York: Springer.

Downs, R., and Stea, D. (1973). *Image and Environment: Cognitive Mapping*. New York: Harper and Row.

Dreyfus, H. (1993). *What Computers Still Can't Do*. Cambridge, Mass.: MIT Press.

Dunbar, R. (1993). Co-evolution of neocortex size, group size and language in humans. *Behavioral and Brain Science* 16(4): 681–735.

Dunbar, R. (1996). *Grooming, Gossip, and the Evolution of Language*. Cambridge, Mass.: Harvard University Press.

Dunbar, R. (1998). The social brain hypothesis. *Evolutionary Anthropology* 6: 178–190.

Ebeling, K. (1982). Experimental investigation of statistical properties of diffuse sound fields in reverberation rooms. *Acustica* 51: 145–153.

Eddins, D. (1993). Amplitude modulation detection of narrow-band noise: Effects of absolute bandwidth and frequency region. *Journal of the Acoustical Society of America* 93: 470–479.

Edwards, G. (1926). Public address system. U.S. Patent 1,587,107. Filed May 22, 1922.

Elbert, T., Candia, V., Altenmuller, E., Rau, H., Sterr, A., Rockstroh, B., Pantev, C., and Taub, E. (1998). Alteration of digit representations in somatosensory cortex in focal hand dystonia. *NeuroReport* 9: 3571–3575.

Elbert, T., Pantov, C, Wienbruch, C., Rockstroh, B., and Taub, E. (1995). Increased cortical representation of the fingers of the left hand in string players. *Science* 270: 305–307.

Elkin, R. (1955). *The Old Concert Rooms of London*. London: Arnold.

Elkins, J. (2001). *Pictures and Tears*. New York: Routledge.

Eliot, T. (1975). Tradition and the individual talent. In F. Kermode (ed.), *Selected Prose of T. S. Eliot*. London: Farber and Farber.

Emmerson, S. (1998). Acoustic/electroacoustic: The relationship with instruments. *Journal of New Music Research* 27: 146–164.

Emmerson, S., and Smalley, D. (2001). Electro-acoustic music. In S. Sadie (ed.), *The New Grove Dictionary of Music and Musicians*. 2nd ed. New York: Macmillan.

Ernst, D. (1977). *The Evolution of Electronic Music*. New York: Schirmer Books.

Ericsson, K., and Simon, H. (1993). *Protocol Analysis: Verbal Reports as Data.* 2nd ed. Cambridge, Mass.: MIT Press.

Etienne, A., Vauclair, J., Emmanuelli, E., Lançon, M., and Stryjenski, J. (1982). Depth perception by means of ambient sounds in small mammals. *Experientia* 38: 553–555.

Fay, R., and Popper, A. (2000). Evolution of hearing in vertebrates: The inner ear and processing. *Hearing Research* 149: 1–10.

Feld, S. (1996). Waterfalls of song: An acoustemology of place resounding in Bosavi, Papua New Guinea. In S. Feld and K. Basso (eds.), *Senses of Place.* New Mexico: School of American Research Press.

Fenton, M. (1985). *Communication in the Chiroptera.* Bloomingdale: Indiana University Press.

Fidi, W. (1970). Delay device particularly for the production of artificial reverberation. U.S. Patent 3,517,344. Issued June 23, 1970. (First filed in Austria December 6, 1966.)

Finkenthal, M. (2001). *Interdisciplinarity: Towards the Definition of a Metadiscipline.* New York: Lang.

Florentine, M., Hunter, W., Robinson, M., Ballou, M., and Buus, S. (1998). On the behavioral characteristics of loud-music listening. *Ear and Hearing* 19: 420–428.

Forgas, J. (2000). Introduction: The role of affect in social cognition. In J. Forgas (ed.), *Feeling and Thinking: The Role of Affect in Social Cognition.* Cambridge: Cambridge University Press.

Forrest, T., Miller, G., and Zagar, J. (1993). Sound propagation in shallow water: Implications for acoustics communications by aquatic animals. *Bioacoustics: The International Journal of Animal Sound and Its Recording* 4: 259–270.

Forsman, K., and Malmquist, M. (1988). Evidence for echolocation in the common shrew, *Sorex araneus. Journal of Zoology* (London) 216: 655–662.

Forsyth, M. (1985). *Buildings for Music: The Architect, the Musician, and the Listener from the Seventeenth Century to the Present Day.* Cambridge, Mass.: MIT Press.

Fouts, R., Fouts, D., and van Cantford, T. (1989). The infant Loulis learns signs from cross-fostered chimpanzees. In R. Gardner, B. Gardner, and T. van Cantford (eds.), *Teaching Sign Language to Chimpanzees.* Albany: State University of New York Press.

Fox, R. (1997). *Conjectures and Confrontations: Science, Evolution, and Social Concern.* New Brunswick, N.J.: Transaction.

Franklin, S. (1995). Science as culture, culture as science. *Annual Review of Anthropology* 24: 163–184.

Freed, D. (1990). Auditory correlates of perceived mallet hardness for set of recorded percussive events. *Journal of the Acoustical Society of America* 87: 311–322.

Frodeman, R., Mitcham, C. and Sacks, A. (2001). Questioning interdisciplinarity. *Science, Technology, and Society Newsletter* 126–127: 1–5.

Fuller, S. (2000). *The Governance of Science: Ideology and the Future of an Open Society*. Philadelphia: Open University Press.

Gade, A. (1989). Investigations of musician's room acoustic conditions in concert halls. *Acustica* 69: 249–262.

Gans, C. (1992). An overview of evolutionary biology of hearing. In D. Webster, R. Fay, and A. Popper (eds.), *The Evolutionary Biology of Hearing*. New York: Springer.

Gardner, H. (1999). *Intelligence Reframed: Multiple Intelligences for the 21st Century*. New York: Basic Books.

Gardner, W. (1992). A real time multichannel room simulator. Paper presented to the 124th Meeting of the Acoustical Society of America, New Orleans, 31 October–4 November.

Gardner, W. (1995). Efficient Convolution without input-output delay. *Journal of the Audio Engineering Society* 43: 127–136.

Gardner, W. (1998). Reverberation algorithms. In M. Kahrs and K. Brandenburg (eds.), *Applications of Digital Signal Processing to Audio and Acoustics*. Boston: Kluwer Academic.

Gardner, W., and Griesinger, D. (1994). Reverberation level matching experiments. In *Proceedings of the Wallace Sabine Centennial Symposium*. Cambridge, Massachusetts, Acoustical Society of America.

Garity, W., and Hawkins, J. (1941). Fantasound. *Journal of the Society of Motion Picture Engineering* 37: 127–146. Available online at http://www.widescreenmuseum.com/sound/Fantasound3.htm (accessed 28 March 2006).

Gazzaniga, M. (2000). Cerebral specialization and interhemispheric communications: Does the corpus callosum enable the human condition? *Brain* 123: 1293–1326.

Gerzon, M. (1973). Periphony: With-height sound reproduction. *Journal of the Audio Engineering Society* 21(1): 2–10.

Gerzon, M. (1976). Unitary (energy preserving) multichannel network with feedback. *Electronic Letters* 12: 278–279.

Gibson, J. (1966). *The Senses Considered as Perceptual Systems*. Boston: Houghton Mifflin.

Gill, S., and Sullivan, I. (1992). *Dictionary of Native American Mythology*. New York: Oxford University Press.

Giménez, A., and Martin, A. (1988). Analysis and assessment of concert halls. *Applied Acoustics* 25: 235–241.

Gish, S., and Morton, E. (1981). Structural adaptations to local habitat acoustics in Carolina wren songs. *Zeitschrift für Tierpsychologie* 56: 74–84.

Gladwell, M. (2005). *Blink: The Power of Thinking without Thinking*. New York: Little, Brown.

Goleman, D. (1997). *Emotional Intelligence*. New York: Bantam Books.

Gordon, M., and Rosenblum, L. (2000). Perception of acoustic occlusion using body-scale judgments. Paper presented at the 139th Meeting of the Acoustical Society of America, Atlanta, May 30–June 3.

Gould, J., and Morgan, C. (1941). Hearing in the rat at high frequencies. *Science* 94: 168.

Gould, S. (1977). *Ontogeny and Phylogeny*. Cambridge, Mass.: Harvard University Press.

Gould, S., and Lewontin, R. (1979). The spandrels of San Marco and the Panglossian program: a critique of the adapationists programme. *Proceedings of the Royal Society of London* 250: 281–288.

Green, G. (1838). On the reflexion and refraction of sound. *Transactions of the Cambridge Philosophical Society* 6, part 3: 403–412.

Griesinger, D. (1989). Practical processors and programs for digital reverberation. In *Proceedings of the Seventh International Conference of the Audio Engineering Society*. Toronto.

Griesinger, D. (1991). Improving room acoustics through time-variant synthetic reverberation. Paper presented to the 90th Convention of the Audio Engineering Society, Paris, February 19–22. Preprint 3014.

Griesinger, D. (1995). How loud is my reverberation? Paper presented to the 98th Convention of the Audio Engineering Society, Paris, February 25–28. Preprint 3943.

Griesinger, D. (1999). Objective measures of spaciousness and envelopment. In *Proceedings of the Nineteenth International Conference of the Audio Engineering Society: Surround Sound: Techniques, Technology, and Perception*.

Griesinger, D. (2000). Recent experiences with electronic acoustical enhancement in concert halls, opera houses, and outdoor venues. Abstract. *Journal of the Acoustical Society of America* 107(5): 2783.

Griesinger, D. (2004). Panel discussion on concert hall acoustics. Boston Audio Society, June 20 at the Massachusetts Institute of Technology.

Griffin, D. (1944). Echolocation by blind men and bats. *Science* 100: 589–590.

Griffin, D. (1958). *Listening in the Dark*. New Haven, Conn.: Yale University Press.

Griffin, D. (1971). The importance of atmospheric attenuation for the echolocation of bats (Chiroptera). *Animal Behavior* 19: 55–61.

Griffin, D. (1986). *Listening in the Dark*. Ithaca, N.Y.: Cornell University Press.

Grose, J., Hall, J., and Buss, E. (1999). Modulation gap detection: effects of modulation rate, carrier separation, and mode of presentation. *Journal of the Acoustical Society of America* 106: 946–952.

Grout, D. (1960). *A History of Western Music*. London: Norton.

Guttman, N., and Julesz, B. (1963). Lower limits of auditory periodicity analysis. *Journal of the Acoustical Society of America* 35: 610.

Haas, H. (1951). Über den Einfluss eines Einfachechos auf die Hösamkeit von Sprache. *Acustica* 1: 49–58.

Hall, E. (1966). *The Hidden Dimension*. New York: Doubleday.

Halmrast, T. (2000). Orchestral timbre: Comb-filter coloration from reflections. *Journal of Sound and Vibration* 232(1): 53–69.

Hammond, L. (1939). Electrical musical instrument. U.S. Patent 2,230,836. Filed July 15, 1939.

Hanson, O. (1932). Planning the NBC studios for Radio City. *Institute of Radio Engineering* 20: 130.

Hardy, H., Telfair, D., and Pielemeier, W. (1942). The velocity of sound in air. *Journal of the Acoustical Society of America* 13: 226–233.

Harley, M. (1994). Space and spatialization in contemporary music: History and analysis, ideas and implementation. Ph.D. diss., McGill University.

Harley, M. (1998). Spatiality of sound and stream segregation in the 20th century instrumental music. *Organised Sound* 3(2): 147–166.

Harris, G. (1963). Periodicity perception using gated noise. *Journal of the Acoustical Society of America* 35: 1229–1233.

Harrison, J. (1998). Sound, space, sculpture: Some thoughts on the "what," "how," and "why" of sound diffusion. *Organised Sound* 3(2): 117–127.

Harrison, J. (2000). Diffusion: Theories and practices, with particular reference to the BEAST system. Paper presented to the Society for Electro-Acoustical Music in Denton, Texas, March 9.

Hauser, M. (1997a). *The Evolution of Communication*. Cambridge, Mass.: MIT Press.

Hauser, M. (1997b). Minding the behavior of deception. In A. Whiten and R. Byrne (eds.), *Machiavellian Intelligence II: Extensions and Evaluations*. New York: Cambridge University Press.

Hausfeld, S., Power, R., Gorta, A., and Harris, P. (1982). Echo perception of shape and texture by sighted subjects. *Perceptual Motor Skills* 55(2): 623–632.

Hawkes, R., and Douglas, H. (1971). Subjective acoustic experience in concert auditoria. *Acustica* 24(5): 235–250.

Hayes, S. (1935). *Facial Vision, or, The Sense of Obstacles*. Watertown, Mass.: Perkins.

Heinz, R. (1993). Binaural room simulation based on an image source model with additional statistical methods to include the diffuse sound scattering of walls to predict the reverberation rail. *Applied Acoustics* 38: 145–159.

Heisenberg, W. (1958). *Physics and Philosophy*. New York: Harper Row.

Helmholtz, H. von (1863). *Die Lehre von dem Tonempfindungen als physiologische Grundlage für die Theorie der Musik*. Braunschweig: Vieweg.

Hertz, B. (1981). One hundred years of stereo: The beginning. *Journal of the Audio Engineering Society* 29(5): 368–372.

Heyer, P. (ed.) (1993). *Architects on Architecture*. New York: Van Nostrand Reinhold.

Hill, W., Hill, A., and Hill, A. (1963). *Antonio Stradivari: His Life and Work*. New York: Dover.

Hiyama, K., Komiyama, S., and Hamasaki, K. (2002). The minimum number of loudspeakers and its arrangement for reproducing the spatial impression of diffuse sound field. Paper presented to the 113th Convention of the Audio Engineering Society, Los Angeles, October 5–8. Preprint 5674.

Hochachka, P., Gunga, H., and Kirsch, K. (1998). Our ancestral physiological phenotype: An adaptation for hypoxia tolerance and endurance performance? *Proceedings of the National Academy of Sciences, U.S.A.* 95: 1915–1920.

Hofman, P., and van Opstal, A. (1998). Spectro-temporal factors in two-dimensional sound localization. *Journal of the Acoustical Society of America* 103: 2634–2648.

Holbrook, R. (2002). VRAS provides acoustical solution—in the electronic domain—at Church of the Living Word. Found at http://www.prosoundweb.com/install/applications/church/vras/livingword.shtml (accessed 28 March 2006).

Holliday, K. (2000). Some American firms and their contributions to the development of the reproducing piano. In M. Saffle (ed.), *Perspectives on American Music, 1900–1950*. New York: Garland.

Holman, T. (2001). Mixing the sound: 2. Perspectives on where do sounds go. *Surround Professional* 4(3): 35–37. Available online at http://www.surroundpro.com/articles/publish/article_176.shtml (accessed 28 March 2006).

Holt, A., Cheek, D., Mellits, E., and Hill, D. (1975). Brain size and the relation of the primate to the nonprimate. In D. Cheek (ed.), *Fetal and Postnatal Cellular Growth: Hormones and Nutrition*. New York: Wiley.

Horace, P. (1977). *Joseph Guarnerius: His Work and His Master*. Boston: Longwood Press.

Horrall, T. (1970). Auditorium acoustics simulator: Form and uses. Paper presented to the 39th Convention of the Audio Engineering Society, New York, 13–15 October. Preprint 761.

Hosler, D. (1994). *The Sounds and Color of Power: The Sacred Metallurgical Technology of Ancient West Mexico*. Cambridge, Mass.: MIT Press.

House, N. (2001). Subjective evaluation of 2-channel vs. surround formats in vehicles. Paper presented to the 110th Convention of the Audio Engineering Society, Amsterdam, May 12–15. Preprint 5284.

Howes, D. (ed.) (1991). *The Varieties of Sensory Experience*. Toronto: University of Toronto Press.

Humphrey, N. (1974). Vision in a monkey with striate cortex: A case study. *Perception* 3: 241–255.

Humphrey, N. (2000). The privatization of sensation. In C. Heyes and L. Huber (eds.), *The Evolution of Cognition*. Cambridge, Mass.: MIT Press.

Humphreys, W. (1940). *Physics of the Air*. New York: McGraw-Hill.

Hunningher, B. (1956). *Acoustics and Acting in the Theatre of Dionysus Eleuthereus*. Amsterdam: North-Holland.

Hunt, F. (1978). *Origins in Acoustics: The Science of Sound from Antiquity to the Age of Newton*. New Haven, Conn.: Yale University Press.

Hunt, K. (1994). The evolution of human bipedality: ecology and functional morphology. *Journal of Human Evolution* 26: 183–202.

Iwamiya, S., and Zhan, M. (1997). A comparison between Japanese and Chinese adjectives which express auditory impressions. *Journal of the Acoustical Society of Japan* (English edition) 18: 319–323.

Jablonski, N., and Chaplin, G. (1993). Origin of habitual terrestrial bipedalism in the ancestor of the Hominidae. *Journal of Human Evolution* 24: 259–280.

Jablonski, N., and Chaplin, G. (2000). The evolution of human skin coloration. *Journal of Human Evolution* 39: 57–106.

Jack, A., and Shallice, T. (2001). Introspective physicalism as an approach to the science of consciousness. *Cognition* 79: 161–196.

Jacob, F. (1982). *The Possible and the Actual*. Seattle: University of Washington Press.

Jacobson, M. (1969). Development of specific neuronal connections. *Science* 163: 543–547.

Jacobson, R. (1998). Cognitive mapping without sight: Four preliminary studies of spatial learning. *Journal of Environmental Psychology* 18: 289–305.

James, J. (1993). *The Music of the Spheres*. New York: Springer.

James, W. (1890). *Principles of Psychology*. New York: Holt.

Jankowicz, D. (2001). Why does subjectivity make us nervous? *Journal of Intellectual Capital* 2(1): 61–73.

Järvilehto, T. (2000). Feeling as knowing: 1. Emotions a reorganization of the organism-environment system. *Consciousness and Emotion* 1(2): 53–65.

Järvilehto, T. (2001). Feeling as knowing: 2. Emotion, consciousness, and brain activity. *Consciousness and Emotion* 2(1): 75–102.

Jaworski, A. (1993). *The Power of Silence: Social and Pragmatic Perspectives*. Newbury Park, Calif.: Sage.

Jaworski, A., (ed.) (1997). *Silence: Interdisciplinary Perspectives*. New York: de Gruyter.

Jay, A. (1994). *Management and Machiavelli*. Oxford: Pfeiffer.

Jellicoe, G. (1989). *International Herald Tribune*, (Paris), November 6.

Jobes, G. (1961). *Dictionary of Mythology, Folklore and Symbols*. New York: Scarecrow Press.

Joffe, T. (1997). Social pressures have selected for an extended juvenile period in primates. *Journal of Human Evolution* 32: 593–605.

Johannes, R. (1981). *Words of the Lagoon: Fishing and Marine Lore in the Palau District of Micronesia*. Berkeley: University of California Press.

Johnson, R., Cole, R., and Ahern, F. (1981). Genetic interpretation of racial/ethnic differences in lactose absorption and tolerance: A review. *Human Biology* 53: 1–14.

Jones, S. (1992). Natural selection in humans. In S. Jones, R. Martin, and D. Pilbeam (eds.), *The Cambridge Encyclopedia of Human Evolution*. New York: Cambridge University Press.

Jot, J. (1992). Étude et réalisation d'un spatialisateur de sons par modèles physiques et perceptifs. Ph.D. diss., École Nationale Supérieure des Télécommunications, Paris.

Jot, J., and Chaigne, A. (1991). Digital delay networks for designing artificial reverberators. Paper presented to the 90th Convention of the Audio Engineering Society, Paris, February 19–22. Preprint 3030.

Jot, J., and Chaigne, A. (1996). Method and system for artificial spatialization of digital audio signals. U.S. Patent 5,491,754. Filed February 19, 1993.

Jowett, B. (1964). *The Dialogues of Plato*. 4th ed. New York: Oxford University Press.

Joyce, W. (1975). Sabine's reverberation time and ergodic auditoriums. *Journal of the Acoustical Society of America* 58(2): 643–655.

Julesz, B., and Hirsh, I. (1972). Visual and auditory perception: An essay of comparison. In E. David and P. Denes (eds.), *Human Communications: A Unified Approach*. New York: McGraw-Hill.

Kaernbach, C., (1993). Temporal and spectral basis of the features perceived in repeated noise. *Journal of the Acoustical Society of America* 94: 91–97.

Kaernbach, C., Schröger, E., and Gunter, T. (1998). Human event-related brain potentials to auditory periodic noise stimuli. *Neuroscience Letters* 242: 17–20.

Kahn, D., and Krubitzer, L. (2002). Massive cross-modal cortical plasticity and the emergence of a new cortical area in developmentally blind mammals. *Proceedings of the National Academy of Sciences, U.S.A.* 99(17): 11429–11434.

Kanada, Y. (2002). How to slice the PI very, very thin. As reported by Associated Press in *New York Times*, December 7, p. 8.

Karamustafaoglu, A., Horbach, U., Pellegrini, R., Mackensen, P., and Theile, G. (1999). Design and application of a data-based auralization system for surround sound. Paper presented at the 106th Convention of the Audio Engineering Society, Munich, May 8–11. Preprint 4976.

Katz, D., and Kahn, R. (1978). *The Social Psychology of Organizations*. 2nd ed. New York: Wiley.

Katz, P. (1999). *The Scalpel's Edge: The Culture of Surgeons*. Boston: Allyn and Bacon.

Kellogg, W. (1962). Sonar system of the blind. *Science* 137: 399–405.

Kelly, G. (1955). *The Psychology of Personal Constructs*. New York: Norton.

Kelvin, W. (1894). *Popular Lectures and Addresses, 1891–1894*. 3 vols. London: Macmillan.

Kendall, G., Martens, W., Freed, D., Ludwig, D., and Karstens, R. (1986). Image model reverberation from recirculating delays. Paper presented to the 81st Convention of the Audio Engineering Society, Los Angeles, November 12–16. Preprint 2408.

King, A., and Schnupp, J. (2000). Sensory convergence in neural function and development. In M. Gazzaniga, (ed.), *The New Cognitive Neurosciences*. 2nd ed. Cambridge, Mass.: MIT Press.

Kish, [D.] (1995). Evaluation of an echo-mobility program for young blind people. Master's thesis, California State University, San Bernadino.

Kish, D. (2001). Echolocation: How humans can "see" without sight. Revised thesis from World Access for the Blind. Available at http://www.worldaccessfortheblind.org/thesis.txt (accessed 3 October 2003).

Kish, D., and Bleier, H. (2000). Echolocation: What it is, and how it can be taught and learned. Paper presented to the California Association of Orientation and Mobility Specialists, Riverside, CA 4 November 2000. Available at http://www.tiresias.org/research/publications/kish.htm (accessed 28 March 2006).

Klapholz, J. (1991). *Fantasia*: Innovations in sound. *Journal of the Audio Engineering Society* 39(1/2): 66–70.

Klein, J. (1990). *Interdisciplinarity, History, Theory, and Practice*. Detroit: Wayne State University Press.

Klein, J. (1996). *Crossing Boundaries: Knowledge, Disciplinarities, and Interdisciplinarities*. Charlottesville: University Press of Virginia.

Kleiner, M., Dalenbäck, B., and Svensson, P. (1993). Auralization: An overview. *Journal of the Audio Engineering Society* 41(11): 861–875.

Kleiner, M., Orlowski, R., and Kirszenstein, J. (1993). A comparison between results from a physical scale model and a computer image source model for architectural acoustics. *Applied Acoustics* 38: 245–265.

Kline, P. (2000). *A Psychometric Primer*. New York: Free Association Books.

Kline, S. (1995). *Conceptual Foundations for Multidisciplinary Thinking*. Stanford, Calif.: Stanford University Press.

Klipsch, P. (1958). Stereophonic sound with two tracks, three channels by means of a phantom circuit (2PH3). *Journal of the Audio Engineering Society* 6(2): 118–123.

KnicKrehm, G. (2004). Panel discussion on concert hall acoustics. Boston Audio Society, June 20 at Massachusetts Institute of Technology.

Knudsen, V. (1932). Resonances in small rooms. *Journal of the Acoustical Society of America* 4: 20–37.

Knudsen, V. (1946). The propagation of sound in the atmosphere: Attenuation and fluctuations. *Journal of the Acoustical Society of America* 18(1): 90–96.

Koelsch, S., Schroger, E., and Tervaniemi, M. (1999). Superior pre-attentive auditory processing in musicians. *NeuroReport* 10: 1309–1313.

Korenaga, Y., and Ando, Y. (1993). A sound-field simulation system and its application to seat-selection system. *Journal of the Audio Engineering Society* 41(11): 920–930.

Kornblith, H. (ed.) (1997). *Naturalizing Epistemology*. 2nd ed. Cambridge, Mass.: MIT Press.

Kövecses, Z. (1990). *Emotional Concepts*. New York: Springer.

Krautheimer, R. (1965). *Early Christian and Byzantine Architecture*. Baltimore: Penguin Books.

Kroodsma, D., and Miller, E. (eds.) (1966). *Ecology and Evolution of Acoustic Communication in Birds*. Ithaca, N.Y.: Cornell University Press.

Krylov, N. (1979). *Works on the Foundation of Statistical Physics*. Princeton, N.J.: Princeton University Press.

Kudo, H., and Dunbar, R. (2001). Neocortex size and social network size in primates. *Animal Behavior* 62: 711–722.

Kuhl, W. (1954). Über Versuche zur Ermittlung der günstigen Nachzeit grosser Musikstudios. *Acustica* 4: 618–634.

Kuhl, W. (1958). The acoustics and technical properties of the reverberation plate. *EBU (European Broadcasting Union) Review*. 49A: 8–14. Also in original German as Über die Akustischen und Technischen Eigenshaften der Nachhallplatte. *Rundfunktechnische Mitteillungen* 2: 111–116.

Kuhl, W. (1970). Reverberation device. U.S. Patent 3,719,905. (First filed in Germany April 29, 1970.)

Kuhn, T. (1996). *The Structure of Scientific Revolutions*. Chicago: University of Chicago Press.

Kukla, A. (2001). *Methods of Theoretical Psychology*. Cambridge, Mass.: MIT Press.

Kurtz, M. (1992). *Stockhausen: A Biography*. London: Farber and Farber.

Kuttner, F. (1990). *Archaeology of Music in Ancient China: Two Thousand Years of Acoustical Experimentation, 1400 B.C.–750 A.D.* New York: Paragon House.

Kuttruff, H. (1954). Über die Frequenzabhängigkeit des Schalldrucks in Räumen. *Acustica* 4(2): 614–617.

Kuttruff, H. (1973). *Room Acoustics*. Reprint, New York: Spon Press, 2000.

Kuttruff, H. (1998). Sound fields in small rooms. In *Proceedings of the Fifteenth International Conference of the Audio Engineering Society*.

Kuttruff, H. (1991). On the audibility of phase distortions in rooms and its significance for sound reproduction and digital simulation in room acoustics. *Acustica* 74: 3–7.

Laakso, T., Välimäki, V., Karjalainen, M., and Laine, U. (1996). Splitting the unit delay: Tools for fractional delay filters. *IEEE Signal Processing Magazine* 13: 30–60.

Lafleur, L., Matese, J., and Spross, R. (1987) Acoustic refraction by a spark discharge in air. *Journal of the Acoustical Society of America* 81: 606–610.

Lakatos, I. (1970). Falsification and the methodology of scientific research programmes. In I. Laskatos and A. Musgrave (eds.), *Criticism and the Growth of Knowledge*. New York: Cambridge University Press.

Laming, D. (1997). *The Measurement of Sensation*. New York: Oxford University Press.

Latour, B., and Woolgar, S. (1979). *Laboratory Life: The Social Construction of Scientific Facts*. Beverly Hills, Calif.: Sage.

Lawson, G., Scarre, C., Cross, I., and Hills, C. (1998). Mounds, megaliths, music and mind: Some thoughts on the acoustical properties *and purposes* of archaeological spaces. *Archaeological Review from Cambridge* 15(1): 111–134.

Lears, R. (2005). Acoustic ecology. *In These Times* 29(3), January 3. Available at http://www.inthesetimes.com/site/main/print/1748 (accessed 28 March 2006).

Le Boeuf, B. and Peterson, R. (1969). Dialects in elephant seals. *Science* 166: 1654–1656.

LeDoux, J. (2000). Emotional circuits in the brain. *Annual Review of Neuroscience* 23: 155–184.

Lee, J. (1989). Note: Why concert halls are not ergodic. *Journal of the Acoustical Society of America* 85(6): 2680–2681.

Lee, R. (1979). *The !Kung San: Men, Women, and Work in a Foraging Society*. Cambridge: Cambridge University Press.

Lehmann, P., and Wilkens, H. (1980). Zusammenhang subjektiver Beurteilung von Konzertsalen mit raumakustishen Kriterien. *Acustica* 45: 256–268.

Letowski, T. (1985). Development of technical listening skills: Timbre solfeggio. *Journal of the Audio Engineering Society* 33: 240–244.

Levinson, S. (1999). Frames of reference and Molyneux's question: Crosslinguistic evidence. In P. Bloom, M. Peterson, L. Nadel, and M. Grant (eds.), *Language of Space*. Cambridge, Mass.: MIT Press.

Lewcock, R., and Rijn, R. (2001). Room acoustics, classical times. In S. Sadie (ed.), *The New Grove Dictionary of Music and Musicians*. 2nd ed. New York: Macmillan.

Lewis-Williams, J., and Dowson, T. (1990). Through the veil: San rock paintings and the rock face. *South African Archaeological Bulletin* 45: 5–16.

Lexicon Corp (2000). *960L Digital Effects System: Owner's Manual.* Bedford, Mass.

Lindskold, S., Albert, K., Baer, R., and Moore, W. (1976). Territorial boundaries of interacting groups and passive audiences. *Sociometry* 39(1): 71–76.

Lippman, E. (1964). *Musical Thought in Ancient Greece.* New York: Columbia University Press.

Litovsky, R. (1997). Developmental changes in the precedence effect: Estimates of minimum audible angle. *Journal of the Acoustical Society of America* 102(3): 1739–1745.

Litovsky, R., Colburn, H., Yost, W., and Guzman, S. (1999). The precedence effect. *Journal of the Acoustical Society of America* 106(4): 1633–1654.

Logan, B., and Schroeder, M. (1963). Artificial reverberation network. U.S. Patent 3,110,771. Issued November 12, 1963.

Loomis, W. (1967). Skin-pigmentation regulation of vitamin-D biosynthesis in man. *Science* 157: 501–506.

Lopez, M., and Gonzales, C. (1987). Experimental study of the acoustics in the Church of the Monastery of "Santo Domingo de Silos." *Acustica* 62(3): 241–248.

López, J., Orduña, F., and González, A. (2000). Modeling and measurement of cross-talk cancellation zones for small displacements for the listener in transaural sound reproduction with different loudspeaker arrangements. Paper presented to the 109th Convention of the Audio Engineering Society, Los Angeles, September 22–25. Preprint 5267.

LoVetri, J., Mardare, D., and Soulodre, G. (1996). Modeling of the seat dip effect using the finite-difference time-domain method. *Journal of the Acoustical Society of America* 100(4): 2204–2212.

Lubman, D. (1998). Archaeological acoustic study of chirped echo from the Mayan pyramid at Chichén Itzá. *Journal of the Acoustical Society of America* 104: 1763. Extended version available at www.ocasa.org/MayanPyramid.htm (accessed 6 June 2006) and www.ocasa.org/MayanPyramid2 .htm (accessed 28 March 2006).

Lubman, D. (2004). Acoustics at the shrine of St. Werburgh. Paper presented to the 148th Meeting of the Acoustical Society of America, San Diego, November 15–19.

Lubman, D., and Kiser, B. (2001). The history of Western civilization told through the acoustics of its worship spaces. Paper presented to the seventeenth International Conference on Acoustics, Rome, September 2–7.

Lucy, J. (1997). Linguistic relativity. *Annual Review of Anthropology* 26: 291–312.

Mace, W. (1977). James J. Gibson's strategy for perceiving: Ask not what's inside your head, but what your head is inside of. In R. Shaw and J. Bransford, (eds.), *Perceiving, Acting, and Knowing.* Hillsdale, N.J.: Erlbaum.

Machiavelli, N. (1958). *The Prince*. Translated by W. Marriot. London: Dent.

MacKinnon, D. (1963). Creativity and the images of self. In R. White (ed.), *The Study of Lives: Essays of Personality in Honor of Henry A. Murray*. New York: Atherton Press.

Malham, D. (1999) Homogeneous and nonhomogeneous surround sound systems. In *Proceedings of the Audio Engineering Society United Kingdom Conference: The Second Century*.

Mann, D., Lu, Z., Hastings, M., and Popper, A. (1998). Detection of ultrasonic tones and simulated dolphin echolocation clicks by a teleost fish, the American shad (*Alosa sapidissima*). *Journal of the Acoustical Society of America* 104: 562–568.

Marks, L. (1982). Bright sneezes and dark coughs, loud sunlight and soft moonlight. *Journal of Experimental Psychology: Human Perception and Performance* 8: 177–193.

Mártinez-Sala, R., Sancho, J., Sanchez, J., Gomez, V., Llinares, J., and Meseguer, F. (1995). Sound attenuation by sculpture. *Nature* 378(6554): 241.

Mason, W. (1976). The architecture of St. Mark's Cathedral and the Venetian polychoral style: A clarification. In J. Pruett (ed.), *Studies in Musicology, Essays in the History, Style, and Bibliography of Music in Memory of Glen Haydon*. Westport, Conn.: Greenwood Press.

McCutchen, B. (1927). Public address system. U.S. Patent 1,642,040. Filed December 12, 1925.

Mehta, V. (1957). A donkey in a world of horses. *Atlantic* 200(1): 24–30.

Mersenne, M. (1644). On the velocity of sound in air. In *Cogitata Physico-Mathematica*. Paris: Bertier.

Merton, R. K. (1976). The ambivalence of scientists. In R. Cohen, P. Feyerabend, and M. Wartofsky (eds.), *Essays in Memory of Imre Lakatos*. Boston: Reidel.

Merzenich, M., and Schreiner, C. (1990). Mammalian auditory cortex: Some comparative observations. In D. Webster, R. Fay, and A. Popper (eds.), *The Evolutionary Biology of Hearing*. New York: Springer.

Messer-Davidow, E., Shumway, D., and Sylvan, D. (eds.) (1993). *Knowledges: Historical and Critical Studies in Disciplinarity*. Charlottesville: University of Virginia Press.

Meyer, D. (2000). Toscanini and the NBC Symphony Orchestra: High, middle, and low culture, 1937–1954. In M. Saffle (ed.), *Perspectives on American Music, 1900–1950*. New York: Garland.

Milgram, S. (1976). Psychological maps of Paris. In H. Proshansky, W. Ittleson, and L. Revlin (eds.), *Environmental Psychology*. New York: Holt, Rinehart, and Winston.

Miller, G., and Taylor, W. (1948). The perception of repeated bursts of noise. *Journal of the Acoustical Society of America* 20: 171–182.

Miller, W. (1993). Silence in the contemporary soundscape. Master's thesis, Simon Fraser University.

Minnaar, P., Olesen, S., Christensen, F., and Møller, H. (2001). Localization with binaural recordings from artificial and human heads. *Journal of the Audio Engineering Society* 49(5): 323–336.

Mithen, S. (2000). Mind, brain and material culture: An archaeological perspective. In P. Carruthers and A. Chamberlain (eds.), *Evolution and the Human Mind*. New York: Cambridge University Press.

Møller, H., Sørensen, M., Jensen, C., and Hammershøi, D. (1996). Binaural technique: Do we need individual recordings? *Journal of the Audio Engineering Society* 44(6): 451–469.

Montgomery, H., Marshall, R., Hemingway, H., Myerson, S., Clarkson, P., Dollery, C., Hayward, M., Holliman, D., Jubb, M., and World, M. (1998). Human gene for physical performance. *Nature* 393: 221–221.

Moore, C. (1981). Time-modulated delay system and improved simulator using same. U.S. Patent 4,268,717. Filed April 19, 1979.

Moorer, J. (1979). About this reverberation business. *Computer Music Journal* 3(2): 13–28.

Morimoto, M., and Iida, K. (1993). A new physical measure for psychological evaluation of sound fields: Front/back energy ratio as a measure of envelopment. Abstract. *Journal of the Acoustical Society of America* 93: 2282.

Morrice, M., Burton, H., and Green, K. (1994). Microgeographic variations and songs in the underwater repertoire of Weddell seal (*Leptonychotes weddellii*) from Vestford Hills, Antarctica. *Polar Biology* 14: 441–446.

Moulton, D. (1993). *Golden Ears*. Groton, Mass.: Moulton Laboratories.

Moulton, D., and Moulton, M. (1998). Codec "transparency," listener "severity," program "intolerance": Suggestive relationships between Rasch measures and some background variables. Paper presented to the 105th Convention of the Audio Engineering Society, San Francisco, September 26–29. Preprint 4843.

Muckel, P., Ensel, L., and Schulte-Fortkamp, B. (1999). Exploration of associated imaginations on sound perception (AISP): A method for helping people to describe and to evaluate sound perceptions. *Journal of the Acoustical Society of America* 105(2, part 2): 1279.

Münte, T., Kohlmetz, C., Nager, W., and Altenmüller, E. (2001). Neuroperception: Superior auditory spatial tuning in conductors. *Nature* 409: 580.

Murphy, G., and Andrew, J. (1993). The conceptual basis of antonymy and synonymy in adjectives. *Journal of Memory and Language* 32: 301–319.

Nader, L. (1996). Introduction. In Nader (ed.), *Naked Science: An Anthropological Inquiry into the Boundaries, Power and Knowledge*. New York: Routledge.

Naguib, M. (1995). Auditory distance assessment of singing conspecifics in Carolina wrens: The role of reverberation and frequency-dependent attenuation. *Animal Behavior* 50: 1297–1307.

Naguib, M., and Wiley, R. (2001). Estimating the distance of a source of sound: mechanism and adaptation for long-distance communication. *Animal Behavior* 62: 825–837.

Nakayama, I. (1984). Preferred time delay of a single reflection for performers. *Acustica* 54: 217–221.

Nelson, M., and Maciver, M. (1999). Prey capture in the weakly electric fish *Apteronotus albifrons*: Sensory acquisition strategies and electrosensory consequences. *Journal of Experimental Biology* 202: 1195–1203.

Newton, I. (1686). *Philosophiae Naturalis Principia Mathematica*. London: Joseph Streater.

Nicol, R., and Emerit, M. (1998). Reproducing 3D sound for videoconferencing: A comparison between holophony and ambisonics. In *Proceedings of the First COST-G6 Workshop on Digital Audio Effects (DAFX98)*.

Nind, T. (2001). Multimedia in cars: The application of Logic 7surround processing to provide surround sound in cars from 2 channel and encoded 5.1 sources. Paper presented to the 110th Convention of the Audio Engineering Society, Amsterdam, May 12–15. Preprint 5286.

Noack, B. (2002). Acoustic upgrades in Prague. *Live Sound International Magazine* March/April.

Noson, D., Sato, S., Sakai, H., and Ando, Y. (2000). Singer responses to sound fields with a simulated reflection. *Journal of Sound and Vibration* 232(1): 39–51.

Novacek, M. (1985). Evidence for echolocation in the oldest known bats. *Nature* 315(9): 140–141.

O'Brien, R., and Iglesias, P. (2001). On the poles and zeros of linear, time-varying systems. *IEEE Transactions on Circuit and Systems-1: Fundamental Theory and Applications* 48(5): 565–577.

O'Donohue, W., and Kitchener, R. (eds.) (1996). *The Philosophy of Psychology*. London: Sage Publications.

Offenhauser, W. (1958). Binaural and stereophonic sound. *Journal of the Audio Engineering Society* 6(2): 67–69.

Okano, T., Beranek, L., and Hidaka, T. (1998). Relations among interaural cross-correlation coefficient (IACC$_E$), lateral fraction (LF$_E$), and apparent source width (ASW) in concert halls. *Journal of the Acoustical Society of America* 104(1): 255–265.

O'Keefe, J., and Nadel, L. (1978). *The Hippocampus as Cognitive Map*. Oxford: Oxford University Press.

Olive, S. (2003). Differences in performance and preference of trained versus untrained listeners in loudspeaker tests: A case study. *Journal of the Audio Engineering Society* 51(9): 806–825.

Ong, W. (1982). *Orality and Literacy: The Technologizing of the Word*. London: Methuen.

Ono, K., Komiyama, S., and Nakabayashi, K. (1996). A method of reproducing concert hall sounds by "loudspeaker walls." *Journal of the Audio Engineering Society* 46(11): 968–995.

Ottman, R. (1991). *Basic Ear Training Skills*. Englewood Cliffs, N.J.: Prentice Hall.

Owsinski, B. (2001). Surround comes of age. *Surround Professional* 4(6): 25–30.

Owsinski, B. (2002). Yamaha SREV1 sampling reverberator. *Surround Professional* 5(2): 40–41.

Pack, A., Herman, L., Hoffman-Kuhnt, M., and Brandstetter, B. (2002). The object behind the echo: Dolphins (*Tursiops truncates*) perceive object shape globally through echolocation. *Behavioural Processes* 58: 1–26.

Paget, V. (1932) [Vernon Lee, pseud.]. *Music and Its Lovers: An Empirical Study of Emotions and Imaginative Responses to Music*. London: Allen and Unwin.

Pallasmaa, J. (1996). *The Eyes of the Skin: Architecture and the Senses*. London: Academy Group.

Pantev, C., Oostenveid, R., Engellen, A., Ross, B., Roberts, L., and Hoke, M. (1998). Increased auditory cortical representation in musicians. *Nature* 392: 811–814.

Pantev, C., Engelien, A., Candia, V., and Elbert, T. (2001a). Representational cortex in musicians: Plastic alterations in response to musical practice. In R. Zattore and I. Peres (eds.), *Biological Foundations of Music*. New York: Annals of the New York Academy of Science. Vol. 930.

Pantev, C., Roberts, L., Shulz, M., Engelien, A., and Ross, B. (2001b). Timbre-specific enhancement of auditory cortical representations in musicians. *NeuroReport* 12(1): 169–174.

Pascual-Leone, A. (2000). Comments on *Scientific American* Frontiers: Changing your mind. Available at www.pbs.og/saf/1101/hotline/index.htm (accessed 28 March 2006).

Passingham, R. (1982). *The Human Primate*. Oxford: Freeman Pres.

Paul, R., and Walker, T. (1979). Arboreal singing in a burrowing cricket, *Anurogryllus arboreus*. *Journal of Comparative Physiology* 132: 217–223.

Pearson, D. (1991). Auditory attention switching: A developmental study. *Journal of Experimental Child Psychology* 51: 320–334.

Peirano, M. (1998). When anthropology is at home: The different contexts of a single discipline. *Annual Review of Anthropology* 27: 105–128.

Peretz, I., Gagnon, L., and Bouchard, B. (1998). Music emotion: Perceptual determinates, immediacy, and isolation after brain damage. *Cognition* 68: 111–141.

Perrott, D. (1993). Auditory and visual location: Two modalities, one world. In *Proceedings of the Twelfth International Conference of the Audio Engineering Society: Perceiving Reproduced Sound*.

Peterson, A. (1985). The philosophy of Niels Bohr. In A. French and P. Kennedy (eds.), *Niels Bohr, A Century Volume*. Cambridge, Mass.: Harvard University Press.

Picasso, P. (1923). Picasso speaks. In A. Barr Jr. (ed.), *Picasso: Fifty Years of His Art*. New York: Museum of Modern Art, 1946.

Picker, J. (2003). *Victorian Soundscapes*. Oxford: Oxford University Press.

Platner, S. (1929). *A Topographical Dictionary of Ancient Rome*. London: Oxford University Press.

Plomin, R., and Thompson, L. (1993). Genetics and high cognitive ability. In G. Bock and K. Ackrill (eds.), *The Origins and Development of High Ability*. Chichester, England: Wiley.

Polack, J. (1989). Digital evaluation of the acoustics of small models: The MIDAS package. *Journal of the Acoustical Society of America* 85: 185–193.

Polack, J. (1992). Modifying chambers to play billiards: The foundation of reverberation theory. *Acustica* 76: 257–272.

Polack, J. (1993). Playing billiards in the concert hall: The mathematical foundations of geometrical room acoustics. *Applied Acoustics* 38: 235–244.

Polack, J., Alrutz, H., and Schroeder, M. (1984). Modulation transfer function of music signals and its application to reverberation measurement. *Acustica* 54: 257–265.

Poletti, M. (1996). An assisted reverberation system for controlling apparent room absorption and volume. Paper presented to the 101st Convention of the Audio Engineering Society, Los Angeles, November 8–11. Preprint 4365.

Poletti, M. (1999). The performance of multichannel sound systems. Ph.D. diss., University of Auckland.

Pollack, G. (1992). Adaptations of basic structures, and mechanisms in the cochlea and central auditory pathways of the mustache bat. In D. Webster, R. Fay, and A. Popper (eds.), *The Evolutionary Biology of Hearing*. New York: Springer.

Popper, K. (1959). *The Logic of Scientific Discovery*. New York: Basic Books.

Popper, K. (1962). *Conjectures and Refutations*. London: Routledge.

Postman, N. (1993). *Technopoly: The Surrender of Culture to Technology*. New York: Random House.

Precoda, K., and Meng, T. (1997). Listener differences in audio compression evaluations. *Journal of the Audio Engineering Society* 45(9): 708–715.

Pressnitzer, D., and McAdams, S. (1999). Two phase effects in roughness perception. *Journal of the Acoustical Society of America* 105: 2773–2782.

Pujol, J., Roset-Llobet, J., Rosinés-Cubells, D., Deus, J., Narberhaus, B., Valls-Solé, J., Capdevila, A., and Pascual-Leone, A. (2000). Brain cortical activation during guitar-induced hand dystonia studied by functional MRI. *NeuroImage* 12(3): 257–267.

Quantz, J. (1966). *On Playing the Flute*. Translated by E. Reilly. London: Faber.

Quiatt, D., and Reynolds, V. (1993). *Primate Behaviour: Information, Social Knowledge, and the Evolution of Culture*. Cambridge: Cambridge University Press.

Raes, A., and Sacerdote, G. (1953). Measurements of the acoustical properties of two Roman basilicas. *Journal of the Acoustical Society of America* 25(5): 954–961.

Rasmussen, S. (1959). *Experiencing Architecture*. Cambridge, Mass.: MIT Press.

Rayleigh, J. (1877). *Theory of Sound*. London: Macmillan.

Recanzone, G., Schreiner, C., and Merzenich, M. (1993). Plasticity in the frequency representation of primary auditory cortex following discrimination training in adult owl monkeys. *Journal of Neuroscience* 13: 87–103.

Reilly, A., and McGrath, D. (1995). Convolution processing for realistic reverberation. Paper presented to the 98th Convention of the Audio Engineering Society, Paris, February 25–28. Preprint 3977.

Rettinger, M. (1957). Reverberation chambers for broadcasting and recording studios. *Journal of the Audio Engineering Society* 5(1): 18–22.

Rettinger, M. (1961). Acoustic considerations in the design of recording studios. *Journal of the Audio Engineering Society* 9(3): 178–183.

Rice, C. (1967). Human echo perception. *Science* 155: 656–664.

Rice, C. (1969). Perceptual enhancement in the early blind. *Psychological Record* 19: 1–14.

Rice, C. (1970). Early blindness, early experience, and perceptual enhancement. *American Foundation for the Blind Research Bulletin* 22: 1–20.

Richards, D., and Wiley, R. (1980). Reverberation and amplitude fluctuations in the propagation of sound in a forest: Implications for animal communications. *American Naturalist* 115: 381–399.

Richardson, R., and Shield, B. (1999). Acoustic measurement of Shakespeare's Globe Theatre. Paper presented at the Forum Acusticum '99, Berlin, March 14–19.

Richerson, P., and Boyd, R. (2005). *Not by Genes Alone: How Culture Transformed Human Evolution*. Chicago: University of Chicago Press.

Riley, D., and Rosenzweig, M. (1957). Echolocation in rats. *Journal of Comparative and Physiological Psychology* 50: 323–328.

Rimell, A. (1999). Immersive Spatial Audio for telepresence applications: System design and implementations. *In Proceedings of the Sixteenth International Conference of the Audio Engineering Society: Spatial Sound Reproduction*.

Riso, D. (1996). *Personality Types*. Boston: Houghton Mifflin.

Ritchie, I. (1991). Fusion of the faculties: A study of the language of the senses in Hausaland. In D. Howes (ed.), *The Varieties of Sensory Experience*. Toronto: University of Toronto Press.

Röder, B., Teder-Sälejärv, W., Sterr, A., Rösler, F., Hillyard, S., and Neville, H. (1999). Improved auditory spatial tuning in blind humans. *Nature* 400(8): 162–166.

Rother, L. (2005). Adventures in opera: A 'Ring' in the rain forest. *New York Times*, May 9, Section A, p. 1.

Rouget, G. (1985). *Music and Trance: A Theory of the Relations between Music and Possession*. Translated by B. Biebuyck. Chicago: University of Chicago Press.

Rumsey, F., Zieliński, S., Kassier, R., and Bech, S. (2005). Relationships between experienced listener ratings of multichannel audio quality and naïve listener preferences. *Journal of the Acoustical Society of America* 117(6): 3832–3840.

Rüsseler, J., Altenmüller, E., Nager, W., Kohlmetz, C., and Münte, T. (2001). Event-related brain potentials to sound omissions differ in musicians and non-musicians. *Neuroscience Letters* 308: 33–36.

Russon, A. (1997). Exploiting the expertise of others. In A. Whiten and R. Byrne (eds.), *Machiavellian Intelligence II: Extensions and Evaluations*. New York: Cambridge University Press.

Ryan, M., and Brenowitz, E. (1985). The role of body size, phylogeny, and ambient noise in the evolution of bird song. *American Naturalist* 126: 87–100.

Rybczynski, W. (1996). Sounds as good as it looks. *Atlantic* 277(6): 108–112.

Sabine, W. (1906). The accuracy of musical taste in regard to architectural acoustics. *Proceedings of the American Academy of Arts and Sciences* 42: 49–84.

Sabine, W. (1922). *Collected Papers on Acoustics*. Reprint, New York: Dover, 1964.

Sacks, O. (2003). The mind's eye: What the blind see. *New Yorker*, July 28, pp. 48–59.

Sadato, N., Okada, T., Honda, M., and Yonekura, Y. (2002). Critical period for cross-modal plasticity in blind humans: A functional MRI study. *NeuroImage* 16: 389–400.

Sánchez-Pérez, J., Caballero, D., Mártinez-Sala, R., Rubio, C., Sánchez-Dehesa, J., Meseguer, F., Llinares, J., and Gálvez, F. (1998). Sound attenuation by a two-dimensional array of rigid cylinders. *Physical Review Letters* 80(24): 5325–5328.

Sato, S., Ando, Y., and Ota, S. (2000). Subjective preferences of cellists for the delay time of a single reflection in a performance. *Journal of Sound and Vibration* 232: 27–37.

Savioja, L., Huopaiemi, J., Lokki, T., and Väänänen, R. (1999). Creating interactive virtual acoustic environment. *Journal of the Audio Engineering Society* 47(9): 675–705.

Schafer, R. (1977). *The Soundscape: Our Sonic Environment and the Tuning of the World*. New York: Knopf.

Schafer, R. (1978). *The Vancouver Soundscape*. Vancouver: ARC.

Schlaug, G., Jäncke, L., Huang, Y., and Steinmetz, H. (1995). In vivo evidence of structural brain asymmetry in musicians. *Science* 267: 699–701.

Schmidt, K., and Cohn, J. (2001). Human facial expressions as adaptations: Evolutionary questions in the facial expression research. *Yearbook of Physical Anthropology* 44: 3–24.

Schooler, J., Ohlsson, S., and Brooks, K. (1993). Thoughts beyond words: When language overshadows insight. *Journal of Experimental Psychology: General* 122(2): 166–183.

Schroeder, M. (1954a). Die statistichen Parameter der Frequenzkurven von grossen Räumen. *Acustica* 4(2): 594–600. Translated as Statistical parameters of the frequency response curves of large rooms. *Journal of the Audio Engineering Society* 35(1987): 299–306.

Schroeder, M. (1954b). Eigenfrequenzstatistik und Anregungsstatistik in Räumen. *Acustica* 4(1): 456–468. Translated as Normal frequency and excitation statistics in rooms: Model experiments with electric waves. *Journal of the Audio Engineering Society* 35(1987): 307–316.

Schroeder, M. (1962a). Frequency-correlation functions of frequency responses in rooms. *Journal of the Acoustical Society of America* 12: 1819–1823.

Schroeder, M. (1962b). Natural sounding artificial reverberation. *Journal of Audio Engineering Society* 10: 219–223.

Schroeder, M. (1970). Digital simulation of sound transmission in reverberant spaces. *Journal of the Acoustical Society of America* 47(2): 424–431.

Schroeder, M. (1975). Diffuse sound reflections by maximum-length sequences. *Journal of the Acoustical Society of America* 57: 149–150.

Schroeder, M. (1979). Integrated-impulse method of measuring sound decay without using impulses. *Journal of the Acoustical Society of America* 66: 497–500.

Schroeder, M. (1996). The "Schroeder frequency" revisited. *Journal of the Acoustical Society of America* 99: 3240–3241.

Schroeder, M., and Atal, B. (1963). Computer simulation of sound transmission in rooms. *IEEE International Convention Record* 7: 150–155.

Schroeder, M., Gottlob, D., and Siebrasse, K. (1975). Comparative study of European concert halls: Correlation of subjective preferences with geometric and acoustic parameters. *Journal of the Acoustical Society of America* 56: 1195–1201.

Schroeder, M., and Kuttruff, H. (1962). On frequency response curves in rooms: Comparison of experimental, theoretical and Monte Carlo results for average frequency spacing between maximum. *Journal of the Acoustical Society of America* 34: 76–80.

Schroeder, M., and Logan, B. (1961). Colorless artificial reverberation. *Journal of the Audio Engineering Society* 9: 192–197.

Schubert, G., and Tzekakis, E. (1999). The ancient Greek theater and its acoustic quality for contemporary performances. Abstract. *Journal of the Acoustical Society of America* 105(2): 1043.

Schultz, T., and Watters, B. (1964). Propagation of sound across audience seating. *Journal of the Acoustical Society of America* 36(5): 885–895.

Seeger, A. (1981). *Nature and Society in Central Brazil: The Suya Indians of Mato Grosso*. Cambridge, Mass.: Harvard University Press.

Sekiguchi, K., and Kimura, S. (1991). Calculation of sound field in a room by finite sound ray integration method. *Applied Acoustics* 32: 121–148.

Seyfarth, R., and Cheney, D. (1986). Vocal development in velvet monkeys. *Animal Behavior* 34: 1640–1658.

Shankland, R. (1973). Acoustics of Greek theater. *Physics Today* 26(10): 30–35.

Shankland, R., and Shankland, H. (1971). Acoustics of St. Peter's and Patriarchal Basilicas in Rome. *Journal of the Acoustical Society of America* 50: 389–395.

Sheridan, T., and van Lengen, K. (2003). Hearing architecture: Exploring and designing the aural environment. *Journal of Architectural Education* 57(2): 37–44.

Shinn-Cunningham, B., Durlach, N., and Held, R. (1998). Adapting to supernormal auditory localization cues: 2. Constraints on adaptation of mean responses. *Journal of the Acoustical Society of America* 103: 3667–3676.

Shively, R. (1998). Subjective evaluation of reproduced sound in automobile spaces. In *Proceedings of the Fifteenth International Conference of the Audio Engineering Society: Audio, Acoustics and Small Spaces.*

Shively, R. (2000). Automotive audio design: A tutorial. Paper presented to the 109th Convention of the Audio Engineering Society, Los Angeles, September 22–25. Preprint 5276.

Shively, R., and House, W. (1996). Perceived boundary effects in an automobile vehicle interior. Paper presented to the 100th Convention of the Audio Engineering Society, Copenhagen, May 11–14. Preprint 4245.

Shlien, S. (2000). Auditory models for gifted listeners. *Journal of the Audio Engineering Society* 48(11): 1032–1044.

Shlien, S., and Soulodre, G. (1996). Measuring the characteristics of "expert" listeners. Paper presented to the 101st Convention of the Audio Engineering Society, Los Angeles, November 8–11. Reprint 4339.

Shuter-Dyson, R., and Gabriel, C. (1981). *The Psychology of Musical Ability.* New York: Methuen.

Sibley, C., and Ahlquist, J. (1984). The phylogeny of the hominid primates, as indicated by DNA-DNA hybridization. *Journal of Molecular Evolution* 20: 2–15.

Sikorav, J. (1986). Implementation of reverberators on digital signal processors. Paper presented to the 80th Convention of the Audio Engineering Society, Montreux, March 4–7. Preprint 2326.

Slawson, D. (1987). *Secret Teachings in the Art of Japanese Gardens: Design Principles and Aesthetic Values.* Tokyo: Kodansha International.

Slonimsky, N. (1965). *Lexicon of Musical Invective: Critical Assaults on Composers since Beethoven's Time.* New York: Coleman-Ross.

Smith, B. (1999). *The Acoustic World of Early Modern England.* Chicago: University of Chicago Press.

Smith, W. (1875). *A Dictionary of Greek and Roman Antiquities.* London: John Murray.

Snell, K., and Hu, H. (1999). The effect of temporal placement on gap-delectability. *Journal of the Acoustical Society of America* 106: 3571–3577.

Sommer, R. (1983). *Social Design: Creating Buildings with People in Mind*. Englewood Cliffs, N.J.: Prentice-Hall.

Sotiropoulou, A., Hawkes, R., and Fleming, D. (1995). Concert hall acoustic evaluations by ordinary concert-goers: I, multi-dimensional descriptions of evaluations. *Acustica* 81: 1–9.

Sousa, D. (2000). *How the Brain Learns*. Thousand Oaks, Calif.: Corwin Press.

Souter, A., et al. (eds.) (1968). *Oxford Latin Dictionary*. Oxford: Clarendon Press.

Spandöck, F. (1934). Akustische Modellversuche. *Annalen der Physik* 20: 345–360.

Stanley, T., and Danko, W. (1998). *The Millionaire Next Door*. Atlanta: Longstreet Press.

Stautner, J., and Pluckette, M. (1982). Designing multi-channel reverberators. *Computer Music Journal* 6(1): 52–65.

Stebbins, R. (2000). *The Making of Symphony Hall, Boston*. Boston: Boston Symphony Orchestra.

Stebbins, W. (1980). The evolution of hearing in mammals. In A. Popper and R. Fay (eds.), *Comparative Studies in Hearing of Vertebrates*. New York: Springer.

Stein, B., and Meredith, M. (1993). *The Merging of the Senses*. Cambridge, Mass.: MIT Press.

Stein, B., Wallace, M., and Stanford, T. (2000). Merging sensory signals in the brain: The development of multisensory integration in the superior colliculus. In M. Gazzaniga (ed.), *The New Cognitive Neurosciences*. 2nd ed. Cambridge, Mass.: MIT Press.

Stills, D. (1986). A note on the origins of "interdisciplinary." *Items: Social Science Research Council*, 48, no. 1 (March): 17–18.

Stockhausen, K. (1959). Musik im Raum. *Die Reihe* 5: 67–72. Translated into English as Music in space. *Die Reihe* 1961, 5: 67–82.

Stoerig, P. (1996). Varieties of vision: From blind responses to conscious recognition. *Trends in Neuroscience* 19: 401–496.

Stone, H., and Sidel, J. (1993). *Sensory Evaluation Practices*. New York: Academic Press.

Supa, M., Cotzin, M., and Dallenbach, K. (1944). "Facial vision": The perception of obstacles by the blind. *American Journal of Psychology* 57: 133–183.

Svensson, U., and Nielsen, J. (1999). Errors in MLS measurements caused by time variance in acoustic systems. *Journal of the Audio Engineering Society* 47: 907–927.

Szuchewycz, B. (1997). Silence in ritual communications. In A. Jaworski (ed.), *Silence: Interdisciplinary Perspectives*. New York: de Gruyter.

Tak, W. (1958). The "electronic poem" performed in the Philips Pavilion at the 1958 Brussels World Fair: The sound effects. *Philips Technical Review* 20: 43–44.

Takahashi, D. (1997). Seat dip effect: The phenomenon and the mechanism. *Journal of the Acoustical Society of America* 102(3): 1326–1334.

Takeuchi, A., and Hulse, S. (1993). Absolute pitch. *Psychological Bulletin* 113: 345–361.

Tannen, D., and Saville-Troike, M. (1985). *Perspectives on Silence*. Norwood, N.J.: Ablex.

Tanner, W., and Swets, J. (1954). A decision-making theory of visual detection. *Psychological Review* 61: 401–409.

Terhardt, E. (1974). On the perception of periodic sound fluctuations (roughness). *Acustica* 30: 201–212.

Tervaniemi, M., Rytkonen, M., Schroger, E., Ilmoniemi, R., and Näätänen, R. (2001). Superior formation of cortical memory traces for melodic patterns in musicians. *Learning and Memory* 8: 295–300.

Thayer, R. (1989). *The Biopyschology of Mood and Arousal*. New York: Oxford University Press.

Thompson, E. (2002). *The Soundscape of Modernity*. Cambridge, Mass.: MIT Press.

Thurlow, W., Mangels, J., and Runge, P. (1967). Head movements during sound localization. *Journal of the Acoustical Society of America* 42(2): 489–493.

Tomasi, T. (1979). Echolocation by the short-tailed shrew *Blarina brevicauda*. *Journal of Mammalogy* 60(4): 751–759.

Torres, R., Svensson, U., and Kleiner, M. (2001). Computation of edge diffraction for more accurate room acoustic auralization. *Journal of the Acoustical Society of America* 109: 600–610.

Traweek, S. (1988). *Beamtimes and Lifetimes: The World of High-Energy Physicists*. Cambridge, Mass.: Harvard University Press.

Treib, M. (1996). *Space Calculated in Seconds*. Princeton, N.J.: Princeton University Press.

Tributsch, H. (1982). *When Snakes Awake*. Cambridge, Mass.: MIT Press.

Tristram, C. (2003). Supercomputing resurrected. *Technology Review* 106(1): 52–60.

Trochimczyk, M. (2001). From circles to nets: On the significance of spatial imagery in new music. *Computer Music Journal* 25(4): 39–56.

Truax, B. (1998). Composition and diffusion: Space in sound in space. *Organised Sound* 3(2): 141–146.

Truax, B. (2001). *Acoustic Communication*. London: Ablex.

Truax, B. (2002). Genres and techniques of soundscape composition as developed at Simon Fraser University. *Organised Sound* 7(1): 5–14.

Tzekakis, E. (1975). Reverberation time of the Rotunda at Thessaloniki. *Journal of the Acoustical Society of America* 57(6): 1207–1209.

Ueda, Y., and Ando, Y. (1997). Effects of air-conditioning on sound propagation in a large space. *Journal of the Acoustical Society of America* 102(5): 2771–2775.

van Duyne, S., and Smith, J. (1995). Multidimensional digital waveguide signal synthesis system and method. U.S. Patent 5,471,007. Issued November 28, 1995.

Van Kirk, W. (2002). The accidental (acoustic) tourist. Abstract. *Journal of the Acoustical Society of America* 112(5): 2284.

Varèse, E. (1998). Spatial music. In E. Schwartz and B. Childs (eds.), *Contemporary Composers on Contemporary Music*. New York: Da Capo Press.

Vassilantonopoulos, S., and Mourjopoulos, J. (2001). Virtual acoustic reconstruction of ritual and public spaces of ancient Greece. *Acustica* 87(5): 604–609.

Vermeulen, R. (1958). Stereo-reverberation. *Journal of the Audio Engineering Society* 6(2): 124–130.

Viemeister, N. (1979). Temporal modulation transfer function based upon modulation thresholds. *Journal of the Acoustical Society of America* 66: 1364–1380.

Vitruvius, M. (30 B.C.). *De Architectura*. Translated by F. Granger. Cambridge, Mass.: Harvard University Press, 1931.

Vogel, S. (1993). Sensation of tone, perception of sound and empiricism. In D. Cahan (ed.), *Hermann von Helmholtz and the Foundation of Nineteenth-Century Science*. Berkeley: University of California Press.

Volkmann, J. (1942). Polycylindrical diffusers in room acoustic design. *Journal of the Acoustical Society of America* 13: 234–243.

von Békésy, G. (1933). Über die Shallfeld verzerrungen in der Nahe von absorbierenden Flächen und ihre Bedeutung für die Raumakustik. *Zeitschrift für Technische Physik* 34: 577–582.

von Békésy, G. (1960). *Experiments in Hearing*. New York: McGraw-Hill.

von Simson, O. (1989). *The Gothic Cathedral*. Princeton, N.J.: Princeton University Press.

Vorländer, M. (1989). Simulation of the transient and steady-state sound propagation in rooms using a new combined ray-tracing/image-source algorithm. *Journal of the Acoustical Society of America* 86: 182–178.

Wakefield, J. (2000). FreeVerb software download at http://www.sonicspot/freeverb/freeverb.html (accessed 28 March 2006).

Walcot, P. (1976). *Greek Drama in Its Theatrical and Social Context*. Cardiff: University of Wales Press.

Waller, S. (1993). Sound and rock art. *Nature* 363(6429): 501.

Waller, S. (1999). Rock art acoustics in the past, present, and future. In *1999 International Rock Art Conference Proceedings*, 2: 11–20.

Waller, S. (2001). Sounds of the spirit world. *American Indian Rock Art* 28: 53–56.

Waller, S. (2002). Psychoacoustic influences of the echoing environments of prehistoric art. Paper presented at the Acoustic Society of America's First Pan-American/Iberian Meeting in Cancún, Mexico, November 19. Abstract published in *Journal of the Acoustical Society of America* 112(5): 2284.

Walton, J. (1984). *The Greek Sense of Theater: Tragedy Reviewed.* New York: Methuen.

Wang, C. (1993). *Sense and Nonsense of Statistical Inference: Controversy, Misuse, and Subtlety.* New York: Dekker.

Ward, D., and Abhayapala, T. (2001). Reproduction of a plane-wave sound field using an array of loudspeakers. *IEEE Transactions on Speech and Audio Processing* 9(6): 697–707.

Warren, R., and Bashford, J. (1981). Perception of acoustic iterance: Pitch and infrapitch. *Perception and Psychophysics* 29: 395–402.

Warren, R., Bashford, J., Cooley, J., and Brubaker, B. (2001). Detection of acoustic repetition for very long stochastic patterns. *Perception and Psychophysics* 63: 175–182.

Waser, P., and Brown, C. (1984). Is there a "sound window" for primate communications? *Behavior Ecology and Sociobiology* 15: 73–76.

Watson, A., and Keating, D. (1999). Architecture and sound: an acoustic analysis of megalithic monuments in prehistoric Britain. *Antiquity* 73: 325–336.

Watson, F. (1926). Optimum conditions for music in rooms. *Science* 64: 209–210.

Watson, J. (1913). Psychology as the behaviorist views it. *Psychological Review* 20: 158–177.

Watson, R. (1973). Psychology: A prescriptive science. In M. Henle, J. Jaynes, and J. Sullivan (eds.), *Historical Conceptions of Psychology.* New York: Springer.

Webster, A. (1991). *Science, Technology, and Society: New Directions.* New Brunswick, N.J.: Rutgers University Press.

Webster, D., Fay, R., and Popper, A. (1992). *The Evolutionary Biology of Hearing.* New York: Springer.

Weeks, R., Horwitz, B., Aziz-Sultan, A., Tian, B., Wessinger, C., Cohen, L., Hallett, M., and Rauschecker, J. (2000). A positron emission tomographic study of auditory localization in the congenitally blind. *Journal of Neuroscience* 20(7): 2664–2672.

Wegel, R. (1932). Wave transmission device. U.S. Patent 1,852,795. Filed October 24, 1928.

Welch, W., and Burt, L. (1994). *From Tinfoil to Stereo: The Acoustic Years of the Recording Industry, 1877–1929.* Miami: University Press of Florida.

Wenzel, E. (1997). Analysis of the role of update rate and system latency in interactive virtual acoustic environments. Paper presented to the 103rd Convention of the Audio Engineering Society, New York, September 26–29. Preprint 4633.

Wenzel, E., Arruda, M., Kistler, D., and Wightman, F. (1993). Localization using nonindividualized head-related transfer functions. *Journal of the Acoustical Society of America* 94(1): 111–123.

West, M., and King, A. (1996). Eco-gen-actics: A systems approach to the ontogeny of avian communications. In D. Kroodsma and E. Miller (eds.), *Ecology and Evolution of Acoustic Communication in Birds*. Ithaca, N.Y.: Cornell University Press.

Westerkamp, H. (1988). Listening and soundmaking: A study of music-as-environment. Master's thesis, Simon Fraser University.

Whiten, A. (1997). The Machiavellian mindreader. In A. Whiten and R. Byrne (eds.), *Machiavellian Intelligence II: Extensions and Evaluations*. New York: Cambridge University Press.

Whiten, A., and Byrne, R. (eds.) (1997). *Machiavellian Intelligence II: Extensions and Evaluations*. New York: Cambridge University Press.

Whorf, B. (1956). *Language, Thought and Reality*. Edited by J. Carroll. Cambridge, Mass.: MIT Press.

Wiegrebe, L., and Patterson, R. (1999). Quantifying the distortion products generated by amplitude modulated noise. *Journal of the Acoustical Society of America* 106: 2709–2718.

Wilkens, H. (1977). Mehrdimensionale Beschreibung subjektiver Beurteilungen der Akustik von Konzertsälen. *Acustica* 38: 10–23.

Wilkes, K. (1984). "External" factors in the development of psychology in the West. In I. Hronszky, M. Fehér, and B. Dajka (eds.), *Scientific Knowledge Socialized: Selected Proceedings of the Fifth Joint International Conference on the History and Philosophy of Science*. Boston: Kluwer Academic.

Williamson, T. (2000). *Knowledge and Its Limits*. Oxford: Oxford University Press.

Wilson, D., Wilczynski, C., Wells, A., and Weiser, L. (2000). Gossip and other aspects of language as group-level adaptations. In C. Heyes and L. Huber, *The Evolution of Cognition*. Cambridge, Mass.: MIT Press.

Wilson, R., and Keil, F. (1999). *The MIT Encyclopedia of the Cognitive Sciences*. Cambridge, Mass.: MIT Press.

Wishart, T. (1996). *On Sonic Art*. Amsterdam: Harwood Academic.

Wisniewski, E. (1993). *Die Berliner Philharmonie und Ihr Kammermusiksaal: Der Konzertsaal als Zentralraum*. Berlin: Gebrüder Mann.

Wissoker, K. (2000). Negotiating a passage between disciplinary borders: A symposium. *Items and Issues: Social Science Research Council* 1(3–4).

Wittkower, R. (1971). *Architectural Principles in the Age of Humanism*. New York: Norton.

Wordsworth, W. (1835). On the power of sound. In D. Nicholson and A. Lee (eds.), *The Oxford Book of English Mystical Verse*. Oxford: The Clarendon Press, 1917.

Worrall, D. (1989). System for a portable multichannel performance space: A technical overview. *Chroma Journal of the Australian Computer Music Association* 1(3): 3–6.

Worrall, D. (1998). Space in sound: sound of space. *Organised Sound* 3(2): 93–99.

Wright, W., and Medendorp, N. (1967). Acoustic radiation from a finite line source with N-wave excitation. *Journal of the Acoustical Society of America* 43: 966–971.

Wundt, W. (1904). *Principles of Physiological Psychology.* London: Swan Sonnenschein.

Yamaha Company (2002). Reverberation for the new millennium: SREV1. Found at http://www.yamaha.co.jp/english/product/proaudio/products/signal_processors/srev1/index.htm (accessed 28 March 2006).

Young, R. (1972). The anthropology of science. *New Humanist* 88(3): 102–105.

Zacharov, N., and Koivuniemi, K. (2001). Unraveling the perception of spatial sound reproduction: Techniques and experimental design. In *Proceedings of the Nineteenth International Conference of the Audio Engineering Society: Surround Sound: Techniques, Technology, and Perception.*

Zaidel, D. (2000). Different organizational concepts and meaning systems in the two cerebral hemispheres. *Psychology of Learning and Motivation* 40: 1–40.

Zheng, W., and Knudsen, E. (1999). Functional selection of adaptive auditory space map by $GABA_A$-mediated inhibition. *Science* 284: 962–965.

Zvonar, R. (1999). A history of spatial music. Available at http://www.zvonar.com/writing/spatial_music/History.html (accessed June 7, 2006).

Index

Absorption of sound
  in anechoic chambers, 18
  in automobiles, 192, 201
  canceling effect of, 201
  creates multiple arenas, 28
  feelings produced by, 146
  from humidity, 245, 338, 339
  idealized, 108, 109, 115, 116
  illuminated, 17
  illusion of infinite, 56
  music responds to, 111
  panels, 51
  in Protestant churches, 101
  in simulations, 243, 263
  statistical assumptions, 253, 263
  suppresses reflections, 28
  by vegetation, 337
  virtual window, 56
Academic values, 278, 280, 296
Accidental acoustics. *See* Acoustic accidents
Acheron Necromancy, 84
Ackerman, Diana, 4
Acousmatic music, 182, 183
Acoustic accidents, 60, 68, 86, 93, 147, 184, 194, 204
Acoustic adaptation in animals, 352
Acoustic amplification, 23, 84, 96, 200
Acoustic archaeologists, 59, 64, 73–77, 82, 85
Acoustic architects, 5, 237. *See also* Acoustic engineers

Acoustic arenas, 22–31. *See also* Acoustic horizons
  and aural architects, 24, 32
  collide, 22
  communities in, 26
  control of, 33
  created by echoes, 94
  created by furnishings, 28
  and culture, 28, 29
  depend on sound barriers, 106
  diverge from visual arenas, 55
  divided by absorption, 28
  and doors, 25, 28, 31
  and electroacoustics, 27, 104, 199
  enlarged by silence, 32
  in factories, 31
  with headphones, 191
  increased by theatric masks, 96
  influence by noise, 22, 28, 32, 34, 65, 103, 106
  integrate citizens, 30
  intentionally shaped, 23, 54
  and large groups, 27, 29
  largest manmade, 97
  limited communications in, 33
  local acoustics, 53
  of natural habitats, 27
  open windows, 28
  ownership and rules, 27, 33, 34
  personal preferences, 25
  and physical boundaries, 23
  private, 26, 29, 31, 59, 106, 191